Convex Analysis in Polynomial Spaces with Applications

Convex Analysis in Polynomial Spaces with Applications is intended to serve a broad audience of undergraduate and graduate students, junior and senior researchers, and as a general self-study guide for anyone who wishes to get acquainted with the geometry of Banach spaces of polynomials with applications. This text is specifically designed to be appealing and accessible to the reader, and provides a general overview on the topic together with new and interesting directions of research. The text also contains original results and material never published before.

Features

- Comprehensive review on the geometry of spaces of polynomials.
- Visually attractive and accessible presentation, with over 75 explanatory figures.
- Contains many examples illustrating the results and techniques appearing in the book.
- Open (and deep!) questions within the area are provided so that the interested reader can begin doing independent research using the techniques presented in the text.
- It also features original results by the authors.

Monographs and Research Notes in Mathematics

Series Editors: John A. Burns, Thomas J. Tucker, Miklos Bona, and Michael Ruzhansky

This series is designed to capture new developments and summarize what is known over the entire field of mathematics, both pure and applied. It will include a broad range of monographs and research notes on current and developing topics that will appeal to academics, graduate students, and practitioners. Interdisciplinary books appealing not only to the mathematical community, but also to engineers, physicists, and computer scientists are encouraged.

This series will maintain the highest editorial standards, publishing well-developed monographs as well as research notes on new topics that are final, but not yet refined into a formal monograph. The notes are meant to be a rapid means of publication for current material where the style of exposition reflects a developing topic.

Direct and Projective Limits of Geometric Banach Structures
Patrick Cabau and Fernand Pelletier

Classical Clifford Algebras
Operator-Algebraic and Free-Probabilistic Approaches
Ilwoo Cho

Perturbed Functional Iterations
A Matrix Free Large-Scale Nonlinear System Solver in Applied Science with an Introduction to D-Mapping
Suhrit Dey

Separate and Joint Continuity
Jiling Cao and Warren B. Moors

Fractional Integrals, Potentials, and Radon Transforms, Second Edition
Boris Rubin

Variational-Hemivariational Inequalities with Applications, Second Edition
Mircea Sofonea and Stanisław Migórski

Classification of Lipschitz Mappings, Second Edition
Torrey M. Gallagher, Víctor Pérez-García, and Łukasz Piasecki

Completely Regular Codes in Distance Regular Graphs
Patrick Solé and Minjia Shi

Non-Newtonian Sequence Spaces with Applications
Feyzi Başar and Bipan Hazarika

Spectral Theory for Linear Operators
Bilel Krichen

Convex Analysis in Polynomial Spaces with Applications
Domingo García, Mingu Jung, Manuel Maestre, Gustavo A. Muñoz Fernández and Juan B. Seoane Sepúlveda

For more information about this series please visit: www.routledge.com/Chapman--HallCRC-Monographs-and-Research-Notes-in-Mathematics/book-series/CRCMONRESNOT

Convex Analysis in Polynomial Spaces with Applications

Domingo García
Mingu Jung
Manuel Maestre
Gustavo A. Muñoz Fernández
Juan B. Seoane Sepúlveda

CRC Press
Taylor & Francis Group
Boca Raton London New York

CRC Press is an imprint of the
Taylor & Francis Group, an **informa** business
A CHAPMAN & HALL BOOK

First edition published 2026
by CRC Press
2385 NW Executive Center Drive, Suite 320, Boca Raton FL 33431

and by CRC Press
4 Park Square, Milton Park, Abingdon, Oxon, OX14 4RN

CRC Press is an imprint of Taylor & Francis Group, LLC

ISBN: 978-1-032-96765-3 (hbk)
ISBN: 978-1-032-96763-9 (pbk)
ISBN: 978-1-003-59063-7 (ebk)

DOI: 10.1201/9781003590637

Typeset in CMR10 font
by KnowledgeWorks Global Ltd.

Publisher's note: This book has been prepared from camera-ready copy provided by the authors.

To our colleague and friend
Richard M. Aron
on occasion of his 80th birthday

Contents

List of Figures

Preface

Any homogeneous polynomial in $\mathbb{K}^n$, where $\mathbb{K}$ stands for either $\mathbb{R}$ or $\mathbb{C}$, of degree $m \in \mathbb{N}$, can be expressed as

$$P(x) = \sum_{|\alpha|=m} a_\alpha x^\alpha, \qquad (\star)$$

where $x = (x_1, \ldots, x_n) \in \mathbb{K}^n$, $\alpha = (\alpha_1, \ldots, \alpha_n) \in (\mathbb{N} \cup \{0\})^n$, with $|\alpha| = \alpha_1 + \cdots + \alpha_n$, and $x^\alpha = x_1^{\alpha_1} \cdots x_n^{\alpha_n}$. Each coefficient a_α belongs to $\mathbb{K}$. The space $\mathcal{P}(^m\mathbb{K}^n)$ represents the finite-dimensional vector space consisting of all homogeneous polynomials of degree m on $\mathbb{K}^n$.

For a given norm $\|\cdot\|$ on $\mathbb{K}^n$, the function

$$\|P\| := \sup\{|P(x)| : x \in \mathsf{B}_X\},$$

for every $P \in \mathcal{P}(^m\mathbb{K}^n)$, where B_X denotes the unit ball of the Banach space $X = (\mathbb{K}^n, \|\cdot\|)$, defines a norm on $\mathcal{P}(^m\mathbb{K}^n)$. This is commonly referred to as the polynomial norm. The space $\mathcal{P}(^m\mathbb{K}^n)$, when endowed with the polynomial norm induced by X, is denoted by $\mathcal{P}(^mX)$.

Apart from the polynomial norm, another frequently used norm on $\mathcal{P}(^m\mathbb{K}^n)$ is the ℓ_p norm of its coefficients. Specifically, for a polynomial P given by $(\star)$ above and any $p \geq 1$, the norm

$$|P|_p := \left(\sum_{|\alpha|=m} |a_\alpha|^p \right)^{\frac{1}{p}},$$

defines an alternative norm on $\mathcal{P}(^m\mathbb{K}^n)$.

The study of the unit ball of a polynomial space has recently motivated a significant volume of publications. However, this problem has much earlier roots. As early as 1966, Konheim and Rivlin [66] provided a characteristic property of the extreme points of the unit ball of the space $\mathcal{P}_n(\mathbb{R})$, which consists of all polynomials on the real line of degree at most n, endowed with the norm

$$\|P\| = \sup\{|P(x)| : x \in [-1, 1]\}.$$

The search for characterizations of the extreme points of the unit ball of a polynomial space has been intensified since the late 1990's. The references [6, 10, 20–23, 39, 45, 46, 49–55, 75–77, 81–83, 88, 89, 94] are just a selection of contributions to the study of the geometry of a number of polynomial spaces of low dimension. The interested reader can find a monograph on this topic in [44].

Perhaps one of the most relevant motivations to study polynomial spaces, and more specifically the extreme points of their unit balls, rests on the fact that an elementary application of the Krein-Milman theorem allows us to obtain optimal constants in a number of polynomial inequalities of interest. For instance, in [2, 4, 62, 66, 71, 72, 84, 85, 87, 106] one can find several applications that obtain optimal Bernstein and Markov type estimates using a combination of geometric results and the so-called Krein-Milman approach. Similarly, several polynomial Bohnenblust-Hille type inequalities (see, e.g., [8, 31, 83]) can be derived (see, for instance, [41, 61, 80]). The same techniques can be applied to obtain sharp estimates of polarization constants and unconditional constants (see [55, 82]) and many other polynomial inequalities of interest (see, for instance, [3, 5]) as it is shown in Chapter 3 of this book.

Interestingly, polynomial norms are frequently non-absolute. Recall that a norm $\|\cdot\|$ on a linear subspace X of $\mathbb{R}^\Lambda$ (here, Λ is any non-empty set) is *absolute* whenever it is complete and satisfies the following conditions:

(a) Given $x, y \in \mathbb{R}^\Lambda$ with $|x(\lambda)| = |y(\lambda)|$ for every $\lambda \in \Lambda$, if $x \in X$, then $y \in X$ with $\|y\| = \|x\|$.

(b) For every $\lambda \in \Lambda$, the function $e_\lambda : \Lambda \to \mathbb{R}$ given by $e_\lambda(\xi) = \delta_{\lambda\xi}$ for $\xi \in \Lambda$, belongs to X with $\|e_\lambda\| = 1$.

One property which can be deduced from the definition is the following: given $x, y \in \mathbb{R}^\Lambda$ with $|y(\lambda)| \leq |x(\lambda)|$ for every $\lambda \in \Lambda$, if $x \in X$, then $y \in X$ with $\|y\| \leq \|x\|$. The ℓ_p-norms together with many other classical norms are clearly absolute in $\mathbb{R}^\Lambda$. Hence polynomial norms can be viewed as a source of natural examples of non-absolute norms. Let us be more precise in this point, consider now $\mathcal{P}_n(\mathbb{R}^k)$ the polynomials on $\mathbb{R}^k$ of degree less than or equal to n and given $m \leq n$, $\mathcal{P}^h_{n,m}(\mathbb{R}^k)$ is the subspace of all homogeneous polynomial of degree m. Given a compact set $K \subset \mathbb{R}^k$ with "enough" numbers of points (that for $k = 1$, it means that K has at least $n + 1$ points. If $k \geq 2$, then a sufficient condition is that K has non-empty interior) it holds that

$$\|P\|_K = \sup\{|P(x)| : x \in K\},$$

is a norm on $\mathcal{P}_n(\mathbb{R}^k)$. If it is denoted by $\Lambda = \{(\lambda_1, \ldots \lambda_k) \in \{0, 1, \ldots, n\}^k : 0 \leq \sum_{j=1}^k \lambda_j \leq n\}$, then a basis of $\mathcal{P}_n(\mathbb{R}^k)$ is $\{x^\lambda = x_1^{\lambda_1} \cdots x_k^{\lambda_k} : \lambda = (\lambda_1, \ldots \lambda_k) \in \Lambda\}$. Now it can be considered the following linear isomorphism

$$L : \mathbb{R}^\Lambda \to \mathcal{P}_n(\mathbb{R}^k),$$

defined by $L((a_\lambda)_{\lambda\in\Lambda}) = \sum_{\lambda\in\Lambda} a_\lambda x^\lambda$.

This isomorphism allows to define the following norm on $\mathbb{R}^\Lambda$ by

$$\|(a_\lambda)_{\lambda\in\Lambda}\| = \sup\{|\sum_{\lambda\in\Lambda} a_\lambda x^\lambda| : x \in K\},$$

that, in general, is not an absolute norm. This can very easily checked in the case of $\mathcal{P}_2(\mathbb{R})$, and $K = [0, 1]$ since

$$\|(0, 0, 1)\| = \|x^2\|_{[0,1]} = 1 > \frac{1}{2} = \|2x^2 - 2x\|_{[0,1]} = \|(0, -2, 2)\|.$$

Analogous way of producing non-absolute norms on certain $\mathbb{R}^\Lambda$ can be obtained by considering $\mathcal{P}^h_{n,m}(\mathbb{R}^k)$ where in this case $\Lambda = \{(\lambda_1, \ldots \lambda_k) \in \{0, 1, \ldots, n\}^k : \sum_{j=1}^k \lambda_j = m\}$.

Although many authors tend to focus on the case where the compact set K is $[-1, 1]^k$, much research has been done for other sets. For example, [2] and [10] studied the case in which K is a $\pi/4$-circle sector, and in [82] where K is a simplex.

The main goal of this book is to provide a full description of the unit ball of some subspaces of polynomials in several variables (more specifically, Chapter 2 is dedicated to the study of m-homogeneous trinomials in two variables). This will require the development of some sophisticated techniques, but some of the key ideas can be traced to the pioneering work of Richard M. Aron (to whom this text is dedicated) and M. Klimek in their 2001 paper [6] (see also [52]) where they defined the following norm in $\mathbb{R}^3$.

$$\|(a,b,c)\|_{\mathbb{R}} = \sup\{|ax^2 + bx + c| \,:\, x \in [-1,1]\},$$

for all $(a,b,c) \in \mathbb{R}^3$.

It is also relevant to mention that they also study an analogous norm in the complex context, i.e. whenever K is the closed unit disk of $\mathbb{C}$. This line of work is not pursued in this book, but let us remark that due to the maximum modulus theorem, the supremum of the absolute value of any complex polynomial on a compact subset K of $\mathbb{C}$ is always attained at some point in the boundary of K. That is not the case, in general, when we are in the real case. Hence, the techniques presented in this book are very different to the ones needed to deal with the complex case.

On top of the previous comments, this text is not only of expository nature, meaning that the whole of Chapter 2 is completely original and contains material never published before, which makes it a little gem for researchers who are interested in getting a state-of-the-art viewpoint of this research topic. The way in which this book is written and the carefulness and rigor in which the authors treat the material in it, make this manuscript an excellent starting point for the interested researcher and, at times, also for the interested graduate student who wishes to begin a mathematical career studying polynomials in Banach spaces and related topics. We have in front of us an excellent, thorough, and self-contained guide to convex analysis in polynomial spaces. A joy to read.

Yun Sung Choi
President, NIMS
South Korea

Authors

Domingo García
Departamento de Análisis Matemático
Universidad de Valencia
Doctor Moliner 50, 46100 Burjasot (Valencia), Spain
Email: domingo.garcia@uv.es

Mingu Jung
Department of Mathematics
Hanyang University
04763 Seoul, Republic of Korea
Email: mingujung@hanyang.ac.kr

Manuel Maestre
Departamento de Análisis Matemático
Universidad de Valencia
Doctor Moliner 50, 46100 Burjasot (Valencia), Spain
Email: manuel.maestre@uv.es

Gustavo A. Muñoz Fernández
Instituto de Matemática Interdisciplinar (IMI)
Departamento de Análisis Matemático y Matemática Aplicada
Facultad de Ciencias Matemáticas
Plaza de Ciencias 3
Universidad Complutense de Madrid
Madrid, 28040 (Spain)
Email: gustavo_fernandez@mat.ucm.es

Juan B. Seoane Sepúlveda
Instituto de Matemática Interdisciplinar (IMI)
Departamento de Análisis Matemático y Matemática Aplicada
Facultad de Ciencias Matemáticas
Plaza de Ciencias 3
Universidad Complutense de Madrid
Madrid, 28040 (Spain)
Email: jseoane@ucm.es

Author Biographies

Domingo García is Professor of mathematics at the University of Valencia, Spain, where he is also Head of Department. He spent two sabbatical years at Universidade Estadual de Campinas, Brazil, in 1993 and Kent State University, USA, in the academic year 2004–2005, respectively. He has extensive research related to Banach space theory, specifically in norm attaining operators, Dirichlet series, and holomorphic functions. He has been both plenary and invited speakers at several international conferences and is the author of more than 100 publications, including several books.

Mingu Jung completed his Ph.D. in 2021 at Pohang University of Science and Technology (POSTECH) under the supervision of Yun Sung Choi and Manuel Maestre. His research focuses primarily on functional analysis and nonlinear analysis. He was a CMC Fellow (Hyo Chul Myung Assistant Professor) at the June E Huh Center for Mathematical Challenges, Korea Institute for Advanced Study (KIAS), and he is currently an Assistant Professor in the Department of Mathematics at Hanyang University, Seoul.

Manuel Maestre got his Ph.D. at the Universidad de Valencia (Spain) in 1982. He is corresponding academician of the Academy of Sciences of Spain and editor of journals of mathematics. His main interests include functional analysis, and complex analysis in finite and infinite dimensions. He is the author of more than 120 scientific publications, including several books. Currently, he is a Professor of Mathematics at Universidad de Valencia (Spain).

Gustavo A. Muñoz Fernández graduated in Mathematics and Physics and obtained a PhD in Mathematics from the Complutense University of Madrid (Spain) in 1999. His main mathematical interests lie in Functional Analysis, particularly in the study of polynomial spaces in normed spaces, the geometry of polynomial spaces, and polynomial inequalities. He has also conducted extensive research on algebraic genericity, specifically the lineability of spaces of rare functions. Currently, he is a Professor at the Complutense University of Madrid and the director of the Interdisciplinary Mathematics Institute (IMI).

Juan B. Seoane Sepúlveda received his first Ph.D. at the Universidad de Cádiz (Spain) jointly with Universität Karlsruhe (Germany) in 2005 under the supervision of Profs. Richard M. Aron and Antonio Aizpuru. His second Ph.D. was earned at Kent State University (Kent, Ohio, USA) in 2006 under the supervision of Profs. Richard M. Aron and Vladimir I. Gurariy. His main interests include Real and Complex Analysis, Operator Theory, Geometry of Banach spaces, History of Mathematics, Lineability and Spaceability, and Mathematical Biology. He is currently a Professor at Universidad Complutense de Madrid (Spain).

Chapter 1

Polynomials on Unbalanced Convex Bodies

1.1 Preliminaries

Let us recall first that given a compact, convex subset C with non-empty interior in a Banach space, or simply a convex body from now on, a point $e \in C$ is said to be *extreme* if $x, y \in C$ and $\lambda x + (1 - \lambda)y = e$, for some $0 < \lambda < 1$, entails $x = y = e$. Equivalently, $e \in C$ is extreme if and only if $C \setminus \{e\}$ is convex. The set of extreme points of C will be denoted by $\mathrm{ext}(C)$. The significance of describing the extreme points of a convex body lies in the Krein-Milman Theorem, which states that a convex body can be characterized by its extreme points. The characterization of the set of extreme points $\mathrm{ext}(C)$ of a convex body $C \subset \mathbb{R}^n$, together with the Krein-Milman Theorem, helps simplify the optimization of a convex function $f : C \to \mathbb{R}$. Instead of searching over the entire set C, we can restrict our attention to its extreme points, using the formula:

$$\max\{f(x) : x \in C\} = \max\{f(e) : e \in \mathrm{ext}(C)\}.$$

We will call this idea the Krein-Milman Approach (see Chapter 3, Section 3.1 for more details). The Krein-Milman Approach can be applied to a wide range of problems. In this book, we are particularly interested in its applications to obtaining sharp inequalities in polynomial spaces of low dimension.

In the last few years, the study of the extreme points in the unit ball of finite dimensional spaces of polynomials has become a fruitful research field. In this section, we will present some significant results on the geometry of polynomial spaces before addressing the main topic of this chapter. One of the first results that appeared in the literature addressing the problem of characterizing the extreme points of the unit ball of a finite-dimensional polynomial space is due to Konheim and Rivlin [66]. They provided a characterization of the extreme polynomials of the unit ball B_n in the space of polynomials of a single real variable (of degree at most n), endowed with the supremum norm on $[-1, 1]$, namely $\mathcal{P}_n(\mathbb{R})$. Konheim and Rivlin's characterization does not provide the extreme polynomials of B_n explicitly, but it can be employed to describe the set $\mathrm{ext}(\mathsf{B}_n)$ with precision.

To understand Konheim and Rivlin's characterization, it will be useful to clarify the notation $N(P)$ for a polynomial $P \in \mathcal{P}_n(\mathbb{R})$, which we will do next.

Definition 1.1.1 *Let $P(x) = a_0 + a_1 x + \cdots + a_n x^n$ be a polynomial in $\mathbb{R}$ of degree less than or equal to n and let $y \in \mathbb{R}$. The multiplicity of P at y, denoted by $N(P, y)$, is defined as the sum of the multiplicities of the roots of the polynomial $P - y$ in $[-1, 1]$. We will write $N(P) := N(P, 1) + N(P, -1)$.*

The characterization of the extreme points of B_n referred to earlier by Konheim and Rivlin (see [66]) is based on the concept introduced in the previous definition.

Theorem 1.1.1 (Konheim and Rivlin [66]) *A polynomial $P \in \mathcal{P}_n(\mathbb{R})$ is an extreme point of B_n if and only if $N(P) > n$.*

DOI: 10.1201/9781003590637-1

The previous result is hardly applicable in certain problems, but it can be used to deduce the extreme points of B_n for small values of n. The case $n = 2$ can be solved as an accessible exercise. Using a different approach, Aron and Klimek characterized the set $\text{ext}(\mathsf{B}_n)$.

Theorem 1.1.2 (Aron and Klimek [6]**)** *If B_2 is the unit ball of $\mathcal{P}_2(\mathbb{R})$, the set of extreme points of B_2 is given by*

$$\text{ext}\,(\mathsf{B}_2) = \left\{\pm\left(t, \pm 2(\sqrt{2t} - t), 1 + t - 2\sqrt{2t}\right) : t \in \left[\frac{1}{2}, 2\right]\right\} \cup \{\pm(0,0,1)\}.$$

An application of Theorem 1.1.1 allows us to obtain a full description of $\text{ext}(\mathsf{B}_3)$ as well. The result is quite far from being trivial and states as follows:

Theorem 1.1.3 (Araújo, Muñoz, Rodríguez, and Seoane [4]**)** *Let $P \in \mathsf{B}_3$ where B_3 is the unit ball of $\mathcal{P}_3(\mathbb{R})$. Then P is an extreme polynomial of B_3 if, and only if, P is of one of the following forms.*

(i) $P(x) = \pm 1$;

(ii) $P(x) = \pm\left(1 - \frac{1}{4}(\pm x + 1)^3\right)$;

(iii) $P(x) = \pm(2x^2 - 1)$;

(iv) $P(x) = \pm\left(1 - \dfrac{4q}{(1-q^2)^2}(x-q)^2\left(x + \dfrac{q^2+1}{2q}\right)\right)$ *or*
$P(x) = \pm\left(1 + \dfrac{4q}{(1-q^2)^2}(x+q)^2\left(x - \dfrac{q^2+1}{2q}\right)\right)$,
for every $q \in \left(-\frac{1}{3}, 0\right)$;

(v) $P(x) = \pm\left(1 + \dfrac{1}{(1+t)^2}(x-t)^2(x-1)\right)$ *or*
$P(x) = \pm\left(1 - \dfrac{1}{(1+t)^2}(x+t)^2(x+1)\right)$,
for every $t \in \left(-\frac{1}{2}, 1\right)$

(vi) $P(x) = \pm\left(1 + \dfrac{4}{(s-r)^3}(x-r)^2\left(x - \dfrac{3s-r}{2}\right)\right)$ *or*
$P(x) = \pm\left(1 - \dfrac{4}{(s-r)^3}(x+r)^2\left(x + \dfrac{3s-r}{2}\right)\right)$,
for every $-1 \leq r < s \leq 1$ *such that* $s \geq \min\left\{3r+2, \frac{r+2}{3}\right\}$.

The procedure employed by Aron and Klimek to derive Theorem 1.1.2 has been adapted many times in the study of other polynomial spaces. In the case of the space $\mathcal{P}_2(\mathbb{R})$, Aron and Klimek's procedure includes the following key steps:

1. Derivation of a formula for $\|ax^2 + bx + c\|$ in terms of the coefficients a, b, and c.
2. Determination of the projection of B_2 onto a coordinate plane. Specifically, Aron and Klimek considered the plane $c = 0$.
3. Calculation of a parametrization for the unit sphere S_2.
4. Characterization of the extreme points of B_2 based on the previous step.

Using Aron and Klimek's procedure, Muñoz and Seoane [88] described the geometry of the spaces of trinomials on the real line with the sup norm on symmetrical intervals.

More specifically, for $m, n \in \mathbb{N}$ with $m > n$, let $\mathcal{P}_{m,n}(\mathbb{R})$ be the space of the trinomials $ax^m + bx^n + c$ with real coefficients a, b, c endowed with the norm

$$\|ax^m + bx^n + c\|_{m,n} = \sup\{|ax^m + bx^n + c| : x \in [-1, 1]\}.$$

The polynomial $ax^m + bx^n + c$ will often be written as (a, b, c). If $\mathsf{B}_{m,n}$ stands for the unit ball of $\mathcal{P}_{m,n}(\mathbb{R})$, then the set $\text{ext}(\mathsf{B}_{m,n})$ has been fully described in [88]. It will be convenient to provide $\text{ext}(\mathsf{B}_{m,n})$ explicitly. The description of $\text{ext}(\mathsf{B}_{m,n})$ depends strongly on whether m and n are even or odd, so several cases need to be addressed. The case where m and n are both odd requires a technical lemma that we state next. This lemma is extremely handy, and it will be used throughout this text on many occasions.

Lemma 1.1.4 (Muñoz and Seoane [88]**)** *If $m, n \in \mathbb{N}$ are such that $m > n$, then the equation*

$$|n + mx| = (m - n)|x|^{\frac{m}{m-n}}$$

has exactly three roots: one at $x = -1$, another at a point $\lambda_0 \in (-\frac{n}{m}, 0)$, and a third at a point $\lambda_1 > 0$. Furthermore, we have

$$|n + mx| < (m - n)|x|^{\frac{m}{m-n}}, \tag{1.1}$$

if and only if $x < \lambda_0$ or $x > \lambda_1$.

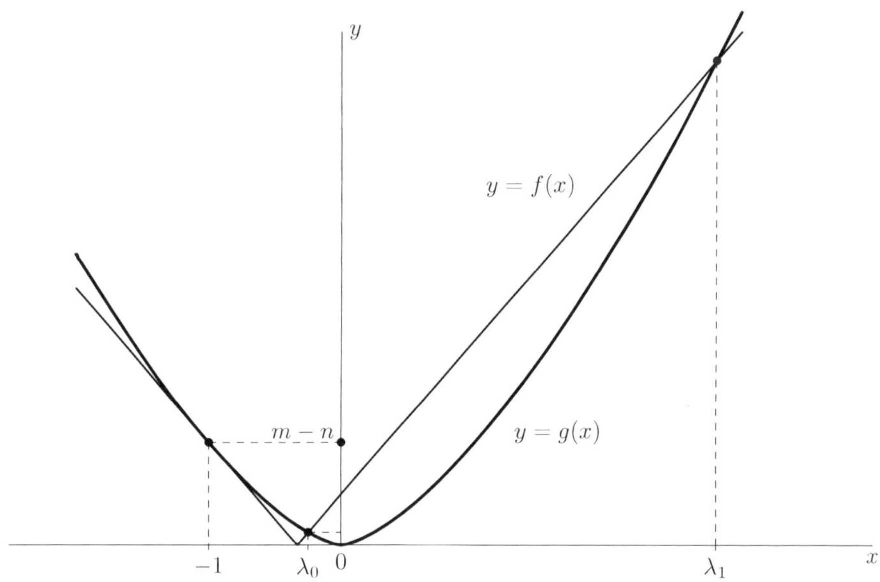

FIGURE 1.1: Illustrating Lemma 1.1.4.

Remark 1.1.5 *The calculations for the previous lemma follow, essentially, a convexity argument. Indeed, let $f(x) = |n + mx|$ and $g(x) = (m - n)|x|^{\frac{m}{m-n}}$ for every $x \in \mathbb{R}$. Keeping in mind that the graph of f is just two straight lines and g is convex, the graphs of f and g can meet at most at four points. Also, the graph of f is tangent to the graph of g at the point $(-1, m - n)$. One can also see that $f(-\frac{n}{m}) < g(-\frac{n}{m})$ and $f(0) > g(0)$, which implies, by continuity, that there exists $\lambda_0 \in (-\frac{n}{m}, 0)$ such that $f(\lambda_0) = g(\lambda_0)$. Moreover, since $\lim_{\lambda \to \infty} [g(\lambda) - f(\lambda)] = \infty$ and $f(0) > g(0)$, there exists $\lambda_1 > 0$ with $f(\lambda_1) = g(\lambda_1)$. To finish, $|n + mx| < (m - n)|x|^{\frac{m}{m-n}}$ follows from the convexity of g (see Figure 1.1).*

In the following results, $\mathsf{S}_{m,n}$ will stand for the unit sphere of $\mathcal{P}_{m,n}(\mathbb{R})$.

Theorem 1.1.6 (Muñoz and Seoane [88]) *Let $m, n \in \mathbb{N}$ be odd. If for every $(a,b) \in \mathbb{R}^2$ we define*

$$f_+(a,b) = \begin{cases} 1 - \frac{(m-n)|a|}{n} \cdot \left|\frac{nb}{ma}\right|^{\frac{m}{m-n}} & \text{if } a \neq 0 \text{ and } -1 < \frac{nb}{ma} < \lambda_0, \\ 1 - |a+b| & \text{otherwise.} \end{cases}$$

and $f_- = -f_+$, then

(a) $\mathsf{S}_{m,n} = \text{graph}\,(f_+|_V) \cup \text{graph}\,(f_-|_W)$.

(b) $\text{ext}(\mathsf{B}_{m,n}) = \left\{\pm\left(t, -\frac{m}{(m-n)^{\frac{m-n}{m}} n^{\frac{n}{m}}} t^{\frac{n}{m}}, 0\right) : \frac{n}{m-n} \le t \le \frac{n}{n+m\lambda_0}\right\} \cup \{\pm(0,0,1)\}$.

Theorem 1.1.7 (Muñoz and Seoane [88]) *If for every $m, n \in \mathbb{N}$ $(m > n)$ with m odd and n even and every $(a,b) \in \mathbb{R}^2$ we define*

$$\begin{aligned} f_+(a,b) &= 1 - |a| - b, & f_-(a,b) &= -f_+(a,-b) = -1 + |a| - b, \\ g_+(a,b) &= 1, & g_-(a,b) &= -g_+(a,b) = -1, \end{aligned}$$

then

(a) $\mathsf{S}_{m,n} = \text{graph}\,(g_+|_U) \cup \text{graph}\,(g_-|_W) \cup \text{graph}\,(f_+|_{V\cup W}) \cup \text{graph}\,(f_-|_{U\cup V})$.

(b) $\text{ext}(\mathsf{B}_{m,n}) = \{\pm(0,2,-1), \pm(1,1,-1), \pm(-1,1,-1), \pm(0,0,-1)\}$.

To understand the remaining cases, we will need implicit functions defined next.

Lemma 1.1.8 *Let $m, n \in \mathbb{N}$ with $m > n$. Then the equation*

$$\frac{(m-n)a}{n}\left(\left|\frac{nb}{ma}\right|\right)^{\frac{m}{m-n}} = 2 - a - b, \tag{1.2}$$

defines implicitly a unique differentiable curve $b = \Gamma_{m,n}(a)$ defined on $(0,\infty)$ such that $\Gamma_{m,n}(2) = 0$ and $\Gamma_{m,n}(n/m) = 1$. Furthermore, although $\Gamma_{m,n}$ cannot be obtained explicitly in general, whenever n is odd, we have

$$\Gamma_{2n,n}(a) = 2(\sqrt{2a} - a).$$

Theorem 1.1.9 (Muñoz and Seoane [88]) *Let $m, n \in \mathbb{N}$ be such that m is even, n is odd and $m > n$. If for every $(a,b) \in \mathbb{R}^2$, we define*

$$f_+(a,b) = 1 - a - |b|, \quad f_-(a,b) = -f_+(-a,b),$$

and for every $(a,b) \in \mathbb{R}^2$ with $a \neq 0$, we define

$$g_+(a,b) = \frac{(m-n)a}{n} \cdot \left(\frac{nb}{ma}\right)^{\frac{m}{m-n}} - 1, \quad g_-(a,b) = -g_+(-a,b),$$

then

(a) $\mathsf{S}_{m,n} = \text{graph}\,(f_+|_{W_{m,n}\cup V_{m,n}}) \cup \text{graph}\,(f_-|_{U_{m,n}\cup V_{m,n}}) \cup \text{graph}\,(g_+|_{W_{m,n}}) \cup \text{graph}\,(g_-|_{U_{m,n}})$.

(b) The extreme points of $\mathsf{B}_{m,n}$ are

$$\pm(0,0,1) \quad \text{and} \quad \pm\,(t, \pm\Gamma_{m,n}(t), 1 - t - \Gamma_{m,n}(t)),$$

where $t \in [n/m, 2]$.

To finish the study of $\mathrm{ext}(\mathsf{B}_{m,n})$, we have to introduce two more functions. For every $a \in \mathbb{R}\backslash\{0\}$ let define

$$\Upsilon_{m,n}(a) = \left(\frac{2m}{m-n}\right)^{\frac{m-n}{m}} \cdot \left(\frac{|a|m}{n}\right)^{\frac{n}{m}}$$

and

$$\Lambda_{m,n}(a) = -\Gamma_{m,n}(|a|),$$

where $b = \Gamma_{m,n}(a)$ for every $a \in (0,\infty)$ is the curve defined in Lemma 1.1.8. Although $\Lambda_{m,n}(a)$ cannot be given explicitly, in the case $m = 2n$ with $n \in \mathbb{N}$ even, we have

$$\Upsilon_{2n,n}(a) = 2\sqrt{2|a|} \text{ and } \Lambda_{2n,n}(a) = 2\left(|a| - \sqrt{2|a|}\right).$$

Theorem 1.1.10 (Muñoz and Seoane [88]) *Let $m, n \in \mathbb{N}$ be even with $m > n$. If for every $(a,b) \in \mathbb{R}^2$ we define*

$$f_+(a,b) = 1 - \left|\frac{a+b}{2}\right| - \frac{a+b}{2}, \quad f_-(a,b) = -f_+(-a,-b),$$

and for every $(a,b) \in \mathbb{R}^2$ with $a \neq 0$ we define

$$g_+(a,b) = \frac{(m-n)a}{n}\left(\left|\frac{nb}{ma}\right|\right)^{\frac{m}{m-n}} - 1, \quad g_-(a,b) = -g_+(-a,-b),$$

then

(a) $\mathsf{S}_{m,n} = \mathrm{graph}\left(f_+|_{W_{m,n}\cup V_{m,n}}\right) \cup \mathrm{graph}\left(f_-|_{U_{m,n}\cup V_{m,n}}\right) \cup \mathrm{graph}\left(g_+|_{W_{m,n}}\right) \cup \mathrm{graph}\left(g_-|_{U_{m,n}}\right)$.

(b) *The extreme points of $\mathsf{B}_{m,n}$ are*

$$\pm(0,0,1), \quad \pm(s, \Lambda_{m,n}(s), -1-s-\Lambda_{m,n}(s)), \quad \text{and} \quad \pm\left(t, -\Upsilon_{m,n}(t), 1\right)$$

where $s \in [\gamma_0, -2]$, $t \in [-\gamma_1, -\gamma_0]$ and

$$\gamma_0 = -\frac{2}{m-n} \cdot \left(\frac{m^m}{n^n}\right)^{\frac{1}{m-n}},$$

$$\gamma_1 = \frac{-2n}{m-n}.$$

Other notable example of geometrical study in a polynomial space of low dimension is due to Choi and Kim [20–22], who characterized the extreme polynomials of the spaces of scalar-valued 2-homogeneous polynomials on the real spaces ℓ_1^2, ℓ_2^2, and ℓ_∞^2. In a similar line, Grecu [51] treated the same problem for scalar-valued 2-homogeneous polynomials on the real spaces ℓ_p^2 with $1 < p < \infty$. See also [49–54] for related questions concerning real or complex homogeneous polynomials of degree 2 or 3. Trigonometric trinomials (real or complex) have also been studied by Aron and Klimek [6] and Neuwirth [89]. In this chapter, our aim is to present a novelty, studied in recent years, by providing a full description of the extreme points of a space of 2-homogeneous polynomials defined on several *non-symmetric* (or unbalanced) convex bodies.

To be more specific, we study the geometry of the unit ball of the space of 2-homogeneous polynomials on $\mathbb{R}^2$ endowed with the norm defined on the simplex, the unit square, or the sectors of the circle.

A complete understanding of the geometry of these spaces – including explicit formulas for the norm and characterization of the extreme points of the unit sphere – is of independent interest and will also be crucial for applications developed in the last chapter. Let us begin with the study of the geometry of the unit ball of the space of 2-homogeneous polynomials on the simplex.

1.2 The unit ball of $\mathcal{P}(^2\Delta)$.

We begin with examining the geometry of the unit ball of the space of 2-homogeneous polynomials on $\mathbb{R}^2$ endowed with the norm defined by

$$\|P\|_\Delta = \sup\{|P(\mathbf{x})| : \mathbf{x} \in \Delta\},$$

where Δ represents the region enclosed by the triangle in $\mathbb{R}^2$ of vertices $(0,0)$, $(0,1)$, and $(1,0)$ (or simplex for short). From now on we will represent the space described just above by $\mathcal{P}(^2\Delta)$. If $L \in \mathcal{L}(^2\mathbb{R}^2)$, we also define

$$\|L\|_\Delta = \sup\{|L(\mathbf{x},\mathbf{y})| : \mathbf{x},\mathbf{y} \in \Delta\}.$$

The mapping that assigns to every $(a,b,c) \in \mathbb{R}^3$ the polynomial $P(x,y) = ax^2 + bxy + cy^2$ allows us to identify $\mathcal{P}(^2\Delta)$ with $(\mathbb{R}^3, \|\cdot\|_\Delta)$, where $\|(a,b,c)\|_\Delta = \|P\|_\Delta$. This representation, together with the standard polynomial representation of $\mathcal{P}(^2\Delta)$, will be used interchangeably throughout this section and, likely in the remainder of the text.

First of all, we derive a formula for $\|\cdot\|_\Delta$ in the following result.

Theorem 1.2.1 (Muñoz, Révész, and Seoane [82]**)** *Let $a,b,c \in \mathbb{R}$ and $P(x,y) = ax^2 + bxy + cy^2$. Then*

$$\|P\|_\Delta = \begin{cases} \max\left\{|a|,|c|,\left|\frac{b^2-4ac}{4(a-b+c)}\right|\right\} & \text{if } a-b+c\neq 0 \text{ and } 0<\frac{2c-b}{2(a-b+c)}<1, \\ \max\{|a|,|c|\} & \text{otherwise.} \end{cases} \tag{1.3}$$

PROOF. *If we let $P(x,y) = ax^2 + bxy + cy^2$, then the norm of P is attained either in $\partial\Delta$, or at a critical point of P in $\mathring{\Delta}$ (the interior of Δ). Let us first evaluate the maximum of P over $\partial\Delta$. To start with, notice that the maximum of $|P|$ over the side $\{(t,1-t) : t \in [0,1]\}$ coincides with the norm of the polynomial $p(t) = (a-b+c)t^2 + (b-2c)t + c$ over the interval $[0,1]$. Performing some calculations, it can be proved that*

$$\|p\|_{[0,1]} = \begin{cases} \max\left\{|a|,|c|,\left|\frac{b^2-4ac}{4(a-b+c)}\right|\right\} & \text{if } a-b+c\neq 0 \text{ and } 0<\frac{2c-b}{2(a-b+c)}<1, \\ \max\{|a|,|c|\} & \text{otherwise.} \end{cases}$$

On the other hand, the maxima of $|P|$ over the sides $\{(t,0) : t \in [0,1]\}$ and $\{(0,t) : t \in [0,1]\}$ are $|a|$ and $|c|$, respectively. Since $\max\{|a|,|c|\} \leq \|p\|_{[0,1]}$, we obviously have that $\|P\|_{\partial\Delta} = \|p\|_{[0,1]}$.

As for the critical points of P in $\mathring{\Delta}$, notice that in case P has a critical point (x,y) other than $(0,0)$ then the restriction of P to the line $L(t) := \{(tx,ty) : t \in \mathbb{R}\}$, always having the form Ct^2 on L, must necessarily have $C = 0$, which implies that $P(x,y) = 0$, and the critical point in question does not qualify for extrema (unless $P \equiv 0$).

Hence, $\|P\|_\Delta = \|P\|_{\partial\Delta} = \|p\|_{[0,1]}$. This concludes the proof.

In order to parametrize S_Δ it will be useful to know what the projection of S_Δ onto any of the coordinate planes looks like.

Theorem 1.2.2 (Muñoz, Révész, and Seoane [82]**)** *The projection of S_Δ onto the ac-plane is $\mathsf{B}_{\ell_\infty^2}$.*

PROOF. *If we take* $(a,b,c) \in \mathsf{S}_\Delta$, *then by* (1.3) *it follows that* $\|(a,c)\|_\infty \le \|(a,b,c)\|_\Delta = 1$. *Therefore,* $\pi_{ac}(\mathsf{S}_\Delta) \subseteq \mathsf{B}_{\ell_\infty^2}$. *However, it is easy to see that* $\pi_{ac}(\mathsf{S}_\Delta)$ *is a convex subset of the* $b = 0$ *plane. Therefore, to conclude the proof it suffices to show that* $\mathsf{S}_{\ell_\infty^2} \subseteq \pi_{ac}(\mathsf{S}_\Delta)$. *But this is easy to check. Indeed, if* $(a,c) \in \mathsf{S}_{\ell_\infty^2}$, *by* (1.3) *we have that*

$$\|(a,0,c)\|_\Delta = \begin{cases} \max\left\{1, \left|\frac{ac}{a+c}\right|\right\} & \text{if } a+c \neq 0 \text{ and } 0 < \frac{c}{a+c} < 1, \\ 1 & \text{otherwise,} \end{cases}$$

and since $\left|\frac{ac}{a+c}\right| < |a| \le 1$ *whenever* $a+c \neq 0$ *and* $0 < \frac{c}{a+c} < 1$, *it follows that* $\|(a,0,c)\|_\Delta = 1$. *Therefore,* $(a,c) = \pi_{ac}(a,0,c) \in \pi_{ac}(\mathsf{S}_\Delta)$.

With the help of Theorems 1.2.1 and 1.2.2, we can now characterize the extreme points of B_Δ.

Theorem 1.2.3 (Muñoz, Révész, and Seoane [82]**)** *If we define the mappings*

$$f_+(a,c) = 2 + 2\sqrt{(1-a)(1-c)}$$

and

$$f_-(a,c) = -f_+(-a,-c) = -2 - 2\sqrt{(1+a)(1+c)},$$

for every $(a,c) \in \mathsf{B}_{\ell_\infty^2}$ *and the set*

$$F = \{(a,b,c) \in \mathbb{R}^3 : (a,c) \in \mathsf{S}_{\ell_\infty^2} \text{ and } f_-(a,c) \le b \le f_+(a,c)\},$$

then

(a) $\mathsf{S}_\Delta = \operatorname{graph}(f_+|_{\mathsf{B}_{\ell_\infty^2}}) \cup \operatorname{graph}(f_-|_{\mathsf{B}_{\ell_\infty^2}}) \cup F$.

(b) $\operatorname{ext}(\mathsf{B}_\Delta) = \{\pm(1, -2-2\sqrt{2(1+t)}, t), \pm(t, -2-2\sqrt{2(1+t)}, 1) : t \in [-1,1]\}$.

PROOF. *Part (a) can be obtained from Theorems 1.2.1 and 1.3.2 through standard calculations. As for part (b), notice that for every* $(a_0,b_0,c_0) \in F$ *such that* $f_-(a_0,c_0) < b_0 < f_+(a_0,c_0)$, *the point* (a_0,b_0,c_0) *is in the interior of the segment* $\{(a,b,c) \in \mathbb{R}^3 : (a,c) \in \mathsf{S}_{\ell_\infty^2}$ *and* $f_-(a_0,c_0) \le b \le f_+(a_0,c_0)\} \subseteq F$. *In addition to that, for every* $(a_0,c_0) \in \mathsf{B}_{\ell_\infty^2}$, *the mapping* f_+ *(respectively* f_-*), is affine on the straight line given by the equation* $(1-c_0)a = (1-a_0)c + a_0 - c_0$ *(respectively* $(1+c_0)a = (1+a_0)c - a_0 + c_0$*). Therefore, the extreme points of* B_Δ *can occur only at the points where the sets* $\operatorname{graph}(f_+|_{\mathsf{B}_{\ell_\infty^2}})$, $\operatorname{graph}(f_-|_{\mathsf{B}_{\ell_\infty^2}})$ *and* F *meet along a non-affine curve. The reader can easily check that this happens only at the points*

$$\pm(1, -2-2\sqrt{2(1+t)}, t) \quad \text{and} \quad \pm(t, -2-2\sqrt{2(1+t)}, 1),$$

where $t \in [-1,1]$.

Figure 1.2 depicts the unit ball of $(\mathbb{R}^3, \|\cdot\|_\Delta)$.

The next section shall be devoted to the study of the geometry of the space of 2-homogeneous polynomials on the square of vertices $(0,0)$, $(0,1)$, $(1,0)$, and $(1,1)$.

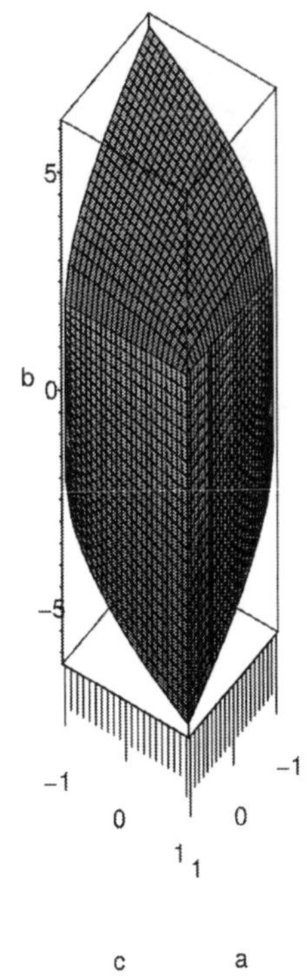

FIGURE 1.2: Unit ball of $(\mathbb{R}^3, \|\cdot\|_\Delta)$.

1.3 The geometry of $\mathcal{P}(^2\square)$

In the following, $\square$ will denote the square of vertices $(0,0)$, $(0,1)$, $(1,0)$, and $(1,1)$. Notice that $\square$ is centrally symmetric but not balanced. Hence the study of polynomials on $\square$ cannot be included in the category of polynomials on a Banach space with the usual sup norm over the unit ball (as we already mentioned at the beginning of this chapter).

If P is a 2-homogeneous polynomial on $\mathbb{R}^2$, we define its norm by

$$\|P\|_\square := \sup\{|P(\mathbf{x})| : \mathbf{x} \in \square\}.$$

Similarly, if L is a bilinear form on $\mathbb{R}^2$, we also define

$$\|L\|_\square := \sup\{|L(\mathbf{x}, \mathbf{y})| : \mathbf{x}, \mathbf{y} \in \square\}.$$

The spaces of 2-homogeneous polynomials, bilinear forms and symmetric bilinear forms on $\mathbb{R}^2$ endowed with the above norms will be represented, respectively, by $\mathcal{P}(^2\square)$, $\mathcal{L}(^2\square)$, and $\mathcal{L}^s(^2\square)$.

As already mentioned in the previous section, the mapping that assigns to every $(a,b,c) \in \mathbb{R}^3$ the polynomial $P(x,y) = ax^2 + bxy + cy^2$ allows us to identify $\mathcal{P}(^2\square)$ with $(\mathbb{R}^3, \|\cdot\|_\square)$, where $\|(a,b,c)\|_\square = \|P\|_\square$.

Again, following the notation used earlier, $\mathsf{S}_\square$ and $\mathsf{B}_\square$ will denote, respectively, the unit sphere and unit ball of the space $\mathcal{P}(^2\square)$. Also, π_{ac} will denote the linear projection given by $\pi_{ac}(a,b,c) = (a,c)$, for every $(a,b,c) \in \mathbb{R}^3$.

First of all we obtain a formula for $\|\cdot\|_\square$:

Theorem 1.3.1 (Gámez, Muñoz, Sánchez, and Seoane [45]) *If $P(x,y) = ax^2 + bxy + cy^2$, then*

$$\|P\|_{\square} = \begin{cases} \max\left\{|a|, |c|, |a+b+c|, \frac{b^2-4ac}{4|c|}\right\} & \text{if } b^2-4ac>0,\ c\neq 0 \text{ and } -\frac{b}{2c}\in(0,1). \\ \max\left\{|a|, |c|, |a+b+c|, \frac{b^2-4ac}{4|a|}\right\} & \text{if } b^2-4ac>0,\ a\neq 0 \text{ and } -\frac{b}{2a}\in(0,1). \\ \max\{|a|, |c|, |a+b+c|\} & \text{otherwise.} \end{cases}$$

PROOF. *Notice that if $(x_0,y_0)\in\mathring{\square}$ (the interior of $\square$) then there exists $\lambda>1$ such that $\lambda(x_0,y_0)\in L=\{(x,1):x\in[0,1]\}\cup\{(1,y):y\in[0,1]\}$. Then $|P(x_0,y_0)|<\lambda^2|P(x_0,y_0)|=|P(\lambda x_0,\lambda y_0)|$, which shows that the norm of P is attained on the set L. Hence, $\|P\|_{\square}=\max\{\|ax^2+bx+c\|_{\square},\|a+by+cy^2\|_{\square}\}$. Considering the zeros of the derivatives of these polynomials ($-\frac{b}{2a}$ and $-\frac{b}{2c}$, respectively), we arrive at*

$$\|P\|_{\square} = \begin{cases} \max\left\{|a|, |c|, |a+b+c|, \left|\frac{4ac-b^2}{4a}\right|, \left|\frac{4ac-b^2}{4c}\right|\right\} & \text{if } -\frac{b}{2a}, -\frac{b}{2c}\in(0,1), \\ \max\left\{|a|, |c|, |a+b+c|, \left|\frac{4ac-b^2}{4a}\right|\right\} & \text{if } -\frac{b}{2a}\in(0,1) \text{ and } -\frac{b}{2c}\notin(0,1), \\ \max\left\{|a|, |c|, |a+b+c|, \left|\frac{4ac-b^2}{4c}\right|\right\} & \text{if } -\frac{b}{2c}\in(0,1) \text{ and } -\frac{b}{2a}\notin(0,1), \\ \max\{|a|, |c|, |a+b+c|\} & \text{otherwise.} \end{cases}$$

If $b^2-4ac>0$ and $-\frac{b}{2a}\in(0,1)$, then $-\frac{b}{2c}\notin(0,1)$. If $b^2-4ac<0$, then

$$\frac{4ac-b^2}{4|a|}\le|a+b+c| \quad \text{and} \quad \frac{4ac-b^2}{4|c|}\le|a+b+c|.$$

Therefore, we obtain the desired formula.

In order to sketch $\mathsf{S}_{\square}$ and obtain the extreme points of $\mathsf{B}_{\square}$, it is important to have a parametrization of $\mathsf{S}_{\square}$. This parametrization can be constructed by projecting $\mathsf{B}_{\square}$ onto the ac-plane. For this matter, we will use the subsets A, B, and C of $[-1,1]^2$ (see Figure 1.3):

$$\begin{aligned} A &:= \{(a,c)\in[-1,1]^2 : \ -1\le a\le 0 \text{ and } a+1\le c\le 1\}, \\ B &:= \left\{(a,c)\in[-1,1]^2 : \ -1\le a\le 1 \text{ and } \max\{-1,a-1\}\le c\le\min\{1,a+1\}\right\}, \\ C &:= \{(a,c)\in[-1,1]^2 : \ 0\le a\le 1 \text{ and } -1\le c\le a-1\}. \end{aligned}$$

Theorem 1.3.2 (Gámez, Muñoz, Sánchez, and Seoane [45]) *The projection of $\mathsf{S}_{\square}$ onto the ac-plane is $[-1,1]^2$.*

PROOF. *If $(a,b,c)\in\mathsf{S}_{\square}$, then it is straightforward from Theorem 1.3.1 that $\max\{|a|,|c|\}\le 1$, or in other words $\pi_{ac}(a,b,c)=(a,c)\in[-1,1]^2$. This shows that $\pi_{ac}(\mathsf{S}_{\square})\subset[-1,1]^2$. Now let $(a,c)\in[-1,1]^2$, then one of the following statements holds:*

1. *$(a,c)\in A$. In this case using Theorem 1.3.1, we have $(a,b,c)\in\mathsf{B}_{\square}$ for all b such that*

$$-2\sqrt{ac+|c|}\le b\le 2\sqrt{ac+|a|}.$$

If $a=-1$ or $c=1$, then all those points are in $\mathsf{S}_{\square}$. However, if $a>-1$ and $c<1$ only $(a,-2\sqrt{ac+|c|},c)$ and $(a,2\sqrt{ac+|a|},c)$ lay in $\mathsf{S}_{\square}$. All this can be easily checked by applying Theorem 1.3.1. In any case, we deduce that $(a,c)\in\pi_{ac}(\mathsf{S}_{\square})$.

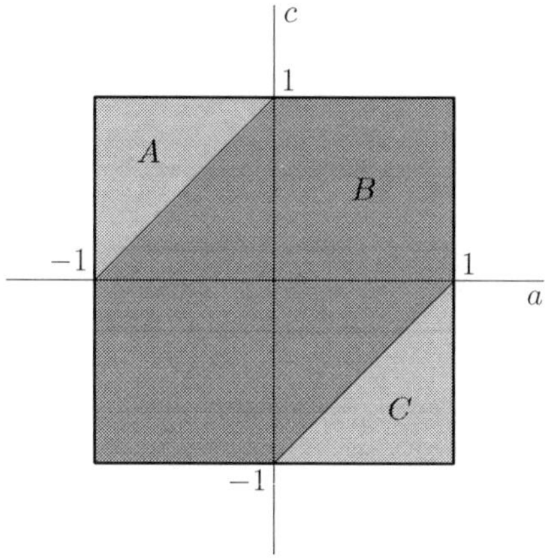

FIGURE 1.3: $\pi_{ac}(\mathsf{B}_\square)$.

2. $(a,c) \in B$. *In this case using Theorem 1.3.1, we see that* $(a,b,c) \in \mathsf{B}_\square$ *for all* b *such that*

$$-1-a-c \le b \le 1-a-c.$$

In fact, if $|a| = 1$ *or* $|c| = 1$*, then all those points are in* $\mathsf{S}_\square$*. However, if* $|a| < 1$ *and* $|c| < 1$*, then only* $(a,-1-a-b,c)$ *and* $(a,1-a-b,c)$ *lay in* $\mathsf{S}_\square$*. Again, all of this can be easily proved using Theorem 1.3.1 once more. Therefore, we deduce that* $(a,c) \in \pi_{ac}(\mathsf{S}_\square)$.

3. $(a,c) \in C$. *Taking into consideration that* $(-a,-c) \in A$*, it follows from case 1 that* $(a,b,c) \in \mathsf{B}_\square$ *for all* b *such that*

$$-2\sqrt{ac+|a|} \le b \le 2\sqrt{ac+|c|}.$$

If $a = 1$ *or* $c = -1$*, we have in fact that those points are in* $\mathsf{S}_\square$*. Otherwise, only the points* $(a, 2\sqrt{ac+|c|}, c)$ *and* $(a, -2\sqrt{ac+|a|}, c)$ *lay in* $\mathsf{S}_\square$*. In any case, we have proved that* $(a,c) \in \pi_{ac}(\mathsf{S}_\square)$.

Theorem 1.3.3 (Gámez, Muñoz, Sánchez, and Seoane [45]**)** *If for every* $(a,c) \in [-1,1]^2$ *we define the mappings*

$$F(a,c) = \begin{cases} 2\sqrt{ac+|a|} & \text{if } (a,c) \in A, \\ 2\sqrt{ac+|c|} & \text{if } (a,c) \in C, \\ 1-a-c & \text{if } (a,c) \in B, \end{cases}$$

$$G(a,c) = -F(-a,-c),$$

where A*,* B*, and* C *are as in Figure 1.3 and the set*

$$H = \{(a,b,c) \in \mathbb{R}^3 : (a,c) \in \partial[-1,1]^2 \text{ and } G(a,c) \le b \le F(a,c)\},$$

then

(a) $\mathsf{S}_\square = \operatorname{graph}(F) \cup \operatorname{graph}(G) \cup H$.

(b) The extreme points of $\mathsf{B}_\square$ *have the form*

$$\pm(t, 2\sqrt{1-t}, -1) \quad \text{and} \quad \pm(-1, 2\sqrt{1-t}, t) \quad \text{with } t \in [0,1]$$

or

$$\pm(1,-1,1),\ \pm(1-3,1),\ \pm(1,0,0),\ \pm(0,0,1).$$

PROOF. *The proof of (a) is outlined in the proof of Theorem 1.3.2. As for the proof of part (b), notice that* $\mathrm{graph}(F|_A)$, $\mathrm{graph}(F|_B)$, $\mathrm{graph}(F|_C)$, $\mathrm{graph}(G|_A)$, $\mathrm{graph}(G|_B)$, $\mathrm{graph}(G|_C)$, *and* H *are all ruled surfaces, therefore, extreme points can only happen at their intersections. By symmetry we just need to focus on the intersections of the surfaces* $\mathrm{graph}(F|_A)$, $\mathrm{graph}(F|_B)$, $\mathrm{graph}(F|_C)$, *and* H.

First we investigate the intersections with H. *The surfaces* $\mathrm{graph}(F|_A)$ *and* H *meet along the curve* $\gamma_1 = \{(-1, 2\sqrt{1-c}, c) : 0 \leq c \leq 1\}$ *and the segment joining the points* $(-1,0,1)$ *and* $(0,0,1)$. *Since* $(-1,0,1) \in \gamma_1$, *the only candidates to be extreme points in this intersection are the points in* γ_1 *and* $(0,0,1)$. *Similarly, the surfaces* $\mathrm{graph}(F|_C)$ *and* H *meet along the curve* $\gamma_2 = \{(a, 2\sqrt{1-a}, -1) : 0 \leq a \leq 1\}$ *and the segment joining the points* $(1,0,-1)$ *and* $(1,0,0)$. *Notice that* $(1,0,-1) \in \gamma_2$, *so the only candidates to be extreme points in this intersection are the points in* γ_2 *and* $(1,0,0)$. *Now the intersection of* $\mathrm{graph}(F|_B)$ *with* H *are the segments joining the point* $(1,-1,1)$ *with the points* $(0,0,1)$ *and* $(1,0,0)$ *on the one hand, and the point* $(-1,3,-1)$ *with the points* $(-1,2,0)$ *and* $(0,2,-1)$ *on the other hand. Clearly, only the endpoints of these segments can be extreme.*

Regarding the other intersections, the surfaces $\mathrm{graph}(F|_A)$ *and* $\mathrm{graph}(F|_B)$ *meet along the segment joining the points* $(-1,2,0)$ *and* $(0,0,1)$ *whereas the surfaces* $\mathrm{graph}(F|_C)$ *and* $\mathrm{graph}(F|_B)$ *meet along the segment joining the points* $(0,2,-1)$ *and* $(1,0,0)$. *Notice that the endpoints of these segments have already appeared above.*

Therefore, by symmetry, the only candidates to be extreme points are the points in the curves $\pm\gamma_k$ *with* $k = 1,2$, *together with the points* $\pm(1,-1,1)$, $\pm(1,-3,1)$, $\pm(1,0,0)$ *and* $\pm(0,0,1)$. *Since the curves* $\pm\gamma_k$ *with* $k = 1,2$ *are nowhere affine, it follows that all their points are extreme in* $\mathsf{B}_\square$. *Also, it is easy to construct support hyperplanes to* $\mathsf{B}_\square$ *at each of the points* $\pm(1,-1,1)$, $\pm(1,-3,1)$, $\pm(1,0,0)$, *and* $\pm(0,0,1)$ *that do not contain any other points of* $\mathsf{B}_\square$. *This concludes the proof.*

Thus, summarizing, the extreme points of $\mathsf{B}_\square$ can be given by

$$\begin{aligned}
P_t(x,y) &= \pm(tx^2 + 2\sqrt{1-t}xy - y^2),\\
Q_t(x,y) &= \pm(-x^2 + 2\sqrt{1-t}xy + ty^2),\\
R_1(x,y) &= \pm(x^2 - 3xy + y^2),\\
R_2(x,y) &= \pm(x^2 - xy + y^2),\\
R_3(x,y) &= \pm x^2,\\
R_4(x,y) &= \pm y^2,
\end{aligned}$$

for $t \in [0,1]$.

1.4 Supremum norms for 2-homogeneous polynomials on circle sectors

Our next aim is to investigate the geometry of the space of 2-homogeneous polynomials on a sector of the unit circle. For each $\alpha, \beta \in [0, 2\pi]$ with $\alpha \leq \beta$ we define the sector $D(\alpha,\beta)$ as

$$D(\alpha,\beta) := \{re^{i\theta} : 0 \leq r \leq 1,\ \alpha \leq \theta \leq \beta\}.$$

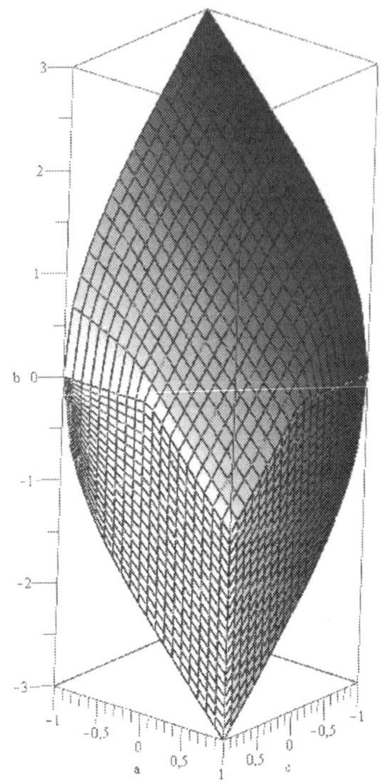

FIGURE 1.4: The unit sphere $\mathsf{B}_{\square}$.

When $\alpha = 0$, we denote $D(0,\beta)$ by $D(\beta)$ for simplicity. If P is a 2-homogeneous polynomial on $\mathbb{R}^2$, we define its norm $\|P\|_{D(\alpha,\beta)}$ by

$$\|P\|_{D(\alpha,\beta)} := \sup\{|P(\mathbf{x})| : \mathbf{x} \in D(\alpha,\beta)\}.$$

The space of 2-homogeneous polynomials on $\mathbb{R}^2$ endowed with the above norm will be represented by $\mathcal{P}(^2D(\alpha,\beta))$. The unit ball and the unit sphere of $\mathcal{P}(^2D(\alpha,\beta))$ will be denoted by $\mathsf{B}_{D(\alpha,\beta)}$ and $\mathsf{S}_{D(\alpha,\beta)}$, respectively, whereas the extreme points of $\mathsf{B}_{D(\alpha,\beta)}$ will be denoted by $\operatorname{ext}(\mathsf{B}_{D(\alpha,\beta)})$. As before, in order to visualize $\mathsf{S}_{D(\alpha,\beta)}$ and $\operatorname{ext}(\mathsf{B}_{D(\alpha,\beta)})$, we will often represent the 2-homogeneous polynomial $P(x,y) = ax^2 + by^2 + cxy$ by $(a,b,c) \in \mathbb{R}^3$, which allows us to identify $\mathcal{P}(^2D(\alpha,\beta))$ with $(\mathbb{R}^3, \|\cdot\|_{D(\alpha,\beta)})$, where $\|(a,b,c)\|_{D(\alpha,\beta)} = \|P\|_{D(\alpha,\beta)}$.

It is worth noting that the unit ball of $\mathcal{P}(^2D(\alpha,\alpha+\beta))$ can be inferred from the unit ball of $\mathcal{P}(^2D(0,\beta))$. Indeed, let

$$\Phi(x,y) = (x\cos\alpha - y\sin\alpha, x\sin\alpha + y\cos\alpha),$$

for all $(x,y) \in \mathbb{R}^2$, i.e., Φ is the rotation of angle α in $\mathbb{R}^2$ with centre at the origin. Notice that Φ transforms $D(0,\beta)$ onto $D(\alpha,\alpha+\beta)$, and hence

$$\|P\|_{D(\alpha,\alpha+\beta)} = \|P \circ \Phi\|_{D(0,\beta)},$$

for all 2-homogeneous polynomial P on $\mathbb{R}^2$. In other words, the mapping

$$\begin{aligned} T : \mathcal{P}(^2D(\alpha,\alpha+\beta)) &\to \mathcal{P}(^2D(\beta)) \\ P &\mapsto P \circ \Phi \end{aligned}$$

defines a linear isometry between $\mathcal{P}(^2D(\alpha, \alpha+\beta))$ and $\mathcal{P}(^2D(\beta))$. It is straightforward to check that T is represented by the matrix

$$\begin{pmatrix} \cos^2\alpha & \sin^2\alpha & \frac{\sin 2\alpha}{2} \\ \sin^2\alpha & \cos^2\alpha & -\frac{\sin 2\alpha}{2} \\ -\sin 2\alpha & \sin 2\alpha & \cos 2\alpha \end{pmatrix}.$$

This isometry enables us to focus our attention on the study of the geometry of $\mathsf{B}_{D(\beta)}$.

The case where $\beta \geq \pi$ is particularly easy to analyze and will be addressed first in Section 1.4.1. When studying the case $0 < \beta < \pi$, one soon realizes that the four cases $0 < \beta \leq \frac{\pi}{4}$, $\frac{\pi}{4} < \beta \leq \frac{\pi}{2}$, $\frac{\pi}{2} < \beta \leq \frac{3\pi}{4}$ and $\frac{3\pi}{4} < \beta < \pi$ must be treated separately. We first explicitly study the limiting cases $\beta = \frac{\pi}{4}, \frac{\pi}{2}, \frac{3\pi}{4}$ due to the high complexity involved in treating the general case in depth. We then address the remaining cases in order to, finally, cover all possibilities for the angle.

1.4.1 The case $\beta \geq \pi$

Observe that for any homogeneous polynomial P on $\mathbb{R}^2$ we have that $|P(-x)| = |P(x)|$ for all $x \in \mathbb{R}^2$. A moment's thought reveals that in order to compute the supremum of $|P|$ over the unit disk $\{(x,y) \in \mathbb{R}^2 : x^2+y^2 \leq 1\}$ we just need to evaluate $|P|$ over one half (any) of the unit disk. Therefore, if we now let $\beta \geq \pi$ then $\mathsf{B}_{D(\beta)} = \mathsf{B}_{\mathcal{P}(^2\ell_2^2)}$, where $\mathsf{B}_{\mathcal{P}(^2\ell_2^2)}$ stands for the closed unit ball of the space $\mathcal{P}(^2\ell_2^2)$ of 2-homogeneous polynomials on $\mathbb{R}^2$ endowed with the sup norm over the unit disk. The extreme points of $\mathsf{B}_{\mathcal{P}(^2\ell_2^2)}$ were described by Choi and Kim in [20]. We provide below an alternative approach:

Let us consider a 2-homogeneous polynomial $P(x,y) = ax^2 + by^2 + cxy$ on $\mathbb{R}^2$. By homogeneity, the supremum of $|P|$ over the closed unit disk is attained on the unit circle, and when restricted to the unit circle,

$$\begin{aligned} f(\theta) := P(\cos\theta, \sin\theta) &= a\cos^2\theta + b\sin^2\theta + c\sin\theta\cos\theta \\ &= a\frac{1+\cos 2\theta}{2} + b\frac{1-\cos 2\theta}{2} + c\frac{\sin 2\theta}{2} \\ &= \frac{1}{2}\left[a+b+(a-b)\cos 2\theta + c\sin 2\theta\right], \end{aligned}$$

for $\theta \in [0, 2\pi]$. Hence, using an elementary duality argument

$$\begin{aligned} \|P\| = \sup_{\theta\in[0,2\pi]} |f(\theta)| &= \frac{1}{2}\left(|a+b| + \|(a-b,c)\|_2\right) \\ &= \frac{1}{2}\left(|a+b| + \sqrt{(a-b)^2+c^2}\right), \end{aligned}$$

where $\|\cdot\|_2$ stands for the Euclidean norm. The latter shows that the unit sphere of $\mathcal{P}(^2\ell_2^2)$ is given by

$$\mathsf{S}_{\mathcal{P}(^2\ell_2^2)} = \{(a,b,c) \in \mathbb{R}^3 : |a+b| + \sqrt{(a-b)^2+c^2} = 2\}.$$

Therefore, if we define

$$f(a,b) = 2\sqrt{1+ab-|a+b|},$$

for all $(a,b) \in [-1,1]^2$, then

$$\mathsf{S}_{\mathcal{P}(^2\ell_2^2)} = \mathrm{graph}(f) \cup \mathrm{graph}(-f).$$

The reader can find in Figure 1.5 a graph of $\mathsf{S}_{\mathcal{P}(^2\ell_2^2)}$.

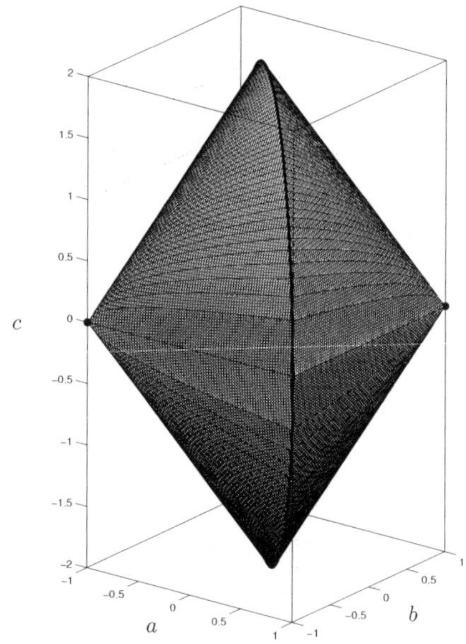

FIGURE 1.5: $\mathsf{S}_{\mathcal{P}(^2\ell_2^2)}$. The extreme points of $\mathsf{B}_{\mathcal{P}(^2\ell_2^2)}$ are drawn with a thicker line and dots

Now let $a \in [-1,1]$ and consider the segment $\{(-1+\lambda(a+1), -1+\lambda(-a+1)) : \lambda \in [0,1]\}$ joining $(-1,-1)$ with the point $(a,-a)$. Then f (and therefore $-f$ too), restricted to that segment, namely

$$g(\lambda) := f(-1+\lambda(a+1), -1+\lambda(-a+1)) = 2\sqrt{1-a^2}\lambda,$$

with $\lambda \in [0,1]$, is linear. Similarly, f (and therefore $-f$ too) restricted to segments joining $(1,1)$ with points of the diagonal $b=-a$ of $[-1,1]^2$ are also linear. In other words, we have seen that graph(f) (and therefore graph$(-f)$ too) is the union of two ruled surfaces. This means that the extreme points of the closed unit ball of $\mathcal{P}(^2\ell_2^2)$, i.e., $\text{ext}(\mathsf{B}_{\mathcal{P}(^2\ell_2^2)})$, must be contained in the set

$$\{(a,-a,\pm f(a,-a)) : a \in [-1,1]\} \cup \{\pm(1,1,0)\}.$$

It is easy to construct supporting planes to each of those points, revealing that they are indeed extreme. We conclude therefore that

$$\begin{aligned}\text{ext}(\mathsf{B}_{\mathcal{P}(^2\ell_2^2)}) &= \{(a,-a,\pm\sqrt{1-a^2}) : a \in [-1,1]\} \cup \{\pm(1,1,0)\} \\ &= \{\pm(a,-a,\sqrt{1-a^2}) : a \in [-1,1]\} \cup \{\pm(1,1,0)\}.\end{aligned}$$

1.4.2 The supremum norm on $D(\beta)$ for $\beta = \frac{\pi}{4}, \frac{\pi}{2}, \frac{3\pi}{4}$.

We obtain first a formula for $\|ax^2+by^2+cxy\|_{D(\beta)}$ where $\beta = \frac{\pi}{4}, \frac{\pi}{2}, \frac{3\pi}{4}$ and $a,b,c \in \mathbb{R}$.

Theorem 1.4.1 (Muñoz, Pellegrino, Seoane, and Weber [81]**)** *If* $a,b,c \in \mathbb{R}$ *and*

$P(x,y)=ax^2+by^2+cxy$ then $\|P\|_{D(\frac{\pi}{4})}$, $\|P\|_{D(\frac{\pi}{2})}$, and $\|P\|_{D(\frac{3\pi}{4})}$ are given, respectively, by

$$\begin{cases}\max\left\{|a|,\frac{1}{2}|a+b+c|,\frac{1}{2}|a+b+\operatorname{sign}(c)\sqrt{(a-b)^2+c^2}|\right\} & \text{if } c(a-b)\geq 0,\\ \max\{|a|,\frac{1}{2}|a+b+c|\} & \text{if } c(a-b)\leq 0,\end{cases}$$

$$\max\left\{|a|,|b|,\frac{1}{2}|a+b+\operatorname{sign}(c)\sqrt{(a-b)^2+c^2}|\right\},\qquad \text{and}$$

$$\begin{cases}\frac{1}{2}\left(|a+b|+\sqrt{(a-b)^2+c^2}\right) & \text{if } c(a-b)\geq 0,\\ \max\left\{|a|,\frac{1}{2}|a+b-c|,\frac{1}{2}|a+b+\operatorname{sign}(c)\sqrt{(a-b)^2+c^2}|\right\} & \text{if } c(a-b)\leq 0.\end{cases}$$

PROOF. *The supremum of $|P|$ over $D(\beta)$ is obviously attained on the set $\{(\cos\theta,\sin\theta):0\leq\theta\leq\beta\}$. When restricted to this set, as shown previously, $P(\cos\theta,\sin\theta)$ is given by*

$$f(\theta)=\frac{1}{2}\left[a+b+(a-b)\cos 2\theta+c\sin 2\theta\right],$$

for $\theta\in[0,\beta]$. Let

$$g(\theta)=\frac{1}{2}\left[a+b+(a-b)\cos\theta+c\sin\theta\right],$$

for $\theta\in[0,2\beta]$. Then it follows that $\|P\|_{D(\beta)}=\sup_{\theta\in[0,2\beta]}|g(\theta)|$. Observe that the two lines appearing in the first and the third formulas coincide when $c(a-b)=0$. This simpler case is left to the reader. Now, if $c(a-b)\neq 0$, then g' vanishes at the points θ_0 such that $\tan\theta_0=\frac{c}{a-b}$.

Assume first that $\beta=\frac{\pi}{4}$. Then the equation $\tan\theta=\frac{c}{a-b}$ has exactly one root $\theta_0\in[0,\frac{\pi}{2}]$ if and only if $\frac{c}{a-b}>0$, in which case

$$\cos\theta_0=\frac{1}{\sqrt{1+\frac{c^2}{(a-b)^2}}}\qquad\text{and}\qquad \sin\theta_0=\frac{\frac{c}{a-b}}{\sqrt{1+\frac{c^2}{(a-b)^2}}}.$$

Hence

$$\begin{aligned}g(\theta_0)&=\frac{1}{2}\left[a+b+(a-b)\frac{1}{\sqrt{1+\frac{c^2}{(a-b)^2}}}+c\frac{\frac{c}{a-b}}{\sqrt{1+\frac{c^2}{(a-b)^2}}}\right]\\ &=\frac{1}{2}\left[a+b+\frac{(a-b)|a-b|}{\sqrt{c^2+(a-b)^2}}+\frac{c^2\operatorname{sign}(a-b)}{\sqrt{c^2+(a-b)^2}}\right]\\ &=\frac{1}{2}\left[a+b+\operatorname{sign}(a-b)\frac{(a-b)^2+c^2}{\sqrt{c^2+(a-b)^2}}\right]\\ &=\frac{1}{2}\left[a+b+\operatorname{sign}(c)\sqrt{c^2+(a-b)^2}\right].\end{aligned}$$

This, together with the fact that $g(0)=a$ and $g(\frac{\pi}{2})=\frac{1}{2}(a+b+c)$, establishes the first formula.

Now, suppose that $\beta=\frac{\pi}{2}$. Then $\tan\theta=\frac{c}{a-b}$ has exactly one solution $\theta_0\in[0,\pi]$. If $c(a-b)>0$ then $\theta_0\in[0,\frac{\pi}{2}]$ and we have already seen that $g(\theta_0)=\frac{1}{2}\left[a+b+\operatorname{sign}(c)\sqrt{c^2+(a-b)^2}\right]$. If $c(a-b)<0$ then $\theta_0\in[\frac{\pi}{2},\pi]$ and

$$\cos\theta_0=-\frac{1}{\sqrt{1+\frac{c^2}{(a-b)^2}}}\qquad\text{and}\qquad \sin\theta_0=-\frac{\frac{c}{a-b}}{\sqrt{1+\frac{c^2}{(a-b)^2}}}.$$

The reader can check, as above, that in this case, we obtain again

$$g(\theta_0) = \frac{1}{2}\left[a + b + \operatorname{sign}(c)\sqrt{c^2 + (a-b)^2}\right].$$

This, together with the fact that $g(0) = a$ *and* $g(\pi) = b$*, proves the second formula.*

Finally, assume that $\beta = \frac{3\pi}{4}$*. If* $c(a-b) > 0$ *then the equation* $\tan\theta = \frac{c}{a-b}$ *has exactly two roots in* $[0, \frac{3\pi}{2}]$*, one* θ_1 *in* $[0, \frac{\pi}{2}]$ *and the other one* θ_2 *in* $[\pi, \frac{3\pi}{2}]$*. Notice that* $\theta_2 - \theta_1 = \pi$*, from which*

$$\begin{aligned}\max\{|g(\theta_1)|, |g(\theta_2)|\} &= \max\left\{\frac{1}{2}\left|a + b + \epsilon \operatorname{sign}(c)\sqrt{c^2 + (a-b)^2}\right| : \epsilon = \pm 1\right\}\\ &= \frac{1}{2}\left(|a+b| + \sqrt{c^2 + (a-b)^2}\right).\end{aligned}$$

The value $\frac{1}{2}\left(|a+b| + \sqrt{c^2 + (a-b)^2}\right)$ *coincides with* $\|P\|_{D(\pi)}$*, as we observed in Section 1.4.1, and therefore* $\|P\|_{D(\frac{3\pi}{4})} = \frac{1}{2}\left(|a+b| + \sqrt{c^2 + (a-b)^2}\right)$ *if* $c(a-b) > 0$*. On the other hand, if* $c(a-b) < 0$*, then* $\tan\theta = \frac{c}{a-b}$ *has exactly one solution* $\theta_0 \in [0, \frac{3\pi}{2}]$*. In fact* $\theta_0 \in [\frac{\pi}{2}, \pi]$*. Following the same computation as in the proof of the second formula,*

$$g(\theta_0) = \frac{1}{2}\left[a + b + \operatorname{sign}(c)\sqrt{c^2 + (a-b)^2}\right].$$

Considering that $g(0) = a$ *and* $g(\frac{3\pi}{2}) = \frac{1}{2}(a+b-c)$*, the proof of the third formula is completed.*

1.4.3 The geometry of $\mathsf{B}_{D(\frac{\pi}{4})}$

The study of this case requires the following technical result:

Lemma 1.4.2 *Let us define* P_1 *and* P_2 *by*

$$\begin{aligned}P_1 &:= \{(a,b) \in \mathbb{R}^2 : \ a \geq -1 \text{ and } 4 + a - 4\sqrt{1+a} \leq b \leq 4 + a + 4\sqrt{1+a}\},\\ P_2 &:= \{(a,b) \in \mathbb{R}^2 : \ a \leq 1 \text{ and } -4 + a - 4\sqrt{1-a} \leq b \leq -4 + a + 4\sqrt{1-a}\}.\end{aligned}$$

Then:

(a) The inequality

$$\frac{1}{2}\left|a + b - \sqrt{2\left[(a-1)^2 + (b-1)^2\right]}\right| \leq 1, \tag{1.4}$$

holds if and only if $(a,b) \in P_1$*.*

(b) The inequality

$$\frac{1}{2}\left|a + b + 2\sqrt{(1-a)(1-b)}\right| \leq 1, \tag{1.5}$$

holds if and only if $(a,b) \in P_2$*.*

PROOF. *In order to prove (a) let* $h(a,b) = \frac{1}{2}\left|a + b - \sqrt{2\left[(a-1)^2 + (b-1)^2\right]}\right|$ *for* $(a,b) \in \mathbb{R}^2$*. Then* $h(a,b) = 1$ *holds either if*

$$a + b - \sqrt{2\left[(a-1)^2 + (b-1)^2\right]} = 2, \tag{1.6}$$

or

$$a+b-\sqrt{2\left[(a-1)^2+(b-1)^2\right]}=-2. \tag{1.7}$$

We have that (1.6) *can only occur if* $a+b-2\geq 0$, *in which case the reader can show easily by squaring the equality* $a+b-2=\sqrt{2\left[(a-1)^2+(b-1)^2\right]}$ *that* $a=b$. *This shows that* $h\equiv 1$ *on the set* $R:=\{(a,a):a\geq 1\}$. *On the other hand,* (1.7) *occurs only if* $a+b+2\geq 0$, *and in that case, by squaring the equality* $a+b+2=\sqrt{2\left[(a-1)^2+(b-1)^2\right]}$ *we obtain easily that* $b=4+a\pm\sqrt{1+a}$ *with* $a\geq -1$. *Observe that the curve described by the equation* $b=4+a\pm\sqrt{1+a}$ *with* $a\geq -1$ *is in the half-plane* $a+b+2\geq 0$. *Hence we have proved so far that the set of zeros of* $h-1$ *is*

$$Z:=R\cup\{(a,b):a\geq -1,\ b=4+a\pm\sqrt{1+a}\}.$$

Now let

$$C_1:=\left\{(a,b):\ a\geq -1,\, b>4+a+4\sqrt{1+a},\, b<4+a-4\sqrt{1+a}\right\}$$

and

$$C_2:=\left\{(a,b):\ a\geq -1,\, 4+a-4\sqrt{1+a}<b<4+a+4\sqrt{1+a}\right\}\setminus R$$

be the two connected components of $\mathbb{R}^2\setminus Z$. *By continuity,* $h-1$ *preserves its sign on both* C_1 *and* C_2. *Since it is straightforward to check that* $h-1$ *is positive on* C_1 *and negative on* C_2, *and* $P_1=C_2\cup Z$, *the result follows.*

One can proceed in exactly the same manner to prove (b), which is spared to the interested reader.

Theorem 1.4.3 *Let* A *and* B *be as in Figure 1.6, namely*

$$\begin{aligned} A&=\{(a,b):\ a\in[-1,1],\ a<b\leq\gamma_1(a)\},\\ B&=\{(a,b):\ a\in[-1,1],\ \gamma_2(a)\leq b\leq a\}, \end{aligned}$$

where γ_1, γ_2 *are defined by*

$$\begin{aligned} \gamma_1(a)&=4+a+4\sqrt{1+a},\\ \gamma_2(a)&=-\gamma_1(-a)=-4+a-4\sqrt{1-a}, \end{aligned}$$

for $a\in[-1,1]$. *Then, the projection of* $\mathsf{S}_{D(\frac{\pi}{4})}$ *over the ab-plane is*

$$\pi_{ab}(\mathsf{S}_{D(\frac{\pi}{4})})=\{(a,b):\ a\in[-1,1],\ \gamma_2(a)\leq b\leq\gamma_1(a)\}.$$

PROOF. *Let* $(a,b)\in A$ *and set* $c=2-a-b$. *First, assume that* $2-a-b>0$, *so that* $c(a-b)<0$. *Therefore, according to the first formula in Theorem 1.4.1, we have*

$$\|(a,b,2-a-b)\|_{D(\frac{\pi}{4})}=\max\left\{|a|,\frac{1}{2}|a+b+c|\right\}=\max\{|a|,1\}=1.$$

On the other hand, if $2-a-b\leq 0$, *then* $c(a-b)\geq 0$. *Applying the first formula in Theorem 1.4.1 again we obtain*

$$\begin{aligned} \|(a,b,2-a-b)\|_{D(\frac{\pi}{4})}&=\max\left\{|a|,1,\frac{1}{2}\left|a+b-\sqrt{(a-b)^2+(2-a-b)^2}\right|\right\}\\ &=\max\left\{|a|,1,\frac{1}{2}\left|a+b-\sqrt{2[(a-1)^2+(b-1)^2]}\right|\right\}. \end{aligned}$$

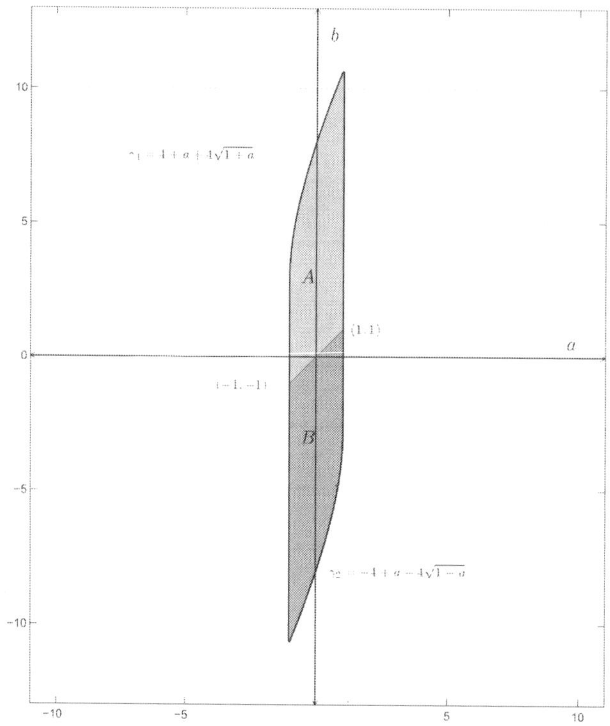

FIGURE 1.6: Projection of $\mathsf{S}_{D(\frac{\pi}{4})}$ onto the ab-plane

Recall that here (a, b) *satisfies* $|a| \le 1$ *and* $2-a \le b \le 4+a+4\sqrt{1+a}$. *Since* $4+a-4\sqrt{1+a} \le 2-a$ *for* $-1 \le a \le 1$, *it follows from Lemma 1.4.2, part (a), that*

$$\frac{1}{2}\left|a + b - \sqrt{2[(a-1)^2 + (b-1)^2]}\right| \le 1,$$

proving that $\|(a, b, 2-a-b)\|_{D(\frac{\pi}{4})} = 1$.

Next, assume that $(a, b) \in B$, *where* B *is defined in Theorem 1.4.5 (see also Figure 1.6), and define* $c = 2\sqrt{(1-a)(1-b)}$. *Notice that if* $(a, b) \in B$ *then* $2-a-b \ge 0$, *from which*

$$\begin{aligned}\frac{1}{2}\left|a + b + \operatorname{sign}(c)\sqrt{(a-b)^2 + c^2}\right| &= \frac{1}{2}\left|a + b + \sqrt{(a-b)^2 + 4(1-a)(1-b)}\right| \\ &= \frac{1}{2}\left|a + b + \sqrt{(2-a-b)^2}\right| = 1.\end{aligned}$$

In this case $c(a-b) \ge 0$, *so using the first formula in Theorem 1.4.1 once again, we arrive at*

$$\left\|\left(a, b, 2\sqrt{(1-a)(1-b)}\right)\right\|_{D(\frac{\pi}{4})} = \max\left\{|a|, \frac{1}{2}\left|a + b + 2\sqrt{(1-a)(1-b)}\right|, 1\right\}.$$

However, under the given conditions, Lemma 1.4.2, part (b) asserts that

$$\frac{1}{2}\left|a + b + 2\sqrt{(1-a)(1-b)}\right| \le 1,$$

which implies that $\left\|\left(a, b, 2\sqrt{(1-a)(1-b)}\right)\right\|_{D(\frac{\pi}{4})} = 1.$

So far, we have shown that

$$\pi_{ab}(\mathsf{S}_{D(\frac{\pi}{4})}) \subset \{(a,b) : \ a \in [-1,1], \ \gamma_2(a) \le b \le \gamma_1(a)\}.$$

On the other hand, if $(a,b) \notin \{(a,b) : \ a \in [-1,1], \ \gamma_2(a) \le b \le \gamma_1(a)\}$ *then either* $|a| > 1$ *or one of the inequalities* $\gamma_2(a) \le b \le \gamma_1(a)$ *fail with* $-1 \le a \le 1$. *In the first case, we would have* $\|(a,b,c)\|_{D(\frac{\pi}{4})} \ge |a| > 1$ *for all* $c \in \mathbb{R}$, *i.e.,* $(a,b) \notin \pi_{ab}(\mathsf{S}_{D(\frac{\pi}{4})})$. *Now if* $|a| \le 1$ *and* $\gamma_1(a) < b$, *then* $a - b < 0$. *If* $c \le 0$, *from the first formula in Theorem 1.4.1 we would have*

$$\|(a,b,c)\|_{D(\frac{\pi}{4})} \ge \max\left\{\frac{1}{2}|a+b+c|, \frac{1}{2}\left|a+b-\sqrt{(a-b)^2+c^2}\right|\right\}.$$

It can be easily checked that $\frac{1}{2}|a+b+c| > 1$ *if* $0 \ge c > 2-a-b$ *and* $\frac{1}{2}\left|a+b-\sqrt{(a-b)^2+c^2}\right| > 1$ *if* $c \le 2-a-b$. *This last statement follows from the fact that* $\frac{1}{2}\left|a+b-\sqrt{(a-b)^2+c^2}\right| > 1$ *is equivalent to the expression* $c^2 > 4(1+a)(1+b)$ *and, in this case,* $c^2 \ge (2-a-b)^2 > 4(1+a)(1+b)$.

Now if $c > 0$, *Theorem 1.4.1 allows us to show that*

$$\|(a,b,c)\|_{D(\frac{\pi}{4})} \ge \frac{1}{2}|a+b+c| = \frac{1}{2}(a+b+c) > 2+a+2\sqrt{1+a} \ge 1,$$

whenever $-1 \le a \le 1$. *In either case,* $(a,b) \notin \pi_{ab}(\mathsf{S}_{D(\frac{\pi}{4})})$.

Finally, using the symmetry of the unit ball and the previous case, if $\gamma_2(a) > b$ *with* $-1 \le a \le 1$ *then* $\|(a,b,c)\|_{D(\frac{\pi}{2})} > 1$ *for all* $c \in \mathbb{R}$, *from which* $(a,b) \notin \pi_{ab}(\mathsf{S}_{D(\frac{\pi}{4})})$. *This concludes the proof.*

Remark 1.4.4 *In the proof of the first part of the following result, it will be useful to observe that*

$$\frac{1}{2}|b+c-1| \le 1 \Leftrightarrow -1-b \le c \le 3-b.$$

Also, if $b \le -1$, *then*

$$\frac{1}{2}\left|b-1+\sqrt{(b+1)^2+c^2}\right| \le 1 \Leftrightarrow |c| \le 3-b.$$

Theorem 1.4.5 (Muñoz, Pellegrino, Seoane, and Weber [81]**)** *Let* A *and* B *be as in Theorem 1.4.3 (see also Figure 1.6) and define*

$$F_1(a,b) = \begin{cases} 2-a-b & \text{if } (a,b) \in A, \\ 2\sqrt{(1-a)(1-b)} & \text{if } (a,b) \in B, \end{cases}$$

and $F_2(a,b) = -F_1(-a,-b)$ *for all* $(a,b) \in \pi_{ab}(\mathsf{S}_{D(\frac{\pi}{4})})$. *If*

$$\Gamma = \{(\pm 1, b, c) \in \mathbb{R}^2 : \ (\pm 1, b) \in \partial\pi_{ab}(\mathsf{S}_{D(\frac{\pi}{4})}), \ F_2(\pm 1, b) \le c \le F_1(\pm 1, b)\},$$

then

(a) $\mathsf{S}_{D(\frac{\pi}{4})} = \operatorname{graph}(F_1) \cup \operatorname{graph}(F_2) \cup \Gamma$.

(b) *The set* $\operatorname{ext}(\mathsf{B}_{D(\frac{\pi}{4})})$ *consists of the elements*

$$\pm\left(t, 4+t+4\sqrt{1+t}, -2-2t-4\sqrt{1+t}\right) \quad \text{for} \quad t \in [-1,1],$$

$$\pm\left(1, s, -2\sqrt{2(1+s)}\right) \quad \text{for} \quad s \in [1, 5+4\sqrt{2}],$$

and

$$\pm(1,1,0).$$

PROOF. *As for the first part of the theorem, notice that* $\mathrm{graph}(F_1) \subset \mathsf{S}_{D(\frac{\pi}{4})}$ *as seen in the proof of Theorem 1.4.3. By symmetry we also have that* $\mathrm{graph}(F_2) \subset \mathsf{S}_{D(\frac{\pi}{4})}$. *Finally* $\Gamma \subset \mathsf{S}_{D(\frac{\pi}{4})}$ *too. Indeed, by symmetry we can focus on the study of the points of* Γ *of the form* $(-1, b, c)$ *with* $-5 - 4\sqrt{2} \leq b \leq 3$ *and* $F_2(-1, b) \leq c \leq F_1(-1, b)$. *Observe that*

$$F_1(-1, b) = \begin{cases} 2\sqrt{2(1-b)} & \text{if } -5 - 4\sqrt{2} \leq b \leq -1, \\ 3 - b & \text{if } -1 \leq b \leq 3, \end{cases}$$

and

$$F_2(-1, b) = \begin{cases} -1 - b & \text{if } -5 - 4\sqrt{2} \leq b \leq -1, \\ 0 & \text{if } -1 \leq b \leq 3. \end{cases}$$

Hence $c \geq F_2(-1, b) \geq 0$ *for all* $b \in [-5 - 4\sqrt{2}, 3]$ *and* $c = 0$ *only holds when* $-3 \leq b \leq 1$. *In that case, since* $\frac{1}{2}|b - 1| \leq 1$, *it follows from Theorem 1.4.1 that*

$$\|(-1, b, 0)\|_{D(\frac{\pi}{4})} = \max\left\{1, \frac{1}{2}|b - 1|\right\} = 1.$$

Otherwise $c > 0$. *First, if* $-5 - 4\sqrt{2} \leq b \leq -1$ *then* $c(-1 - b) \geq 0$. *Therefore, using Theorem 1.4.1 we have*

$$\|(-1, b, c)\|_{D(\frac{\pi}{4})} = \max\left\{1, \frac{1}{2}|b + c - 1|, \frac{1}{2}|b - 1 + \sqrt{(b+1)^2 + c^2}|\right\}.$$

Since $-5 - 4\sqrt{2} \leq b \leq -1$ *and* $-3 - b \leq c \leq 2\sqrt{2(1-b)} \leq 3 - b$, *from Remark 1.4.4 it follows that* $\|(-1, b, c)\|_{D(\frac{\pi}{4})} = 1$. *Finally, if* $-1 < b \leq 3$ *then* $c(-1 - b) < 0$, *which implies, from Theorem 1.4.1 that*

$$\|(-1, b, c)\|_{D(\frac{\pi}{4})} = \max\left\{1, \frac{1}{2}|b + c - 1|\right\}.$$

Since now $-1 \leq b \leq 3$ *and* $-3 - b \leq 0 < c \leq -1 - b$, *from Remark 1.4.4 we have that* $\|(-1, b, c)\|_{D(\frac{\pi}{4})} = 1$ *too.*

We have proved so far that $\mathrm{graph}(F_1) \cup \mathrm{graph}(F_2) \cup \Gamma \subset \mathsf{S}_{D(\frac{\pi}{4})}$. *On the other hand, suppose* $(a, b, c) \notin \mathrm{graph}(F_1) \cup \mathrm{graph}(F_2) \cup \Gamma$. *Obviously,* $(0, 0, 0) \notin \mathsf{S}_{D(\frac{\pi}{4})}$, *so we may assume that* $(a, b, c) \neq (0, 0, 0)$. *The straight line* $\{\lambda(a, b, c) : \lambda \in \mathbb{R}\}$ *certainly meets the set* $\mathrm{graph}(F_1) \cup \mathrm{graph}(F_2) \cup \Gamma$ *as the reader may easily check. Put,* $(a, b, c) = \lambda_0(a_0, b_0, c_0)$ *with* $\lambda_0 \neq 0, 1$ *and* $(a_0, b_0, c_0) \in \mathrm{graph}(F_1) \cup \mathrm{graph}(F_2) \cup \Gamma$. *Then* $\|(a, b, c)\|_{D(\frac{\pi}{4})} = \lambda_0 \|(a_0, b_0, c_0)\|_{D(\frac{\pi}{4})} = \lambda_0 \neq 1$, *and therefore* $(a, b, c) \notin \mathsf{S}_{(\frac{\pi}{4})}$. *This concludes part (a).*

Part (b) is an easy consequence of the fact that the sets $\mathrm{graph}(F_1)$, $\mathrm{graph}(F_2)$, *and* Γ *are ruled surfaces in* $\mathbb{R}^3$. *Actually* Γ *is contained in the two planes* $a = \pm 1$. *Hence the only possible extreme points are contained in the intersections* $\mathrm{graph}(F_1) \cap \Gamma$, $\mathrm{graph}(F_2) \cap \Gamma$, *and* $\mathrm{graph}(F_1) \cap \mathrm{graph}(F_2)$. *Once we have removed the interior of all straight lines contained in those intersections, we end up with the following candidates to be extreme points of* $\mathsf{B}_{D(\frac{\pi}{4})}$:

$$\pm\left(t, 4 + t + 4\sqrt{1+t}, -2 - 2t - 4\sqrt{1+t}\right) \quad \text{for} \quad t \in [-1, 1],$$

$$\pm\left(1, s, -2\sqrt{2(1+s)}\right) \quad \text{for} \quad s \in [1, 5 + 4\sqrt{2}],$$

and

$$\pm(1, 1, 0).$$

Constructing a supporting hyperplane to each of the previous points is left as a straightforward exercise for the interested reader, which finishes the proof.

The reader can find a sketch of $\mathsf{S}_{D(\frac{\pi}{4})}$ in Figure 1.7.

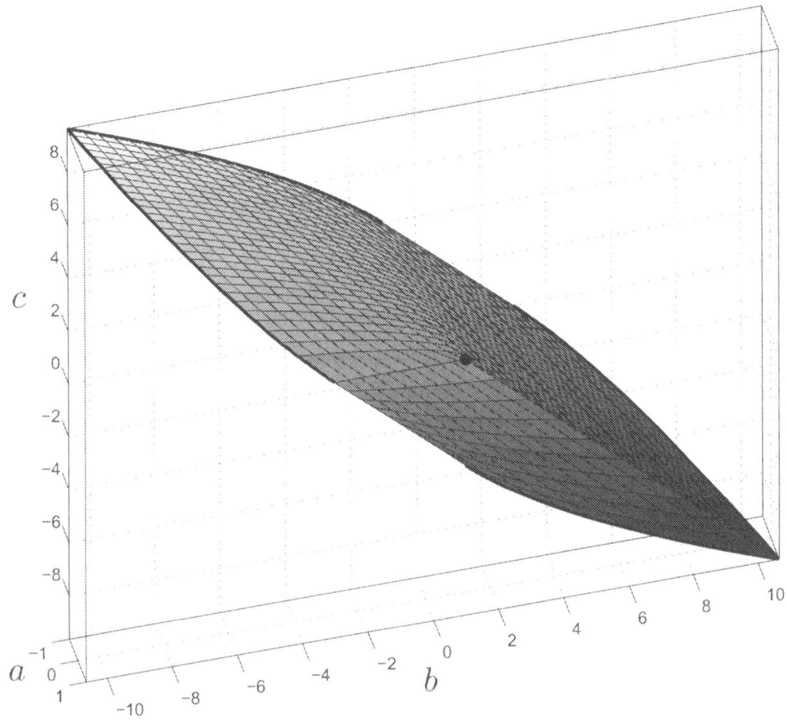

FIGURE 1.7: $\mathsf{S}_{D(\frac{\pi}{4})}$. The extreme points of $\mathsf{B}_{D(\frac{\pi}{4})}$ are drawn with a thicker line and dots

1.4.4 The geometry of $\mathsf{B}_{D(\frac{\pi}{2})}$

Theorem 1.4.6 (Muñoz, Pellegrino, Seoane, and Weber [81]**)** *The projection of* $\mathsf{S}_{D(\frac{\pi}{2})}$ *over the ab-plane is* $[-1,1]^2$.

PROOF. *Using the second formula in Theorem 1.4.1 we have that*

$$\|(a,b,c)\|_{D(\frac{\pi}{2})} \geq \max\{|a|,|b|\},$$

for all $(a,b,c) \in \mathbb{R}^3$. *This shows right away that* $\pi_{ab}(\mathsf{S}_{D(\frac{\pi}{2})}) \subset [-1,1]^2$. *In order to prove that* $[-1,1]^2 \subset \pi_{ab}(\mathsf{S}_{D(\frac{\pi}{2})})$ *take* $(a,b) \in \mathbb{R}^2$ *so that* $|a| \leq 1$ *and* $|b| \leq 1$. *Then, using again the same formula, one shows easily that*

$$\begin{aligned}
&\left\|\left(a,b,2\sqrt{(1-a)(1-b)}\right)\right\|_{D(\frac{\pi}{2})} \\
&\quad = \max\left\{|a|,|b|,\frac{1}{2}\left|a+b+\sqrt{(a-b)^2+4(1-a)(1-b)}\right|\right\} \\
&\quad = \max\left\{|a|,|b|,\frac{1}{2}\left|a+b+\sqrt{(2-a-b)^2}\right|\right\} \\
&\quad = \max\{|a|,|b|,1\} = 1.
\end{aligned}$$

Observe that, similarly,

$$\left\|\left(a,b,-2\sqrt{(1+a)(1+b)}\right)\right\|_{D(\frac{\pi}{2})} = 1.$$

This concludes the proof.

Theorem 1.4.7 (Muñoz, Pellegrino, Seoane, and Weber [81]) *If we define the mappings*

$$G_1(a,b) = 2\sqrt{(1-a)(1-b)}$$

and

$$G_2(a,b) = -f_+(-a,-b) = -2\sqrt{(1+a)(1+b)},$$

for every $(a,b) \in [-1,1]^2$ and the set

$$\Upsilon = \{(a,b,c) \in \mathbb{R}^3 : (a,b) \in \partial[-1,1]^2 \text{ and } G_2(a,b) \leq c \leq G_1(a,b)\},$$

where $\partial[-1,1]^2$ is the boundary of $[-1,1]^2$, then

(a) $\mathsf{S}_{D(\frac{\pi}{2})} = \text{graph}(G_1) \cup \text{graph}(G_2) \cup \Upsilon$.

(b) *The set $\text{ext}(\mathsf{B}_{D(\frac{\pi}{2})})$ consists of the elements*

$$\pm\left(1, t, -2\sqrt{2(1+t)}\right), \; \pm\left(t, 1, -2\sqrt{2(1+t)}\right), \quad \text{and} \quad \pm(1,1,0),$$

with $t \in [-1,1]$.

PROOF. *Regarding part (a), we have already seen in the proof of Theorem 1.4.6 that*

$$\text{graph}(G_1) \cup \text{graph}(G_2) \subset \mathsf{S}_{D(\frac{\pi}{2})}.$$

Also, if $(a,b,c) \in \Upsilon$, then $G_2(a,b) \leq c \leq G_1(a,b)$ with $\max\{|a|,|b|\} = 1$. Suppose $a = -1$. Then $0 \leq c \leq 2\sqrt{2(1-b)}$ with $|b| \leq 1$ and

$$\begin{aligned}\frac{1}{2}\left|a + b + \text{sign}(c)\sqrt{(a-b)^2 + c^2}\right| &\leq \frac{1}{2}\left|-1 + b + \sqrt{(1+b)^2 + 8(1-b)}\right| \\ &= \frac{1}{2}\left|-1 + b + \sqrt{(3-b)^2}\right| = 1.\end{aligned}$$

Hence using the second formula in Theorem 1.4.1 we see that $\|(-1,b,c)\|_{D(\frac{\pi}{2})} = 1$. It is left to the reader, as a simple exercise, to prove that the same holds if $a = 1$ or $|b| = 1$. Therefore,

$$\text{graph}(G_1) \cup \text{graph}(G_2) \cup \Upsilon \subset \mathsf{S}_{D(\frac{\pi}{2})}.$$

On the other hand, suppose $(a,b,c) \notin \text{graph}(G_1) \cup \text{graph}(G_2) \cup \Upsilon$. Obviously, $(0,0,0) \notin \mathsf{S}_{D(\frac{\pi}{2})}$, so we may assume that $(a,b,c) \neq (0,0,0)$. The straight line $\{\lambda(a,b,c) : \lambda \in \mathbb{R}\}$ certainly meets the set $\text{graph}(G_1) \cup \text{graph}(G_2) \cup \Upsilon$ as the reader can easily check. Put $(a,b,c) = \lambda_0(a_0,b_0,c_0)$ with $\lambda_0 \neq 0,1$ and $(a_0,b_0,c_0) \in \text{graph}(G_1) \cup \text{graph}(G_2) \cup \Upsilon$. Then $\|(a,b,c)\|_{D(\frac{\pi}{2})} = \lambda_0\|(a_0,b_0,c_0)\|_{D(\frac{\pi}{2})} = \lambda_0 \neq 1$, and therefore $(a,b,c) \notin \mathsf{S}_{D(\frac{\pi}{2})}$. This concludes part (a).

Part (b) is an easy consequence of the fact that the sets $\text{graph}(G_1)$, $\text{graph}(G_2)$, and Υ are ruled surfaces in $\mathbb{R}^3$. Actually Υ is contained in the four planes $a = \pm 1$ and $b = \pm 1$. Using this with the aid of Figure 1.8, the reader will quickly conclude that the only possible extreme points are contained in the intersections $\text{graph}(G_1) \cap \Upsilon$, $\text{graph}(G_2) \cap \Upsilon$ and $\text{graph}(G_1) \cap \text{graph}(G_2)$. The union of these three intersections consists of the elements

$$\pm\left(1, t, -2\sqrt{2(1+t)}\right), \; \pm\left(t, 1, -2\sqrt{2(1+t)}\right), \; (\pm 1, t, 0), \quad \text{and} \quad (t, \pm 1, 0),$$

with $t \in [-1,1]$. Observe that $(\pm 1, t, 0)$ and $(t, \pm 1, 0)$ with $|t| < 1$ cannot obviously be extreme and the point $\pm(1,-1,0)$ are already considered in the curves $\left(1, t, -2\sqrt{2(1+t)}\right)$,

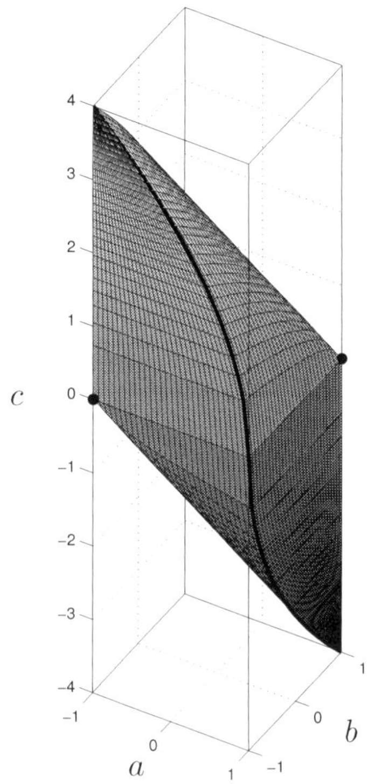

FIGURE 1.8: $\mathsf{S}_{D(\frac{\pi}{2})}$. The extreme points of $\mathsf{B}_{D(\frac{\pi}{2})}$ are drawn with a thicker line and dots.

$\pm\left(t, 1, -2\sqrt{2(1+t)}\right)$ *with* $t \in [-1, 1]$. *Hence the candidates to be extreme points of* $\mathsf{B}_{D(\frac{\pi}{2})}$ *reduce to*

$$\pm\left(1, t, -2\sqrt{2(1+t)}\right), \ \pm\left(t, 1, -2\sqrt{2(1+t)}\right), \quad \textit{and} \quad \pm(1, 1, 0),$$

with $t \in [-1, 1]$. *Constructing a supporting hyperplane to each of the previous points is a straightforward exercise that is left to the interested reader, which finishes the proof.*

A sketch of $\mathsf{S}_{D(\frac{\pi}{2})}$ is shown in Figure 1.8.

1.4.5 The geometry of $\mathsf{B}_{D(\frac{3\pi}{4})}$

Let C, D and F as in Figure 1.9, namely

$$\begin{aligned} C &:= \{(a,b) : \ a \in [-1,0], \ 0 < b \leq \ \delta(a)\}, \\ D &:= \{(a,b) : \ a \in [-1,1], \ -1 \leq b \leq -|a|\}, \\ F &:= [-1,1]^2 \backslash (C \cup D), \end{aligned}$$

where $\delta(a) = -4 + a + 4\sqrt{1-a}$ for $a \in [-1, 0]$. We show first that $\pi_{ab}(\mathsf{S}_{D(\frac{3\pi}{4})})$ is the union of C, D and F.

Theorem 1.4.8 (Muñoz, Pellegrino, Seoane, and Weber [81]**)** *The projection of* $\mathsf{S}_{D(\frac{3\pi}{4})}$ *over the ab-plane is* $[-1, 1]^2$.

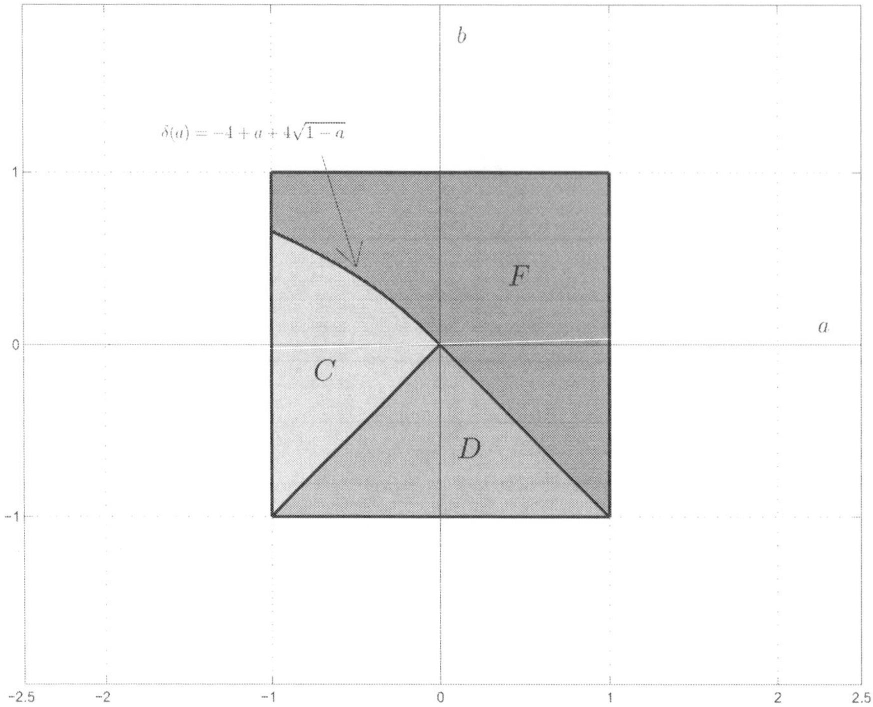

FIGURE 1.9: Projection of $\mathsf{S}_{D(\frac{3\pi}{4})}$ on the ab plane.

PROOF. *Let $(a, b) \in C$, and consider the point $(a, b, 2 + a + b)$. Using Theorem 1.4.1 along with the fact that $2 + a + b > 0$ and $(a - b)(2 + a + b) < 0$, we have*

$$\|(a, b, 2 + a + b)\|_{D(\frac{3\pi}{4})} = \max\left\{|a|, 1, \frac{1}{2}\left|a + b + \sqrt{2[(1 + a)^2 + (1 + b)^2]}\right|\right\}.$$

Now notice that $-C \subset P_1$, where P_1 is defined in Lemma 1.4.2. Hence

$$\frac{1}{2}\left|a + b + \sqrt{2[(1 + a)^2 + (1 + b)^2]}\right| \le 1,$$

which implies that $\|(a, b, 2 + a + b)\|_{D(\frac{3\pi}{4})} = 1$.

If $(a, b) \in D$ then we put $c = 2\sqrt{(1 + a)(1 + b)}$. Observe that $(a - b)c \ge 0$ and $a + b \le 0$, so by applying Theorem 1.4.1 again, we have

$$\begin{aligned}\|(a, b, c)\|_{D(\frac{3\pi}{4})} &= \frac{1}{2}\left(|a + b| + \sqrt{(a - b)^2 + c^2}\right)\\ &= \frac{1}{2}\left(-a - b + \sqrt{(2 + a + b)^2}\right) = 1.\end{aligned}$$

Finally, let $(a, b) \in F$ and define $c = 2\sqrt{(1 - a)(1 - b)}$. If $a - b \ge 0$ then Theorem 1.4.1 yields

$$\begin{aligned}\|(a, b, c)\|_{D(\frac{3\pi}{4})} &= \frac{1}{2}\left(|a + b| + \sqrt{(a - b)^2 + c^2}\right)\\ &= \frac{1}{2}\left(a + b + \sqrt{(2 - a - b)^2}\right) = 1.\end{aligned}$$

If, on the contrary, $a-b<0$*, then, applying Theorem 1.4.1 once more yields*

$$\|(a,b,c)\|_{D(\frac{3\pi}{4})} = \max\left\{|a|, \frac{1}{2}|a+b-c|, \frac{1}{2}\left|a+b+\text{sign}(c)\sqrt{(a-b)^2+c^2}\right|\right\}.$$

However

$$\frac{1}{2}\left|a+b+\text{sign}(c)\sqrt{(a-b)^2+c^2}\right| = \frac{1}{2}\left|a+b+\sqrt{(2-a-b)^2}\right| = 1$$

and

$$\frac{1}{2}|a+b-c| = \frac{1}{2}\left|a+b-2\sqrt{(1-a)(1-b)}\right| \leq 1,$$

proving that $\|(a,b,c)\|_{D(\frac{3\pi}{4})} = 1$.

So far, we have shown that $[-1,1]^2 = C \cup D \cup F \subset \pi_{ab}(\mathsf{S}_{D(\frac{3\pi}{4})})$. *On the other hand, if* $(a,b) \notin [-1,1]^2$ *then either* $|a|>1$*, in which case*

$$\|(a,b,c)\|_{D(\frac{3\pi}{4})} \geq |a| > 1,$$

for all $c \in \mathbb{R}$*, or* $|b|>1$ *and* $|a|\leq 1$*. In the latter situation, due to symmetry, we just need to focus on the case* $b<-1$ *and* $|a|\leq 1$*. Since now* $a-b>0$*, if* $c \geq 0$*, then by Theorem 1.4.1*

$$\begin{aligned}\|(a,b,c)\|_{D(\frac{3\pi}{4})} &= \frac{1}{2}\left(|a+b| + \sqrt{(a-b)^2+c^2}\right)\\ &\geq \frac{1}{2}\left(|a+b|+|a-b|\right) = |b| > 1,\end{aligned}$$

and if $c<1$*, by Theorem 1.4.1 again*

$$\begin{aligned}\|(a,b,c)\|_{D(\frac{3\pi}{4})} &\geq \frac{1}{2}\left|a+b+\text{sign}(c)\sqrt{(a-b)^2+c^2}\right|\\ &= \frac{1}{2}\left(|a+b| + \sqrt{(a-b)^2+c^2}\right) \geq |b| > 1.\end{aligned}$$

Therefore, if $(a,b)\notin[-1,1]^2$ *then* $(a,b)\notin \pi_{ab}(\mathsf{S}_{D(\frac{3\pi}{4})})$*, which concludes the proof.*

Theorem 1.4.9 (Muñoz, Pellegrino, Seoane, and Weber [81]**)** *Define*

$$H_1(a,b) = \begin{cases} 2+a+b & \text{if } (a,b)\in C,\\ 2\sqrt{(1+a)(1+b)} & \text{if } (a,b)\in D,\\ 2\sqrt{(1-a)(1-b)} & \text{if } (a,b)\in F,\end{cases}$$

where C*,* D *and* F *were introduced at the beginning of Section 1.4.5, and* $H_2(a,b) = -H_1(-a,-b)$ *for all* $(a,b)\in[-1,1]^2$*. If*

$$\Omega = \{\pm(-1,b,c)\in\mathbb{R}^3 : \ -1\leq b\leq 1, \ 0\leq c\leq H_1(-1,b)\},$$

then

(a) $\mathsf{S}_{D(\frac{3\pi}{4})} = \text{graph}(H_1)\cup\text{graph}(H_2)\cup\Omega.$

(b) The set $\operatorname{ext}(\mathsf{B}_{D(\frac{3\pi}{4})})$ *consists of the elements*

$$\pm\left(t,-4+t+4\sqrt{1-t},-2+2t+4\sqrt{1-t}\right) \quad \textit{for} \quad t\in[-1,0],$$

$$\pm\left(s,-s,2\sqrt{1-s^2}\right) \quad \textit{for} \quad s\in[0,1],$$

$$\pm\left(-1,r,2\sqrt{2(1-r)}\right) \quad \textit{for} \quad r\in[-5+4\sqrt{2},1],$$

and

$$\pm(1,1,0).$$

PROOF. *In the proof of Theorem 1.4.8 we saw already that*

$$\operatorname{graph}(H_1)\cup\operatorname{graph}(H_2)\subset \mathsf{S}_{D(\frac{3\pi}{4})}.$$

On the other hand, let us consider a typical element of Ω*, namely* $\pm(-1,b,c)$ *with* $|b|\leq 1$ *and* $0\leq c\leq H_1(-1,b)$*. Since* $-(1+b)c\leq 0$*, using Theorem 1.4.1 we have*

$$\|\pm(-1,b,c)\|_{D(\frac{3\pi}{4})}=\max\left\{1,\frac{1}{2}\left|1-b+c\right|,\frac{1}{2}\left|1-b-\sqrt{(1+b)^2+c^2}\right|\right\}.$$

Observe that

$$\begin{aligned}\frac{1}{2}\left|1-b+c\right| &= \frac{1}{2}(1-b+c)\leq\frac{1}{2}(1-b+H_1(-1,b))\\ &=\begin{cases}1 & \textit{if } -1\leq b\leq -5+4\sqrt{2},\\ \frac{1}{2}(1-b+2\sqrt{2(1-b)}) & \textit{if } -5+4\sqrt{2}\leq b\leq 1,\end{cases}\end{aligned}$$

and that $\frac{1}{2}(1-b+2\sqrt{2(1-b)})\leq 1$ *for* $-5+4\sqrt{2}\leq b\leq 1$*. Also, notice that*

$$\frac{1}{2}\left|1-b-\sqrt{(1+b)^2+c^2}\right|\leq 1\Leftrightarrow |c|\leq 2\sqrt{2(1-b)},$$

for $|b|\leq 1$*. Since*

$$0\leq c\leq H_1(-1,b)=\begin{cases}1-b & \textit{if } -1\leq b\leq -5+4\sqrt{2},\\ 2\sqrt{2(1-b)} & \textit{if } -5+4\sqrt{2}\leq b\leq 1,\end{cases}$$

and $1-b\leq 2\sqrt{2(1-b)}$ *for* $-1\leq b\leq -5+4\sqrt{2}$*, we conclude that*

$$\frac{1}{2}\left|1-b-\sqrt{(1+b)^2+c^2}\right|\leq 1.$$

Hence $\|\pm(-1,b,c)\|_{D(\frac{3\pi}{4})}=1$.

This shows that $\operatorname{graph}(H_1)\cup\operatorname{graph}(H_2)\cup\Omega\subset\mathsf{S}_{D(\frac{3\pi}{4})}$*. In order to prove the reverse inclusion, we follow the same strategy as in the proof of Theorems 1.4.5 and 1.4.7, noting that any straight line through the origin intersects the set* $\operatorname{graph}(H_1)\cup\operatorname{graph}(H_2)\cup\Omega$*.*

For the proof of the second part of the statement, notice that the set Ω *and the surfaces defined by the formulas* $c=2\sqrt{1-a-b+ab}$*,* $c=2\sqrt{1+a+b+ab}$ *and* $c=2+a+b$*, are ruled surfaces in* $\mathbb{R}^3$*. Therefore, the extreme points can only lie in the non-straight intersections of the four surfaces* Ω*,* $c=2\sqrt{1-a-b+ab}$*,* $c=2\sqrt{1+a+b+ab}$ *and* $c=2+a+b$ *– that is, in the set of the points described in the statement. As the reader can check easily by constructing supporting hyperplanes, all those points are indeed extreme, concluding the proof.*

The reader can find a sketch of $\mathsf{S}_{D(\frac{3\pi}{4})}$ from different viewpoints in Figure 1.10.

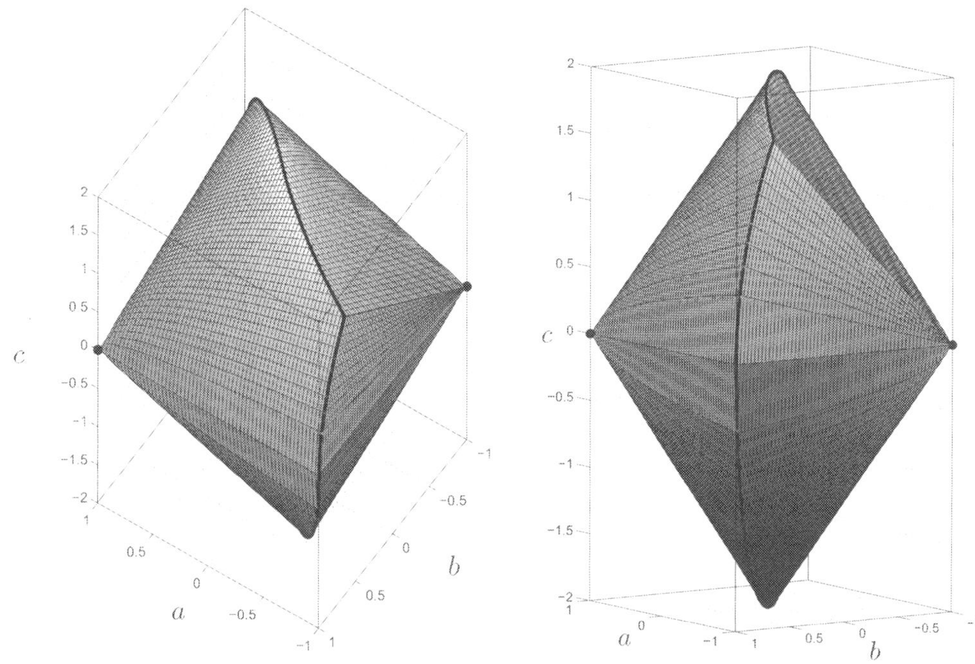

FIGURE 1.10: Two different perspectives of $\mathsf{S}_{D(\frac{3\pi}{4})}$. The extreme points of $\mathsf{B}_{D(\frac{3\pi}{4})}$ have been drawn with a thicker line and dots

1.4.6 The remaining cases

Let $P(x,y) = ax^2 + by^2 + cxy$ be a 2-homogeneous polynomial on $\mathbb{R}^2$. By the homogeneity property of P, the supremum of $|P|$ over $D(\beta)$ is attained on the set $B := \{(\cos(\theta), \sin(\theta)) : 0 \leq \theta \leq \beta\}$ and when restricted to B, P is given by:

$$\begin{aligned} f(\theta) :&= P(\cos(\theta), \sin(\theta)) = a\cos^2(\theta) + b\sin^2(\theta) + c\sin(\theta)\cos(\theta) \\ &= a\frac{1+\cos(2\theta)}{2} + b\frac{1-\cos(2\theta)}{2} + c\frac{\sin(2\theta)}{2} \\ &= \frac{1}{2}[a + b + (a-b)\cos(2\theta) + c\sin(2\theta)], \end{aligned}$$

for all $\theta \in [0, \beta]$.

In this section, we obtain a formula for the norm $\|ax^2 + by^2 + cxy\|_{D(\beta)}$, where β is in one of the following open intervals: $\left(0, \frac{\pi}{4}\right), \left(\frac{\pi}{4}, \frac{\pi}{2}\right), \left(\frac{\pi}{2}, \frac{3\pi}{4}\right)$ or $\left(\frac{3\pi}{4}, \pi\right)$. We define the values M and N as follows and use them throughout the rest of this section:

$$M := \cos(2\beta) \quad \text{and} \quad N := \sin(2\beta).$$

Theorem 1.4.10 (Bernal, Muñoz, Rodríguez, and Seoane [10]**)** *If $a, b, c \in \mathbb{R}$ and $P(x,y) = ax^2 + by^2 + cxy$, then $\|P\|_{D(\beta)}$ where $0 < \beta < \frac{\pi}{4}, \frac{\pi}{4} < \beta < \frac{\pi}{2}, \frac{\pi}{2} < \beta < \frac{3\pi}{4}$*

or $\frac{3\pi}{4} < \beta < \pi$ *is given, respectively, by*

$$\begin{cases} \max\left\{|a|, \frac{1}{2}\left|(1+M)a+(1-M)b+cN\right|, \frac{1}{2}\left|a+b+\operatorname{sign}(c)\sqrt{(a-b)^2+c^2}\right|\right\} & (1), \\ \max\{|a|, \frac{1}{2}\left|(1+M)a+(1-M)b+cN\right|\} & (2), \end{cases}$$

(1) if $c(a-b) > 0$ *and* $\tan(2\beta) > \dfrac{c}{a-b}$,

(2) if $\left(c(a-b) > 0 \text{ and } \tan(2\beta) \le \dfrac{c}{a-b}\right)$ *or* $(c(a-b) \le 0)$,

$$\begin{cases} \max\left\{|a|, \frac{1}{2}\left|(1+M)a+(1-M)b+cN\right|, \frac{1}{2}\left|a+b+\operatorname{sign}(c)\sqrt{(a-b)^2+c^2}\right|\right\} & (3), \\ \max\{|a|, \frac{1}{2}\left|(1+M)a+(1-M)b+cN\right|\} & (4), \end{cases}$$

(3) if $\left(c(a-b) < 0 \text{ and } \tan(2\beta) > \dfrac{c}{a-b}\right)$ *or* $(c(a-b) > 0)$ *or* $(a = b)$,

(4) if $\left(c(a-b) < 0 \text{ and } \tan(2\beta) \le \dfrac{c}{a-b}\right)$ *or* $(c = 0)$,

$$\begin{cases} \frac{1}{2}\left(|a+b| + \sqrt{(a-b)^2+c^2}\right) & (5), \\ \max\left\{|a|, \frac{1}{2}\left|(1+M)a+(1-M)b+cN\right|, \frac{1}{2}\left|a+b+\operatorname{sign}(c)\sqrt{(a-b)^2+c^2}\right|\right\} & (6), \end{cases}$$

(5) if $\left(c(a-b) > 0 \text{ and } \tan(2\beta) > \dfrac{c}{a-b}\right)$ *or* $(c = 0)$,

(6) if $\left(c(a-b) > 0 \text{ and } \tan(2\beta) \le \dfrac{c}{a-b}\right)$ *or* $(c(a-b) < 0)$ *or* $(a = b)$,

$$\begin{cases} \frac{1}{2}\left(|a+b| + \sqrt{(a-b)^2+c^2}\right) & (7), \\ \max\left\{|a|, \frac{1}{2}\left|(1+M)a+(1-M)b+cN\right|, \frac{1}{2}\left|a+b+\operatorname{sign}(c)\sqrt{(a-b)^2+c^2}\right|\right\} & (8), \end{cases}$$

(7) if $\left(c(a-b) < 0 \text{ and } \tan(2\beta) > \dfrac{c}{a-b}\right)$ *or* $(c(a-b) \ge 0)$,

(8) if $\left(c(a-b) < 0 \text{ and } \tan(2\beta) \le \dfrac{c}{a-b}\right)$,

PROOF. *Let*

$$g(\theta) = \frac{1}{2}\left[a+b+(a-b)\cos\theta + c\sin\theta\right],$$

for $\theta \in [0, 2\beta]$. *Then clearly* $\|P\|_{D(\beta)} = \sup_{\theta\in[0,2\beta]} |g(\theta)|$, *and the derivative of* g *is given by*

$$g'(\theta) = \frac{1}{2}\left[(b-a)\sin\theta + c\cos\theta\right].$$

Assume that $0 < \beta < \frac{\pi}{4}$. *Suppose first that* $c(a-b) = 0$. *Then, there are three different cases:*

Case (1): Suppose that $c = 0$ *and* $a - b \neq 0$. *Put* $g(\theta) = \frac{1}{2}[a+b+(a-b)\cos\theta]$. *Therefore,* $g'(\theta) = \frac{b-a}{2}\sin(\theta) = 0$ *if and only if* $\theta = k\pi$ *for all* $k \in \mathbb{Z}$. *However,* $\theta \in (0, 2\beta)$ *where* $0 < 2\beta < \frac{\pi}{2}$. *Then there are no critical points in the open interval* $(0, 2\beta)$. *Therefore, the maximum and minimum of* g *are attained at the endpoints of the interval, that is, at* $\theta = 0,\ 2\beta$ *with* $g(0) = a$ *and* $g(2\beta) = \frac{1}{2}[a+b+(a-b)\cos(2\beta)]$.

Case (2): Suppose that $c \neq 0$ *and* $a = b$. *Put* $g(\theta) = \frac{1}{2}[a + b + c \sin \theta]$. *As in case (1), the maximum and minimum are attained at the endpoints since* $g'(\theta) = \frac{c}{2} \cos(\theta) \neq 0$ *because* $\theta \in (0, 2\beta)$ *with* $0 < 2\beta < \frac{\pi}{2}$, *and* $g(0) = \frac{1}{2}[a + b] = a$, $g(2\beta) = \frac{1}{2}[a + b + c \sin(2\beta)]$.

Case (3): Suppose that $c = 0$ *and* $a = b$. *Then* $g(\theta) := \frac{1}{2}[a + b] = a$ *is constant in* $[0, 2\beta]$.

Now suppose that $c(a - b) \neq 0$. *There exists only one* $\theta_0 \in (0, 2\beta)$ *such that* $g'(\theta_0) = 0$ *i.e.* $\tan(\theta_0) = \frac{c}{a-b}$ *if and only if* $\frac{c}{a-b} > 0$ *and* $\tan(2\beta) > \frac{c}{a-b}$.

If $\frac{c}{a-b} > 0$ *and* $\tan(2\beta) > \frac{c}{a-b}$, *then*

$$\cos \theta_0 = \frac{1}{\sqrt{1 + \frac{c^2}{(a-b)^2}}} \qquad \textit{and} \qquad \sin \theta_0 = \frac{\frac{c}{a-b}}{\sqrt{1 + \frac{c^2}{(a-b)^2}}}.$$

Hence the maximum and minimum of g are attained at $\theta = 0,\ 2\beta$ *or* θ_0 *with*

$$\begin{aligned}
g(\theta_0) &= \frac{1}{2}\left[a + b + (a - b)\frac{1}{\sqrt{1 + \frac{c^2}{(a-b)^2}}} + c\frac{\frac{c}{a-b}}{\sqrt{1 + \frac{c^2}{(a-b)^2}}}\right] \\
&= \frac{1}{2}\left[a + b + \frac{(a - b)|a - b|}{\sqrt{c^2 + (a - b)^2}} + \frac{c^2 \operatorname{sign}(a - b)}{\sqrt{c^2 + (a - b)^2}}\right] \\
&= \frac{1}{2}\left[a + b + \operatorname{sign}(a - b)\frac{(a - b)^2 + c^2}{\sqrt{c^2 + (a - b)^2}}\right] \\
&= \frac{1}{2}\left[a + b + \operatorname{sign}(c)\sqrt{c^2 + (a - b)^2}\right].
\end{aligned}$$

This, together with the fact that

$$g(2\beta) = \frac{1}{2}\left[a + b + (a - b)\cos(2\beta) + c \sin(2\beta)\right]$$

and $g(0) = a$, *prove the first part of the first formula. If* $c(a - b) < 0$ *or if* $c(a - b) > 0$ *and* $\tan(2\beta) \leq \frac{c}{a-b}$, *then the maximum and minimum of g are attained at the extreme points of the interval, this proves the second part of the first formula.*

Assume that $\frac{\pi}{4} < \beta < \frac{\pi}{2}$. *Suppose first that* $c(a - b) = 0$. *Then, as in the previous part, there are three different cases: The cases (1) and (3) are analogous, but the case (2) needs some work, because* $g'(\theta) = \frac{c}{2} \cos(\theta) = 0$ *in* $(0, 2\beta)$ *when* $\theta = \frac{\pi}{2}$ *and* $\frac{\pi}{2} < 2\beta < \pi$. *But* $g''(\theta) = -\frac{c}{2} \sin(\theta)$ *and therefore* $g''\left(\frac{\pi}{2}\right) = -\frac{c}{2} \neq 0$. *So there is a maximum or a minimum at* $\theta = \frac{\pi}{2}$. *Thus, the maximum and minimum are attained at* $\theta = 0,\ 2\beta$ *or* $\frac{\pi}{2}$ *with* $g(2\beta) = \frac{1}{2}[a + b + c \sin(2\beta)] = \frac{1}{2}[2a + c \sin(2\beta)]$, $g(0) = a$ *and* $g\left(\frac{\pi}{2}\right) = \frac{1}{2}[2a + c]$.

Suppose that $c(a - b) \neq 0$. *If* $c(a - b) > 0$, *then there exits only one* $\theta_0 \in \left(0, \frac{\pi}{2}\right)$ *such that* $\tan(\theta_0) = \frac{c}{a-b}$ *and, as we have already seen,* $g(\theta_0) = \frac{1}{2}\left[a + b + \operatorname{sign}(c)\sqrt{c^2 + (a - b)^2}\right]$.

If $c(a - b) < 0$, *then there exists only one* $\theta_0 \in \left(\frac{\pi}{2}, 2\beta\right)$ *such that* $\tan(\theta_0) = \frac{c}{a-b}$ *if and only if* $\tan(2\beta) > \frac{c}{a-b}$. *If* $\tan(2\beta) > \frac{c}{a-b}$, *then we proceed as above using*

$$\cos \theta_0 = -\frac{1}{\sqrt{1 + \frac{c^2}{(a-b)^2}}} \qquad \textit{and} \qquad \sin \theta_0 = -\frac{\frac{c}{a-b}}{\sqrt{1 + \frac{c^2}{(a-b)^2}}}.$$

Hence

$$\begin{aligned} g(\theta_0) &= \frac{1}{2}\left[a+b-(a-b)\frac{1}{\sqrt{1+\frac{c^2}{(a-b)^2}}} - c\frac{\frac{c}{a-b}}{\sqrt{1+\frac{c^2}{(a-b)^2}}}\right] \\ &= \frac{1}{2}\left[a+b-\frac{(a-b)|a-b|}{\sqrt{c^2+(a-b)^2}} - \frac{c^2\operatorname{sign}(a-b)}{\sqrt{c^2+(a-b)^2}}\right] \\ &= \frac{1}{2}\left[a+b-\operatorname{sign}(a-b)\frac{(a-b)^2+c^2}{\sqrt{c^2+(a-b)^2}}\right] \\ &= \frac{1}{2}\left[a+b+\operatorname{sign}(c)\sqrt{c^2+(a-b)^2}\right]. \end{aligned}$$

This, together with the fact that $g(2\beta) = \frac{1}{2}[a+b+(a-b)\cos(2\beta)+c\sin(2\beta)]$ *and* $g(0)=a$, *proves the first part of the second formula. If* $c(a-b)<0$ *and* $\tan(2\beta)\le\frac{c}{a-b}$, *then the maximum and minimum of* g *are attained at the endpoints of the interval, which prove the second part of the second formula.*

Assume that $\frac{\pi}{2}<\beta<\frac{3\pi}{4}$. *If* $c(a-b)=0$, *then we proceed as in the second part of the proof, but in case (1) notice that* $g'(\pi)=0$, *which gives us* $g(\pi)=b$. *Suppose that* $c(a-b)>0$; *then there exist* $\theta_1,\ \theta_2\in(0,2\beta)$ *with* $\theta_1\in\left(0,\frac{\pi}{2}\right]$ *and* $\theta_2\in[\pi,2\beta]$ *such that* $\tan(\theta_1)=\tan(\theta_2)=\frac{c}{a-b}$ *if and only if* $\tan(2\beta)>\frac{c}{a-b}$. *Otherwise, only* θ_1 *exists.*

We know that $g(\theta_1)=\frac{1}{2}\left[a+b+\operatorname{sign}(c)\sqrt{c^2+(a-b)^2}\right]$. *In the first case, since* $\theta_2-\theta_1=\pi$, *we have*

$$\begin{aligned} \max\{|g(\theta_1)|,|g(\theta_2)|\} &= \max\left\{\frac{1}{2}\left|a+b+\epsilon\operatorname{sign}(c)\sqrt{c^2+(a-b)^2}\right| : \epsilon=\pm1\right\} \\ &= \frac{1}{2}\left(|a+b|+\sqrt{c^2+(a-b)^2}\right). \end{aligned}$$

In the second case, the maximum and minimum of g *are attained at* $\theta=0,\ 2\beta$ *or* θ_1.

Now suppose that $c(a-b)<0$. *Then there exists exactly one* $\theta_0\in(0,2\beta)$ *such that* $\tan(\theta_0)=\frac{c}{a-b}$. *Applying the same reasoning as in the previous parts of the proof, we deduce that* $g(\theta_0)=\frac{1}{2}\left[a+b+\operatorname{sign}(c)\sqrt{c^2+(a-b)^2}\right]$. *This together with* $g(0)=a$ *and*

$$g(2\beta)=\frac{1}{2}\left[a+b+(a-b)\cos(2\beta)+c\sin(2\beta)\right],$$

proves the third formula.

Finally, assume that $\frac{3\pi}{4}<\beta<\pi$. *If* $c(a-b)=0$, *then we use the same arguments as above, but notice that in case (2) we have* $g'(\frac{3\pi}{2})=0$, *which gives us* $g(\frac{3\pi}{2})=\frac{1}{2}[2a-c]$. *Suppose that* $c(a-b)\ne0$. *If* $c(a-b)>0$, *then there exist only one* $\theta_1\in\left(0,\frac{\pi}{2}\right]$ *and only one* $\theta_2\in\left[\pi,\frac{3\pi}{2}\right]$ *such that* $\tan(\theta_1)=\tan(\theta_2)=\frac{c}{a-b}$. *Then, using the previous calculations, we see that*

$$\max\{|g(\theta_1)|,|g(\theta_2)|\}=\frac{1}{2}\left(|a+b|+\sqrt{c^2+(a-b)^2}\right).$$

If $c(a-b)<0$, *then there exist* $\theta_1,\ \theta_2\in(0,2\beta)$ *with* $\theta_1\in\left[\frac{\pi}{2},\pi\right]$ *and* $\theta_2\in\left[\frac{3\pi}{2},2\beta\right)$ *such that* $\tan(\theta_1)=\tan(\theta_2)=\frac{c}{a-b}$ *if and only if* $\tan(2\beta)>\frac{c}{a-b}$. *Otherwise, only* θ_1 *exists. In the first case, we get*

$$\max\{|g(\theta_1)|,|g(\theta_2)|\}=\frac{1}{2}\left(|a+b|+\sqrt{c^2+(a-b)^2}\right),$$

where we took

$$\cos\theta_1 = -\frac{1}{\sqrt{1+\frac{c^2}{(a-b)^2}}} \qquad \textit{and} \qquad \sin\theta_1 = -\frac{\frac{c}{a-b}}{\sqrt{1+\frac{c^2}{(a-b)^2}}}$$

and

$$\cos\theta_2 = \frac{1}{\sqrt{1+\frac{c^2}{(a-b)^2}}} \qquad \textit{and} \qquad \sin\theta_2 = \frac{\frac{c}{a-b}}{\sqrt{1+\frac{c^2}{(a-b)^2}}}.$$

In the second case, the maximum or minimum value of g is attained at 0, θ_1 or 2β with

$$g(\theta_1) = \frac{1}{2}\left[a+b+\operatorname{sign}(c)\sqrt{c^2+(a-b)^2}\right],$$

$g(2\beta) = \frac{1}{2}[a+b+(a-b)\cos(2\beta)+c\sin(2\beta)]$ *and* $g(0)=a$, *which proves the last formula.*

1.4.6.1 The geometry of $\mathcal{P}(^2D(\beta))$ when $0<\beta\le\frac{\pi}{2}$

In this section, we derive a projection of the unit sphere of $\mathcal{P}(^2D(\beta))$ onto the ab-plane for values of β satisfying $0<\beta<\frac{\pi}{4}$ or $\frac{\pi}{4}<\beta<\frac{\pi}{2}$. Additionally, we present a parametrization of the unit sphere and identify the extreme points of the unit ball of $\mathcal{P}(^2D(\beta))$. The special cases $\beta=\frac{\pi}{4},\frac{\pi}{2}$ have already been discussed in earlier sections of this text (see also [81]).

To proceed, we first establish a generalization of Lemma 4.1 from [81], which will play a crucial role in subsequent results.

Lemma 1.4.11 *Let us define the sets P_1, P_2 and P_3 by*

$$P_1 := \left\{(a,b)\in\mathbb{R}^2\colon a\ge -1,\ \frac{(1+M)a+2(2+M)-4\sqrt{(1+M)(1+a)}}{1-M} \le b \le \frac{(1+M)a+2(2+M)+4\sqrt{(1+M)(1+a)}}{1-M}\right\},$$

where $0<\beta<\pi$ and $\beta\neq\frac{\pi}{2}$;

$$P_2 := \left\{(a,b)\in\mathbb{R}^2\colon |a|\le 1,\ \frac{(1+M)a-2(2+M)-4\sqrt{(1+M)(1-a)}}{1-M}\le b\le 1\right\},$$

where $0<\beta<\frac{\pi}{2}$; and

$$P_3 := \left\{(a,b)\in\mathbb{R}^2\colon |a|\le 1,\ \frac{(1+M)a-2(2+M)+4\sqrt{(1+M)(1-a)}}{1-M}\le b\le 1\right\},$$

where $\frac{\pi}{2}<\beta<\pi$. Then, we have:

(a) *If $0<\beta<\pi$ and $\beta\neq\frac{\pi}{2}$, the inequality*

$$\frac{1}{2}\left|a+b-\sqrt{2\left[\frac{1}{1-M}(a-1)^2+\frac{1}{1+M}(b-1)^2\right]}\right|\le 1,$$

holds if and only if $(a,b)\in P_1$.

(b) If $(a,b) \in P_2$ (when $0<\beta<\frac{\pi}{2}$) or $(a,b)\in P_3$ (when $\frac{\pi}{2}<\beta<\pi$), then

$$\frac{1}{2}\left|(1+M)a+(1-M)b+2N\sqrt{(1-a)(1-b)}\right| \le 1,$$

PROOF. *To prove part (a), we define*

$$h(a,b)=\frac{1}{2}\left(a+b-\sqrt{2\left[\frac{1}{1-M}(a-1)^2+\frac{1}{1+M}(b-1)^2\right]}\right)$$

for $(a,b)\in\mathbb{R}^2$. Then $|h(a,b)|=1$ holds if either

$$a+b-\sqrt{2\left[\frac{1}{1-M}(a-1)^2+\frac{1}{1+M}(b-1)^2\right]}=2, \tag{1.8}$$

or

$$a+b-\sqrt{2\left[\frac{1}{1-M}(a-1)^2+\frac{1}{1+M}(b-1)^2\right]}=-2. \tag{1.9}$$

Equation (1.8) can only happen when $a+b-2\ge 0$; in this case, by squaring (1.8) we arrive at

$$b=\frac{(1+M)a-2M}{1-M},$$

with $a+b-2\ge 0$. It is easy to prove that the line described by the equation $b=\frac{(1+M)a-2M}{1-M}$ is in the half-plane $a+b-2\ge 0$ if $a\ge 1$. This shows that $h\equiv 1$ on the set $R_h := \left\{\left(a,\frac{(1+M)a-2M}{1-M}\right) : a\ge 1\right\}$. On the other hand, (1.9) occurs only if $a+b+2\ge 0$, in which case, by squaring (1.9) we obtain

$$b=\frac{(1+M)a+2(2+M)\pm 4\sqrt{(1+M)(1+a)}}{1-M},$$

with $a\ge -1$. Observe that the curves described by the equation

$$b=\frac{(1+M)a+2(2+M)\pm 4\sqrt{(1+M)(1+a)}}{1-M},$$

with $a\ge -1$, are in the half-plane $a+b+2\ge 0$. Hence we have proved so far that the set of zeros of $|h|-1$, namely Z_h, is given by

$$R_h\cup\left\{(a,b)\colon a\ge -1,\ b=\frac{(1+M)a+2(2+M)\pm 4\sqrt{(1+M)(1+a)}}{1-M}\right\}.$$

Now define

$$C_1^1 := \left\{(a,b)\colon a\ge -1,\, b>\frac{(1+M)a+2(2+M)+4\sqrt{(1+M)(1+a)}}{1-M}\right\},$$

$$C_1^2 := \left\{(a,b)\colon a\ge -1,\, b<\frac{(1+M)a+2(2+M)-4\sqrt{(1+M)(1+a)}}{1-M}\right\},$$

and

$$C_2 := \left\{ (a,b)\colon a \geq -1,\ \frac{(1+M)a + 2(2+M) - 4\sqrt{(1+M)(1+a)}}{1-M} < b \right.$$
$$\left. < \frac{(1+M)a + 2(2+M) + 4\sqrt{(1+M)(1+a)}}{1-M} \right\} \setminus R_h$$

be the three connected components of $\{(a,b) : a \geq -1\} \setminus Z_h$. *By continuity,* $|h| - 1$ *preserves its sign on* C_1^1, C_1^2 *and* C_2 *(see Figure 1.11). It is easy to see that* $|h| - 1$ *is positive on* C_1^1 *and* C_1^2, *and negative on* C_2, *and* $P_1 = C_2 \cup Z_h$. *Therefore, the result follows.*

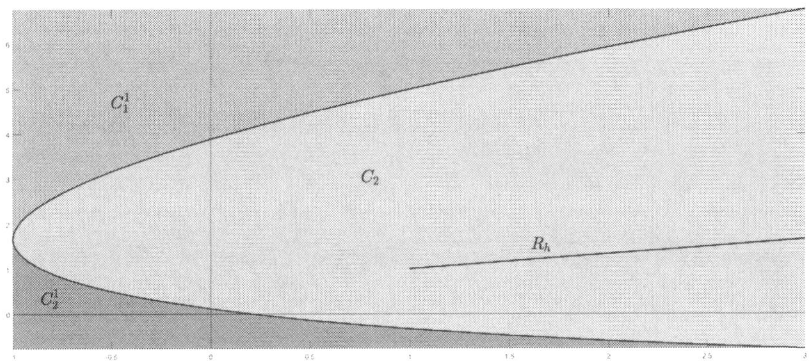

FIGURE 1.11: Sketch of C_1^1, C_1^2, C_2, and R_h for $\beta = 2\pi/3$.

In order to prove (b), we assume that $0 < \beta < \frac{\pi}{2}$ *(the proof in the case* $\frac{\pi}{2} < \beta < \pi$ *can be done in a similar way). Let us define* $g\colon P_2 \to \mathbb{R}$ *as*

$$g(a,b) = \frac{1}{2}\left[(1+M)a + (1-M)b + 2N\sqrt{(1-a)(1-b)}\right].$$

Since $(a,b) \in P_2$, *we have that* $(1-a)(1-b) \geq 0$, *which proves that* g *is well defined. Now, take* $(a,b) \in P_2$ *such that* $|g(a,b)| = 1$. *Thus, either*

$$(1+M)a + (1-M)b + 2N\sqrt{(1-a)(1-b)} = 2, \tag{1.10}$$

or

$$(1+M)a + (1-M)b + 2N\sqrt{(1-a)(1-b)} = -2. \tag{1.11}$$

On one hand, since $N > 0$, *equation* (1.10) *holds only if* $2 - (1+M)a - (1-M)b \geq 0$. *By squaring* (1.10), *we arrive at*

$$b = \frac{(1+M)a - 2M}{1-M},$$

with $2 - (1+M)a - (1-M)b \geq 0$. *It is easy to prove that, since* $a \leq 1$, *the curve described by the equation* $b = \frac{(1+M)a-2M}{1-M}$ *is in the half-plane* $2 - (1+M)a - (1-M)b \geq 0$. *This shows that* $g \equiv 1$ *on the set* $R_g := \left\{\left(a, \frac{(1+M)a-2M}{1-M}\right) : |a| \leq 1\right\}$. *On the other hand, since*

$N>0$, (1.11) occurs only if $2+(1+M)a+(1-M)b\le 0$. In this case, by squaring (1.11) we obtain

$$b=\frac{(1+M)a-2(2+M)\pm 4\sqrt{(1+M)(1-a)}}{1-M}.$$

It is easy to prove that, since $(a,b)\in P_2$, the curve described by the equation

$$b=\frac{(1+M)a-2(2+M)-4\sqrt{(1+M)(1-a)}}{1-M}$$

is in the half-plane $2+(1+M)a+(1-M)b\le 0$. However, the curve described by the equation

$$b=\frac{(1+M)a-2(2+M)+4\sqrt{(1+M)(1-a)}}{1-M}$$

with $(a,b)\in P_2$ is in the half-plane $2+(1+M)a+(1-M)b\le 0$ if $a=1$. Hence we have proved so far that the set of zeros of $|g|-1$, namely Z_g, is given by

$$R_g\cup\left\{(a,b)\colon |a|\le 1,\ b=\frac{(1+M)a-2(2+M)-4\sqrt{(1+M)(1-a)}}{1-M}\right\}.$$

Next, define

$$D_1:=\left\{(a,b)\colon |a|\le 1, \frac{(1+M)a-2M}{1-M}<b\le 1\right\},$$

$$D_2:=\left\{(a,b)\colon |a|\le 1,\right.$$
$$\left.\frac{(1+M)a-2(2+M)-4\sqrt{(1+M)(1-a)}}{1-M}<b<\frac{(1+M)a-2M}{1-M}\right\}$$

and

$$D_3:=\left\{(a,b)\colon |a|\le 1,\ b<\frac{(1+M)a-2(2+M)-4\sqrt{(1+M)(1-a)}}{1-M}\right\}.$$

These form the three connected components of $\{(a,b)\colon |a|\le 1,\ b\le 1\}\setminus Z_g$. By continuity, $|g|-1$ preserves its sign on D_1, D_2, and D_3 (see Figure 1.12). Since $|g|-1$ is clearly positive on D_3 and negative on $D_1\cup D_2$, and $P_2=D_1\cup D_2\cup Z_g$, the result follows.

Next, we obtain the projection of the unit sphere of $\mathcal{P}(^2D(\beta))$ onto the ab-plane.

Let $\beta\in(0,\frac{\pi}{2})$. We need to define the following three functions $\gamma_1,\gamma_2,\delta:[-1,1]\to\mathbb{R}$ as well as the following two sets $A,B\subset\mathbb{R}^2$ (see Figure 1.14 for a picture of the sets A and B):

$$\gamma_1(a):=\frac{(1+M)a+2(2+M)+4\sqrt{(1+M)(1+a)}}{1-M},$$
$$\gamma_2(a):=-\gamma_1(-a)=\frac{(1+M)a-2(2+M)-4\sqrt{(1+M)(1-a)}}{1-M},$$
$$\delta(a):=\frac{(1+M)a-2M}{1-M},$$
$$A:=\{(a,b)\in\mathbb{R}^2:\ a\in[-1,1],\ \delta(a)\le b\le\gamma_1(a)\},$$
$$B:=\{(a,b)\in\mathbb{R}^2:\ a\in[-1,1],\ \gamma_2(a)\le b<\delta(a)\}.$$

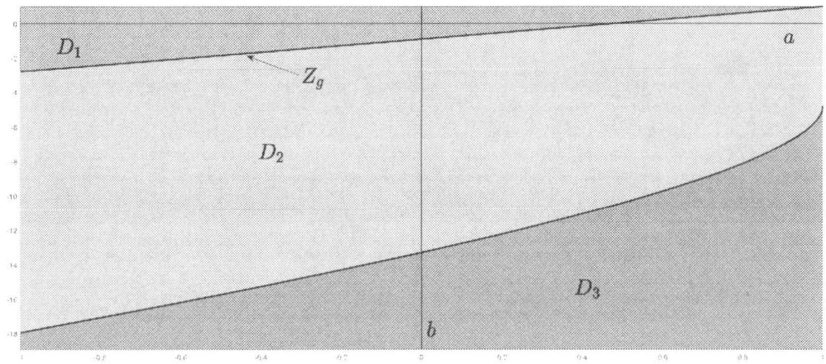

FIGURE 1.12: Sketch of D_1, D_2, D_3 and R_g for $\beta = \pi/5$.

Theorem 1.4.12 (Bernal, Muñoz, Rodríguez, and Seoane [10]) *Let $0 < \beta < \frac{\pi}{4}$ or $\frac{\pi}{4} < \beta < \frac{\pi}{2}$. Then, the projection of $\mathsf{S}_{D(\beta)}$ onto the ab-plane (see Figures 1.13 and 1.14 for a sketch of the projection) is*

$$\pi_{ab}(\mathsf{S}_{D(\beta)}) = \{(a, b) \in \mathbb{R}^2 : \ a \in [-1, 1], \ \gamma_2(a) \le b \le \gamma_1(a)\}.$$

PROOF. *Assume that $0 < \beta < \frac{\pi}{4}$ (the case $\frac{\pi}{4} < \beta < \frac{\pi}{2}$ follows by using similar arguments). First, we will begin by making some calculations that will tell us when the formulas of Theorem 1.4.10 are equal. This equality must occur by continuity. Since $0 < \beta < \frac{\pi}{4}$, observe that for $-1 \le a \le 1$ we have $\delta(a) \le a$, and also*

$$\delta(-1) = \frac{-3M - 1}{1 - M} > \gamma_2(-1) = \frac{-3M - 5 - 4\sqrt{2(1 + M)}}{1 - M}.$$

Now, by solving b in terms of a in the next equation (where each part of the equality comes from solving c in the formulas of Theorem 1.4.10)

$$\frac{1}{N}\left[2 - (1 + M)a - (1 - M)b\right] = 2\sqrt{(1 - a)(1 - b)}$$

for $-1 \le a \le 1$, we see that this equality is attained along the straight line $b = \delta(a)$. Basically what this tells us is that the equations

$$\tan(2\beta) = \frac{2\sqrt{(1 - a)(1 - b)}}{a - b}$$

and

$$\tan(2\beta) = \frac{\frac{1}{N}\left[2 - (1 + M)a - (1 - M)b\right]}{a - b}$$

are both satisfied simultaneously along the straight line $b = \delta(a)$. It can be checked that: if $b < \delta(a)$, then

$$\tan(2\beta) > \frac{2\sqrt{(1 - a)(1 - b)}}{a - b}; \tag{1.12}$$

if $\delta(a) \le b < a \le 1$, then

$$\tan(2\beta) \le \frac{\frac{1}{N}\left[2 - (1 + M)a - (1 - M)b\right]}{a - b}; \tag{1.13}$$

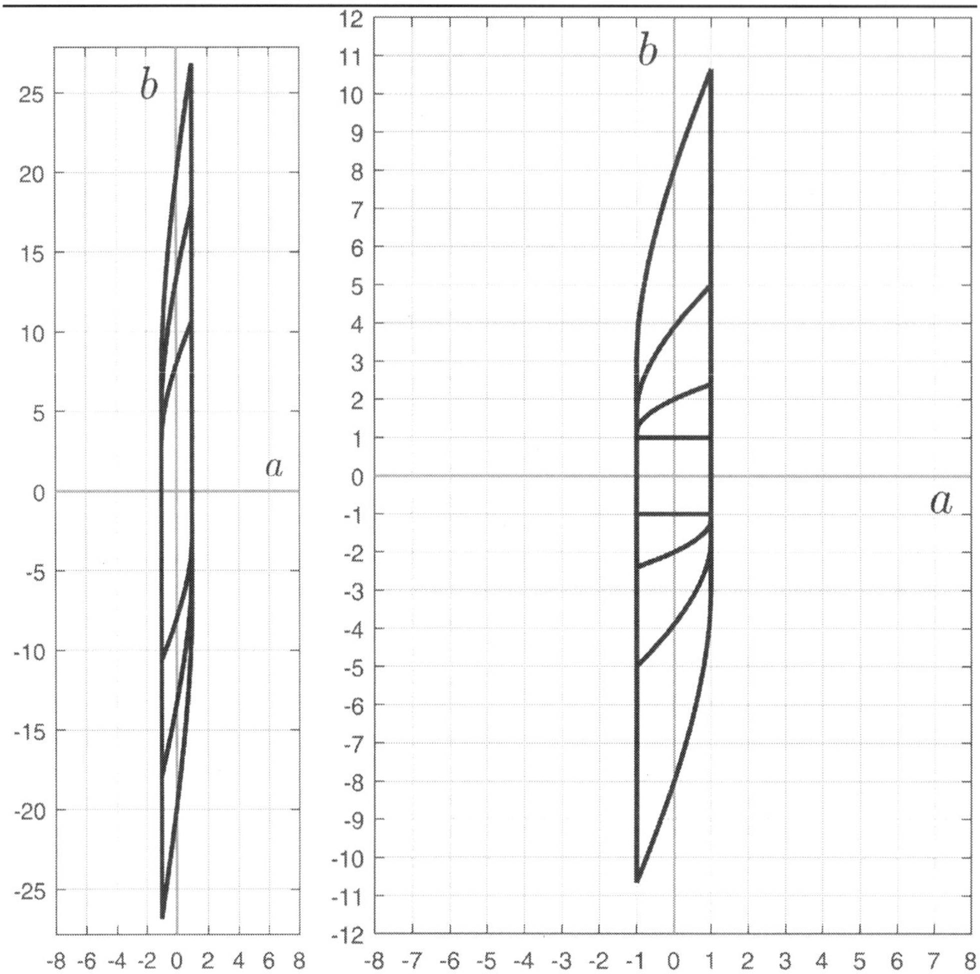

FIGURE 1.13: Scaled pictures of the projection onto the ab-plane of $\mathsf{B}_{D(\beta)}$ with $\beta = \frac{\pi}{k}$ for $k = 4, 5, 6$ on the left and $\beta = \frac{\pi}{4}\left(1 + \frac{j}{3}\right)$ with $j = 0, 1, 2, 3$ on the right. Observe that the projections expand along the b axis as β approaches to 0 and to the unit square $[-1, 1]^2$ as β approaches to $\frac{\pi}{2}$.

and if $b > a$, then

$$\tan(2\beta) > \frac{\frac{1}{N}\left[2 - (1+M)a - (1-M)b\right]}{a - b}. \tag{1.14}$$

Let $(a, b) \in A$ with $b > a$ and set $c = \frac{1}{N}\left[2 - (1+M)a - (1-M)b\right]$. Suppose first that $c \geq 0$, since $(a, b) \in A$ with $b > a$, we have $c(a - b) \leq 0$, therefore using Theorem 1.4.10 we obtain

$$\left\|\left(a, b, \frac{1}{N}\left[2 - (1+M)a - (1-M)b\right]\right)\right\|_{D(\beta)} = \max\{|a|, 1\} = 1.$$

Suppose now that $c < 0$; then since $(a, b) \in A$ with $b > a$ we have $c(a - b) > 0$. Also, since $b > a$, we have inequality (1.14), i.e., $\tan(2\beta) > \frac{c}{a-b}$, therefore according to the first

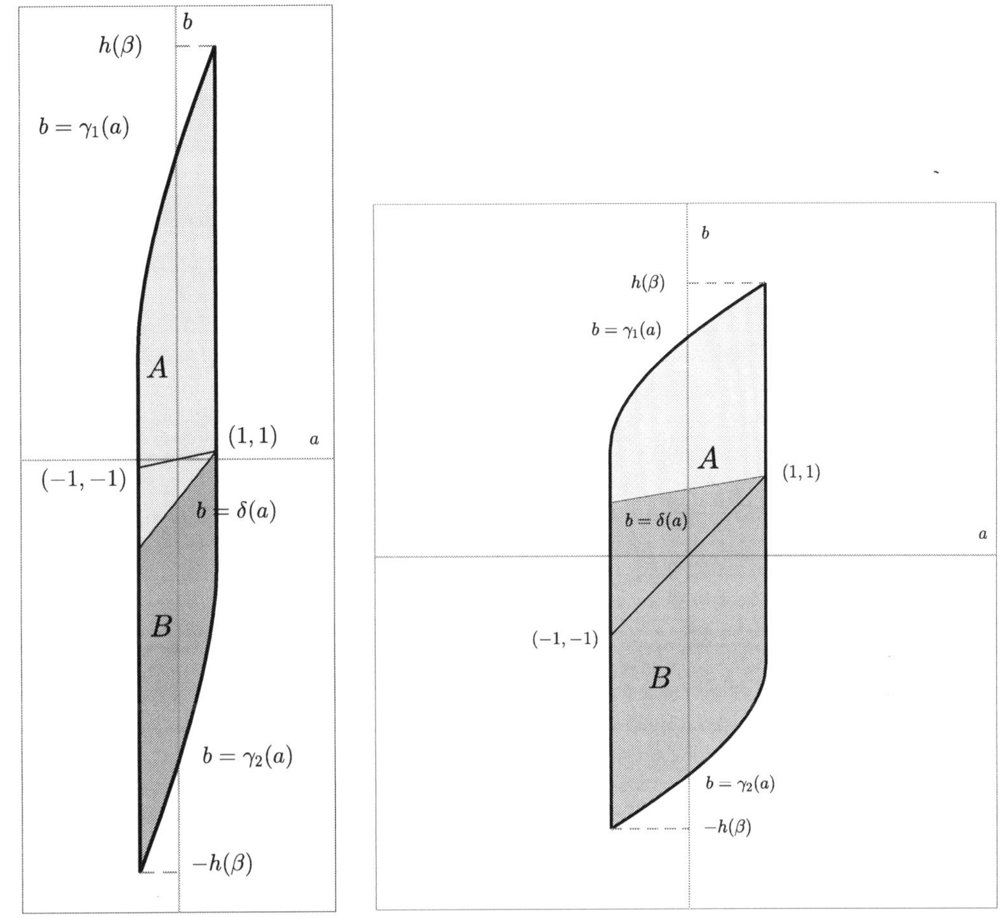

FIGURE 1.14: General view of the projection of $\mathsf{B}_{D(\beta)}$ onto the ab-plane with $0 < \beta \leq \pi/2$. On the left we have the case $\beta = \frac{\pi}{4} - 0.4$. The b axis has been rescaled by a factor 0.2. On the right we have represented the case $\beta = \frac{\pi}{4} + 0.4$. The axes here are scaled. In both cases $h(\beta) = \frac{5+3M+4\sqrt{2(1+M)}}{1-M}$.

formula in Theorem 1.4.10 we have

$$\begin{aligned}
&\left\|\left(a, b, \frac{1}{N}\left[2-(1+M)a-(1-M)b\right]\right)\right\|_{D(\beta)} \\
&= \max\left\{|a|, 1, \frac{1}{2}\left|a+b-\sqrt{(a-b)^2+\frac{1}{N^2}\left[2-(1+M)a-(1-M)b\right]^2}\right|\right\} \\
&= \max\left\{|a|, 1, \frac{1}{2}\left|a+b-\sqrt{2\left[\frac{1}{1-M}(a-1)^2+\frac{1}{1+M}(b-1)^2\right]}\right|\right\},
\end{aligned}$$

where in the last equation we used the known equality: $N^2 + M^2 = 1$. *Recall here that* (a, b) *satisfies* $|a| \leq 1$ *and*

$$2-(1+M)a < (1-M)b \leq (1+M)a + 2(2+M) + 4\sqrt{(1+M)(1+a)}, \tag{1.15}$$

where the first inequality comes from the assumption that $c < 0$ and the second inequality comes from the hypothesis that $b \leq \gamma_1(a)$. Since it can be checked that

$$(1+M)a + 2(2+M) - 4\sqrt{(1+M)(1+a)} \leq 2 - (1+M)a$$

for $|a| \leq 1$, we can apply Lemma 1.4.11, part (a), and therefore

$$\left\|\left(a, b, \frac{1}{N}\left[2-(1+M)a-(1-M)b\right]\right)\right\|_{D(\beta)} = 1.$$

Let $(a,b) \in A$ with $b < a$ and set $c = \frac{1}{N}\left[2-(1+M)a-(1-M)b\right]$. Suppose that $c \leq 0$, so that $c(a-b) \leq 0$. Applying now Theorem 1.4.10 we have

$$\left\|\left(a, b, \frac{1}{N}\left[2-(1+M)a-(1-M)b\right]\right)\right\|_{D(\beta)} = \max\{|a|, 1\} = 1.$$

Suppose that $c > 0$. Since $\delta(a) \leq b < a \leq 1$, from inequality (1.13), *we know that* $\tan(2\beta) \leq \frac{c}{a-b}$, *therefore using Theorem 1.4.10 we get*

$$\left\|\left(a, b, \frac{1}{N}\left[2-(1+M)a-(1-M)b\right]\right)\right\|_{D(\beta)} = \max\{|a|, 1\} = 1.$$

Let $(a,b) \in B$ and set $c = 2\sqrt{(1-a)(1-b)}$. Since $(a,b) \in B$ we see that $2-a-b > 0$, and therefore

$$\begin{aligned}\frac{1}{2}\left|a+b+\operatorname{sign}(c)\sqrt{(a-b)^2+c^2}\right| &= \frac{1}{2}\left|a+b+\sqrt{(a-b)^2+4(1-a)(1-b)}\right| \\ &= \frac{1}{2}\left|a+b+\sqrt{(2-a-b)^2}\right| = 1.\end{aligned}$$

In this case, since $b < a$ and $c > 0$, we have $c(a-b) > 0$; and since $b < \delta(a)$, we know from inequality (1.12) *that* $\tan(2\beta) > \frac{c}{a-b}$. *By Lemma 1.4.11, part (b), and applying Theorem 1.4.10 we have*

$$\begin{aligned}&\left\|\left(a, b, 2\sqrt{(1-a)(1-b)}\right)\right\|_{D(\beta)} \\ &= \max\left\{|a|, \frac{1}{2}\left|(1+M)a+(1-M)b+2N\sqrt{(1-a)(1-b)}\right|, 1\right\} = 1.\end{aligned}$$

We have proved so far that

$$\pi_{ab}(\mathsf{S}_{D(\beta)}) \supset \{(a,b) : \ a \in [-1,1], \ \gamma_2(a) \leq b \leq \gamma_1(a)\}.$$

On the other hand, suppose that $(a,b) \notin \{(a,b) : \ a \in [-1,1], \ \gamma_2(a) \leq b \leq \gamma_1(a)\}$. We then have three separate cases:

Case (1): If $|a| > 1$, then $\|(a,b,c)\| \geq |a| > 1$ for all $c \in \mathbb{R}$. Therefore, $(a,b) \notin \pi_{ab}(\mathsf{S}_{D(\beta)})$.

Case (2): Assume $|a| \leq 1$ and $\gamma_1(a) < b$, then $a-b < 0$. If $c < 0$ and $\tan(2\beta) > \frac{c}{a-b}$, *then*

$$\begin{aligned}\|(a,b,c)\|_{D(\beta)} \geq \max\Bigg\{&\frac{1}{2}\left|(1+M)a+(1-M)b+Nc\right|, \\ &\frac{1}{2}\left|a+b-\sqrt{(a-b)^2+c^2}\right|\Bigg\}.\end{aligned}$$

In this case, the value of c must satisfy exactly one of the following inequalities:

$$0 > c > \frac{1}{N}\left[2 - (1+M)a - (1-M)b\right]$$

or

$$c \leq \frac{1}{N}\left[2 - (1+M)a - (1-M)b\right].$$

If the first inequality is satisfied by c, then it can be easily checked that

$$\frac{1}{2}\left|(1+M)a + (1-M)b + Nc\right| > 1.$$

On the other hand, if c satisfies the second inequality, then

$$\frac{1}{2}\left|a + b - \sqrt{(a-b)^2 + c^2}\right| > 1. \tag{1.16}$$

Indeed, notice that $\frac{1}{2}\left|a + b - \sqrt{(a-b)^2 + c^2}\right| > 1$ *if and only if either*

$$a + b - 2 > \sqrt{(a-b)^2 + c^2}, \tag{1.17}$$

or

$$a + b + 2 < \sqrt{(a-b)^2 + c^2}. \tag{1.18}$$

Assume that inequality (1.17) is satisfied, then $a+b-2 > 0$, which implies that $b > 1$ due to the assumption $a < b$. Now, by squaring (1.17), we have that $c^2 < 4(1-a)(1-b)$. Since $a \leq 1$ and $c^2 > 0$ we have that $b \leq 1$ which leads to a contradiction. Thus, inequality (1.17) is not satisfied under these conditions. Thus, assume that inequality (1.18) is satisfied. Notice that $a + b + 2 > 0$, therefore we can square inequality (1.18) and we obtain $c^2 > 4(1+a)(1+b)$. Now, assume that $\frac{1}{N}\left[2 - (1+M)a - (1-M)b\right] \geq 0$, then $b \leq \frac{2-(1+M)a}{1-M}$. Furthermore, we have $\gamma_1(a) < b \leq \frac{2-(1+M)a}{1-M}$, however this implies that $a < -1$ and we have a contradiction. Therefore,

$$c \leq \frac{1}{N}\left[2 - (1+M)a - (1-M)b\right] < 0,$$

and we can square the last inequality. Finally, since $b > \gamma_1(a)$, it can be proved that

$$c^2 \geq \frac{1}{N^2}\left[2 - (1+M)a - (1-M)b\right]^2 > 4(1+a)(1+b),$$

which implies inequality (1.16).

Now, if $c < 0$ and $\tan(2\beta) \leq \dfrac{c}{a-b}$, then

$$\|(a,b,c)\|_{D(\beta)} \geq \frac{1}{2}\left|(1+M)a + (1-M)b + Nc\right|.$$

However, in this case c has a more restricted form, because if we assume that

$$c \geq \frac{1}{N}\left[-2 - (1+M)a - (1-M)b\right],$$

then, since $0 > \tan(2\beta)(a-b) \geq c$, we have

$$b \leq \frac{(1+M)a + 2M}{1-M}. \tag{1.19}$$

However, one can check that $\frac{(1+M)a+2M}{1-M} < \gamma_1(a)$. *Therefore, inequality* (1.19) *implies* $b < \gamma_1(a)$ *and we have a contradiction. This means that* c *satisfies the inequality*

$$c < \frac{1}{N}\left[-2-(1+M)a-(1-M)b\right]$$

and therefore

$$\frac{1}{2}\left|(1+M)a+(1-M)b+Nc\right| > 1.$$

Assume that $c \geq 0$. *Using the assumptions that* $b > \gamma_1(a)$ *and* $c \geq 0$, *it can be checked that* $\frac{1}{2}\left[(1+M)a+(1-M)b+Nc\right] > 0$. *Then, applying Theorem 1.4.10 we have*

$$\begin{aligned}\|(a,b,c)\|_{D(\beta)} &\geq \frac{1}{2}|(1+M)a+(1-M)b+Nc| = \frac{1}{2}[(1+M)a+(1-M)b+Nc] \\ &> (1+M)a+2+M+2\sqrt{(1+M)(1+a)} > 1.\end{aligned}$$

It follows in any case that $(a,b) \notin \pi_{ab}(\mathsf{S}_{D(\beta)})$.

Case (3): Using the symmetry of the unit ball and the previous case, if $\gamma_2(a) > b$ *with* $|a| \leq 1$, *then* $\|(a,b,c)\|_{D(\beta)} > 1$ *for all* $c \in \mathbb{R}$, *from which* $(a,b) \notin \pi_{ab}(\mathsf{S}_{D(\beta)})$. *This concludes the proof.*

Remark 1.4.13 *Let* $0 < \beta < \frac{\pi}{4}$. *The following two straightforward equivalences will be useful in the proof of the first part of the next result*

$$\begin{aligned}&\frac{1}{2}|-1-M+(1-M)b+Nc| \leq 1 \\ &\iff \frac{1}{N}[-1+M-(1-M)b] \leq c \leq \frac{1}{N}[3+M-(1-M)b].\end{aligned}$$

Also, if $b \leq -1$, *then*

$$\frac{1}{2}\left|-1+b+\sqrt{(1+b)^2+c^2}\right| \leq 1 \iff |c| \leq 2\sqrt{2(1-b)}.$$

Theorem 1.4.14 (Bernal, Muñoz, Rodríguez, and Seoane [10]) *Let* $0 < \beta < \frac{\pi}{4}$ *or* $\frac{\pi}{4} < \beta < \frac{\pi}{2}$, *and let* A *and* B *be defined as before Theorem 1.4.12. Define*

$$F_1(a,b) = \begin{cases}\frac{1}{N}[2-(1+M)a-(1-M)b] & \text{if } (a,b) \in A, \\ 2\sqrt{(1-a)(1-b)} & \text{if } (a,b) \in B,\end{cases}$$

and $F_2(a,b) = -F_1(-a,-b)$ *for all* $(a,b) \in \pi_{ab}(\mathsf{S}_{D(\beta)})$. *If*

$$\begin{aligned}\Gamma_F = &\left\{(1,b,c) \in \mathbb{R}^3 : \ (1,b) \in \partial\pi_{ab}(\mathsf{S}_{D(\beta)}), \ F_2(1,b) \leq c \leq F_1(1,b)\right\} \\ &\cup \left\{(-1,b,c) \in \mathbb{R}^3 : \ (-1,b) \in \partial\pi_{ab}(\mathsf{S}_{D(\beta)}), \ F_2(-1,b) \leq c \leq F_1(-1,b)\right\},\end{aligned}$$

then

(a) $\mathsf{S}_{D(\beta)} = \operatorname{graph}(F_1) \cup \operatorname{graph}(F_2) \cup \Gamma_F$ *(see Figures 1.15 and 1.16 for a picture of* $\mathsf{S}_{D(\beta)}$ *with* $\beta = \pi/6$ *and* $\beta = 3\pi/8$, *respectively).*

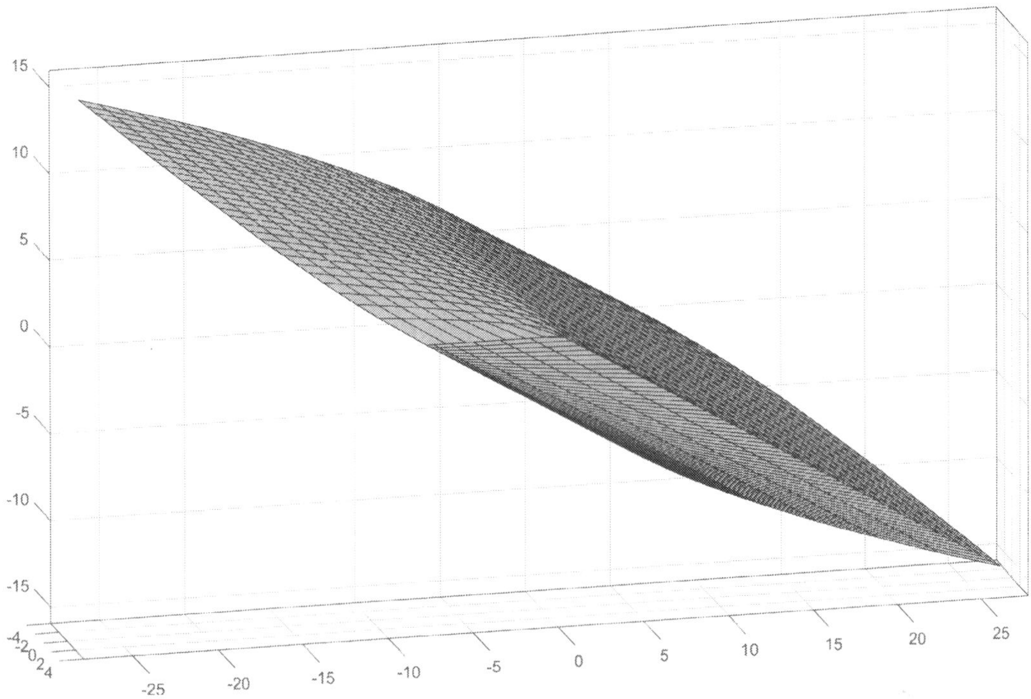

FIGURE 1.15: Representation of $\mathsf{S}_{D(\beta)}$ with $\beta = \pi/6$.

(b) The set $\mathrm{ext}(\mathsf{B}_{D(\beta)})$ *consists of the elements*

$$\pm\Bigg(t, \frac{(1+M)t + 2(2+M) + 4\sqrt{(1+M)(1+t)}}{1-M},$$
$$\frac{1}{N}\left[-2(1+M)(1+t) - 4\sqrt{(1+M)(1+t)}\right]\Bigg) \quad \textit{for} \quad t \in [-1,1],$$

$$\pm\left(1, s, -2\sqrt{2(1+s)}\right) \quad \textit{for} \quad s \in \left[\frac{1+3M}{1-M}, \frac{5+3M+4\sqrt{2(1+M)}}{1-M}\right],$$

and

$$\pm(1,1,0).$$

PROOF. *Assume that* $0 < \beta < \frac{\pi}{4}$ *(if* $\frac{\pi}{4} < \beta < \frac{\pi}{2}$*, then the result follows by using similar arguments). Notice that* $\mathrm{graph}(F_1) \subset \mathsf{S}_{D(\beta)}$ *as seen in Theorem 1.4.12. By symmetry,* $\mathrm{graph}(F_2) \subset \mathsf{S}_{D(\beta)}$*. Furthermore,* $\Gamma_F \subset \mathsf{S}_{D(\beta)}$*. Indeed, by symmetry we can focus on the points of* Γ_F *of the form* $(-1, b, c)$ *with*

$$\gamma_2(-1) = \frac{-5-3M-4\sqrt{2(1+M)}}{1-M} \leq b \leq \gamma_1(-1) = \frac{3+M}{1-M},$$

and $F_2(-1,b) \leq c \leq F_1(-1,b)$*. Notice that*

$$F_1(-1,b) = \begin{cases} \frac{1}{N}[3+M-(1-M)b] & \textit{if } \frac{-1-3M}{1-M} \leq b \leq \frac{3+M}{1-M}, \\ 2\sqrt{2(1-b)} & \textit{if } \frac{-5-3M-4\sqrt{2(1+M)}}{1-M} \leq b \leq \frac{-1-3M}{1-M}, \end{cases}$$

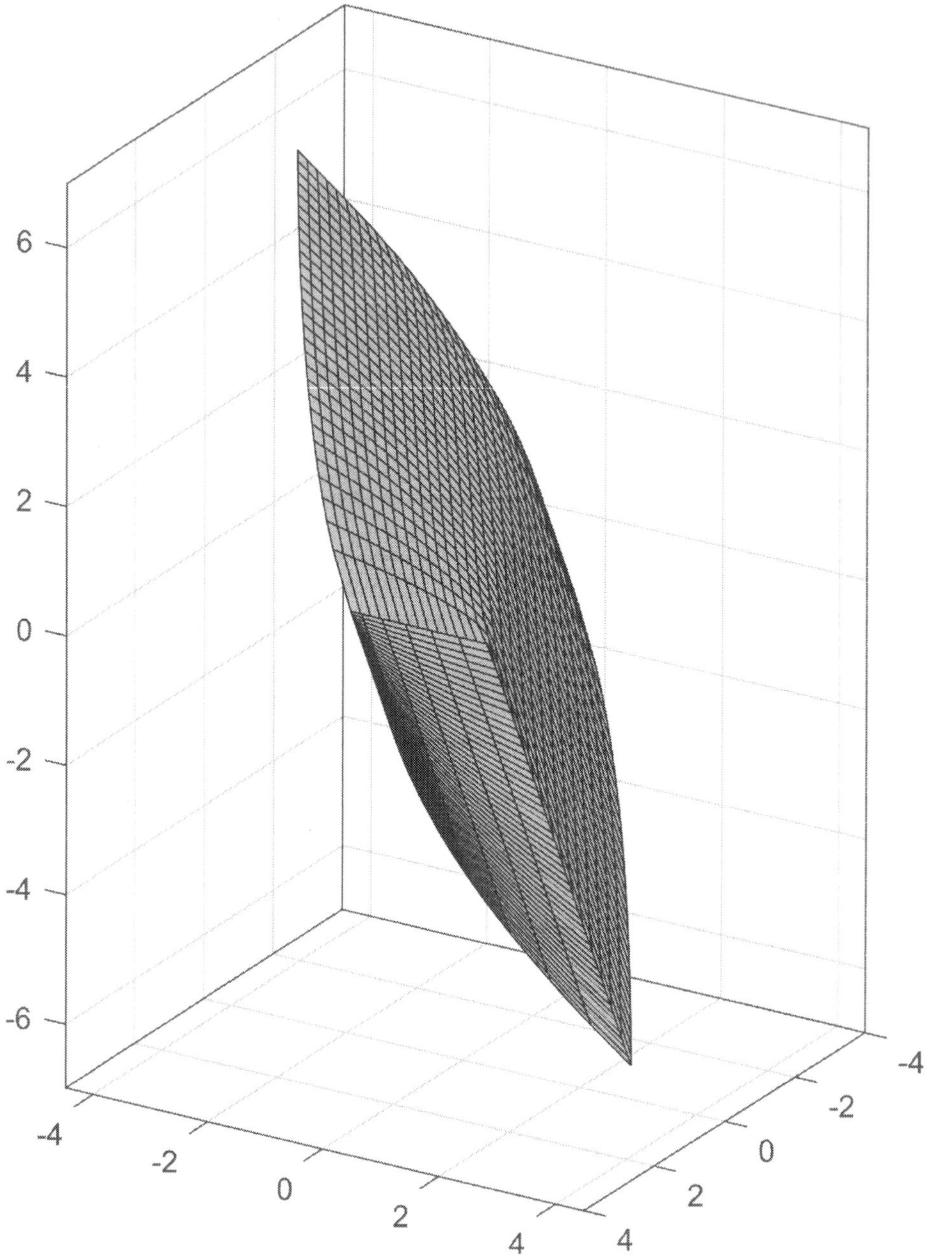

FIGURE 1.16: Representation of $\mathsf{S}_{D(\beta)}$ with $\beta = 3\pi/8$.

and

$$F_2(-1,b) = \begin{cases} \frac{1}{N}[-1+M-(1-M)b] & \text{if } \frac{-5-3M-4\sqrt{2(1+M)}}{1-M} \le b \le -1, \\ 0 & \text{if } -1 \le b \le \frac{3+M}{1-M}. \end{cases}$$

Hence, $c \ge F_2(-1,b) \ge 0$ *for all* $b \in \left[\frac{-5-3M-4\sqrt{2(1+M)}}{1-M}, \frac{3+M}{1-M}\right]$*, since it can be checked that* $\frac{1}{N}[-1+M-(1-M)b] \ge 0$ *when* $b \le -1$*. Moreover,* c *can only be* 0 *if* $-1 \le b \le \frac{3+M}{1-M}$*.*

Assume first that $c = 0$. *Then, by Remark 1.4.13, we obtain*

$$\frac{1}{2}|-1-M+(1-M)b| \leq 1.$$

Therefore, since $c(-1-b) = 0$,

$$\|(-1,b,c)\|_{D(\beta)} = \max\left\{1, \frac{1}{2}|-1-M+(1-M)b|\right\} = 1.$$

Assume now that $c > 0$. *In this case,* $\frac{-5-3M-4\sqrt{2(1+M)}}{1-M} \leq b < -1$, *which implies, for instance, that* $c(-1-b) > 0$. *If* $\tan(2\beta) \leq \frac{c}{-1-b}$, *then*

$$\|(-1,b,c)\|_{D(\beta)} = \max\left\{1, \frac{1}{2}|-1-M+(1-M)b+Nc|\right\}.$$

We know that $\frac{-5-3M-4\sqrt{2(1+M)}}{1-M} \leq b < -1$. *Now, one can check that*

$$0 < \frac{1}{N}[-1-M-(1-M)b] \leq c \leq 2\sqrt{2(1-b)}.$$

Therefore, we can apply Remark 1.4.13 to obtain $\|(-1,b,c)\|_{D(\beta)} = 1$. *If* $\tan(2\beta) > \frac{c}{-1-b}$, *by using the same arguments, we have*

$$\begin{aligned}
&\|(-1,b,c)\|_{D(\beta)}\\
&= \max\left\{1, \frac{1}{2}|-1-M+(1-M)b+Nc|, \frac{1}{2}\left|-1+b+\sqrt{(1+b)^2+c^2}\right|\right\}\\
&= 1.
\end{aligned}$$

Thus after the cascade of previous calculations, we have proved so far that

$$\text{graph}(F_1) \cup \text{graph}(F_2) \cup \Gamma_F \subset \mathsf{S}_{D(\beta)}.$$

Now, suppose $(a,b,c) \notin \text{graph}(F_1) \cup \text{graph}(F_2) \cup \Gamma_F$. *Since* $(0,0,0) \notin \mathsf{S}_{D(\beta)}$, *we may assume that* $(a,b,c) \neq (0,0,0)$. *It can be checked that the straight line* $\{\lambda(a,b,c) : \lambda \in \mathbb{R}\}$ *certainly meets the set* $\text{graph}(F_1) \cup \text{graph}(F_2) \cup \Gamma_F$. *Put* $(a,b,c) = \lambda_0(a_0,b_0,c_0)$ *with* $\lambda_0 \neq 0, 1$ *and* $(a_0,b_0,c_0) \in \text{graph}(F_1) \cup \text{graph}(F_2) \cup \Gamma_F$. *Then* $\|(a,b,c)\|_{D(\beta)} = \lambda_0\|(a_0,b_0,c_0)\|_{D(\beta)} = \lambda_0 \neq 1$, *and therefore* $(a,b,c) \notin \mathsf{S}_{D(\beta)}$. *We have proved (a).*

Part (b) of the theorem follows from the observation that the sets $\text{graph}(F_1)$, $\text{graph}(F_2)$, *and* Γ_F *are ruled surfaces in* $\mathbb{R}^3$. *Indeed, the restriction of* F_1 *to the straight lines passing through* $(1,1)$ *with positive slope are affine functions. A similar argument can be applied to* F_2, *whereas* Γ_F *is contained in the two planes* $a = \pm 1$. *Hence the only possible extreme points are contained in the intersections* $\text{graph}(F_1) \cap \Gamma_F$, $\text{graph}(F_2) \cap \Gamma_F$ *and* $\text{graph}(F_1) \cap \text{graph}(F_2)$. *Removing the interiors of all line segments within those intersections, we are left with the following candidates for extreme points of* $\mathsf{B}_{D(\beta)}$:

$$\begin{aligned}
\pm\Bigg(&t, \frac{(1+M)t+2(2+M)+4\sqrt{(1+M)(1+t)}}{1-M},\\
&\frac{1}{N}\left[-2(1+M)(1+t)-4\sqrt{(1+M)(1+t)}\right]\Bigg) \quad \textit{for} \quad t \in [-1,1],
\end{aligned}$$

$$\pm\left(1, s, -2\sqrt{2(1+s)}\right) \quad \textit{for} \quad s \in \left[\frac{1+3M}{1-M}, \frac{5+3M+4\sqrt{2(1+M)}}{1-M}\right],$$

and

$$\pm(1,1,0).$$

To confirm that these are indeed the extreme points, it suffices to construct a supporting hyperplane at each of the previous points. This is left for the reader to check, and the proof for $0<\beta<\frac{\pi}{4}$ *is complete.*

Remark 1.4.15 *The set* Γ_F *in Theorem 1.4.14 is nothing but the union of the intersection of* $\mathsf{S}_{D(\beta)}$ *with the planes* $a=-1$ *and* $a=1$*. The reader may find interesting Figure 1.17, where we can see the intersection of* $\mathsf{S}_{D(\beta)}$ *with the plane* $a=-1$*.*

More explicitly we have

$$\begin{aligned}\Gamma_F = &\left\{(-1,b,c)\in\mathbb{R}^3 : b_0\le b\le b_2,\ 0\le c\le F_1(-1,b)\right\}\\ &\cup\left\{(1,b,c)\in\mathbb{R}^3 : -b_2\le b\le -b_0,\ F_2(1,b)\le c\le 0\right\}\end{aligned}$$

where $b_0=-\frac{5+3M+4\sqrt{2(1+M)}}{1-M}$ *and* $b_2=\frac{3+M}{1-M}$ *(see Figure 1.17).*

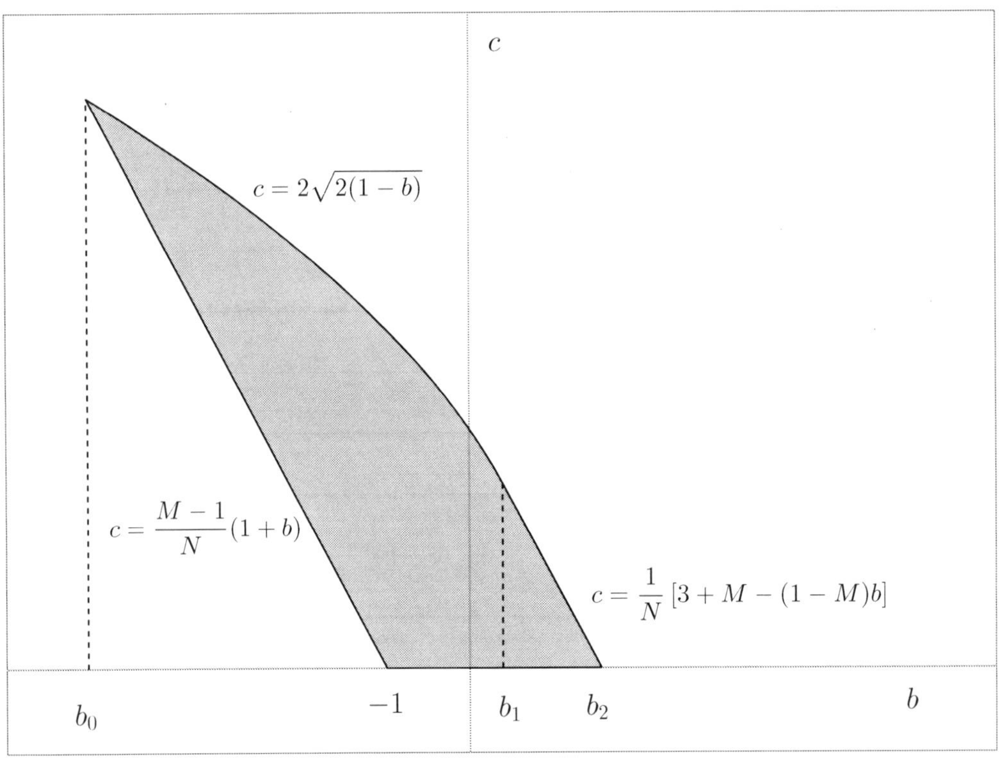

FIGURE 1.17: Representation of the intersection of the plane $a=-1$ and $\mathsf{S}_{D(\beta)}$ for $0<\beta<\pi/2$ and where $b_0=-\frac{5+3M+4\sqrt{2(1+M)}}{1-M}$, $b_1=-\frac{1+3M}{1-M}$ and $b_2=\frac{3+M}{1-M}$. In particular, in this figure $\beta=\pi/2-0.5$.

1.4.6.2 The geometry of $\mathcal{P}(^2D(\beta))$ when $\frac{\pi}{2} < \beta \leq \pi$

In this section, we describe the projection of the unit sphere of $\mathcal{P}(^2D(\beta))$ onto the ab-plane for $\frac{\pi}{2} < \beta < \frac{3\pi}{4}$ and $\frac{3\pi}{4} < \beta < \pi$. We also provide a parametrization of the unit sphere and identify the extreme points of the unit ball of $\mathcal{P}(^2D(\beta))$. The special cases $\beta = \frac{3\pi}{4}$ and $\beta = \pi$ have already been treated in previous sections of this part of the text (see, also, [81]).

Let us define the sets

$$\begin{aligned} C &:= \left\{(a,b) \in \mathbb{R}^2 : \ a \in [-1,1], \ b \leq \gamma_3(a) \text{ and } b \geq \widehat{\delta}(a)\right\}, \\ D &:= \left\{(a,b) \in \mathbb{R}^2 : \ a \in [-1,1], \ b \leq -a \text{ and } b \leq \widehat{\delta}(a)\right\}, \\ E &:= [-1,1]^2 \backslash (C \cup D), \end{aligned}$$

where

$$\begin{aligned} \gamma_3(a) &:= \frac{(1+M)a - 2(2+M) + 4\sqrt{(1+M)(1-a)}}{1-M}, \\ \widehat{\delta}(a) &:= \frac{(1+M)a + 2M}{1-M}. \end{aligned}$$

See Figure 1.18 for an illustration of the sets C, D, and E.

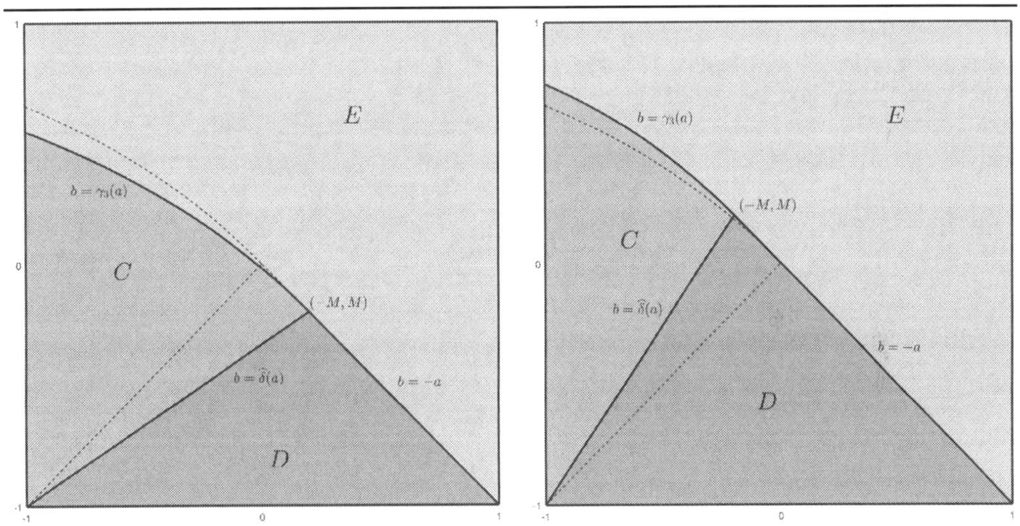

FIGURE 1.18: Scaled pictures of the projection onto the ab-plane of $\mathsf{B}_{D(\beta)}$ with $\beta \in [\pi/2, \pi]$. On the left, we depict the sets C, D, and E for $\beta = 3\pi/4 - 0.1$. On the right, we have C, D, and E for $\beta = 3\pi/4 + 0.1$. In both cases, we draw the case $\beta = 3\pi/4$ with a dashed line.

Remark 1.4.16 *Notice that the function $\widehat{\delta}(a)$ is the same as $-\delta(-a)$, where the angles for $\widehat{\delta}$ and δ are taken over the intervals $\left(\frac{\pi}{2}, \frac{3\pi}{4}\right)$ and $\left(0, \frac{\pi}{4}\right)$, respectively.*

Theorem 1.4.17 (Bernal, Muñoz, Rodríguez, and Seoane [10]**)** *Let $\frac{\pi}{2} < \beta < \frac{3\pi}{4}$ or $\frac{3\pi}{4} < \beta < \pi$. The projection of $\mathsf{S}_{D(\beta)}$ onto the ab-plane is $[-1,1]^2$.*

PROOF. *Assume that* $\frac{\pi}{2} < \beta < \frac{3\pi}{4}$ *(the proof for* $\frac{3\pi}{4} < \beta < \pi$ *can be done by using similar arguments). Let* $(a,b) \in C$*. First of all, notice that* $-C \subset P_1$*, where* P_1 *is the set defined in Lemma 1.4.11. Since* $N < 0$, $-\frac{1}{N}[2+(1+M)a+(1-M)b] > 0$ *if and only if* $2 + (1+M)a+(1-M)b > 0$*. Then, since* $1+M,\ 1-M > 0$*, we have that the minimum of* $2+(1+M)a+(1-M)b$ *is attained at* $a = b = -1$ *and in this case is* 0*. Thus,* $-\frac{1}{N}[2+(1+M)a+(1-M)b] > 0$*.*

Set $c = -\frac{1}{N}[2+(1+M)a+(1-M)b]$*. First assume that* $b > a$ *(we denote by* C_1 *the set* $\{(a,b) \in C : b > a\}$*), then* $c(a-b) < 0$*. Therefore, using Theorem 1.4.10 we have*

$$\begin{aligned}&\left\|\left(a,b,-\frac{1}{N}[2+(1+M)a+(1-M)b]\right)\right\|_{D(\beta)}\\&=\max\left\{|a|,1,\frac{1}{2}\left|a+b+\sqrt{(a-b)^2+\frac{1}{N^2}[2+(1+M)a+(1-M)b]^2}\right|\right\}\\&=\max\left\{|a|,1,\frac{1}{2}\left|a+b+\sqrt{2\left[\frac{1}{1-M}(1+a)^2+\frac{1}{1+M}(1+b)^2\right]}\right|\right\}.\end{aligned}$$

Since $-C \subset P_1$*, we have that* $-C_1 \subset P_1$*. Hence*

$$\left\|\left(a,b,-\frac{1}{N}[2+(1+M)a+(1-M)b]\right)\right\|_{D(\beta)} = 1.$$

Assume now that $b < a$ *(we denote by* C_2 *the set* $\{(a,b) \in C : b < a\}$*), then* $c(a-b) > 0$*. It can be proved that*

$$\frac{N}{M} = \frac{-\frac{1}{N}[2+(1+M)a+(1-M)b]}{a-b}$$

if and only if

$$b = \widehat{\delta}(a).$$

In fact, notice that $\tan(2\beta) \le \frac{c}{a-b}$ *in* C_2*. Therefore, using Theorem 1.4.10 we have*

$$\begin{aligned}&\left\|\left(a,b,-\frac{1}{N}[2+(1+M)a+(1-M)b]\right)\right\|_{D(\beta)}\\&=\max\left\{|a|,1,\frac{1}{2}\left|a+b+\sqrt{2\left[\frac{1}{1-M}(1+a)^2+\frac{1}{1+M}(1+b)^2\right]}\right|\right\}.\end{aligned}$$

Since $-C \subset P_1$*, it follows that* $-C_2 \subset P_1$ *as well. Therefore, we can apply Lemma 1.4.11 and we have that* $\|(a,b,c)\|_{D(\beta)} = 1$*.*

Let $(a,b) \in D$ *and set* $c = 2\sqrt{(1+a)(1+b)}$*. Since* $(a,b) \in D$*, we have* $c(a-b) \ge 0$*. In fact,* $c(a-b) = 0$ *if and only if* $a = b = -1$*. Therefore,* $c = 0$ *and, using Theorem 1.4.10, we have*

$$\|(-1,-1,0)\|_{D(\beta)} = \max\{1,|M|\} = 1.$$

Assume that $c(a-b) > 0$*. It can be proved that*

$$\frac{N}{M} = \frac{2\sqrt{(1+a)(1+b)}}{a-b} \tag{1.20}$$

implies

$$b = \widehat{\delta}(a)$$

or

$$b = \frac{(1-M)a - 2M}{1+M}.$$

Observe that the previous two equations were obtained by squaring (1.20)*. Recall here that we have*

$$\frac{N}{M} = \frac{-\frac{1}{N}[2 + (1+M)a + (1-M)b]}{a-b}$$

if and only if $b = \widehat{\delta}(a)$. *Therefore, we just need to focus on* $b = \widehat{\delta}(a)$, *because in this half-line, for* $b \geq -1$, *we have*

$$\tan(2\beta) = \frac{-\frac{1}{N}[2 + (1+M)a + (1-M)b]}{a-b} = \frac{2\sqrt{(1+a)(1+b)}}{a-b}.$$

Notice that $\tan(2\beta) > \frac{c}{a-b}$ *holds throughout D. Therefore, making use of Theorem 1.4.10 we have*

$$\begin{aligned}\left\|\left(a, b, 2\sqrt{(1+a)(1+b)}\right)\right\|_{D(\beta)} &= \frac{1}{2}\left(|a+b| + \sqrt{(a-b)^2 + 4(1+a)(1+b)}\right)\\ &= \frac{1}{2}\left(-a - b + \sqrt{(2+a+b)^2}\right) = 1.\end{aligned}$$

Let $(a, b) \in E$, *and define* $c = 2\sqrt{(1-a)(1-b)}$. *Firstly, suppose that* $b > a$; *then* $c(a-b) < 0$. *Therefore, from Theorem 1.4.10 we arrive at*

$$\begin{aligned}&\left\|\left(a, b, 2\sqrt{(1-a)(1-b)}\right)\right\|_{D(\beta)}\\ &= \max\left\{|a|, \frac{1}{2}\left|(1+M)a + (1-M)b + 2N\sqrt{(1-a)(1-b)}\right|, 1\right\}.\end{aligned}$$

Thus, after Lemma 1.4.11, part (b), we have $\left\|\left(a, b, 2\sqrt{(1-a)(1-b)}\right)\right\|_{D(\beta)} = 1$.

Also, notice the following property: if $b = -a$ *and* $b \leq \widehat{\delta}(a)$, *then*

$$\left\|\left(a, -a, 2\sqrt{1-a^2}\right)\right\|_{D(\beta)} = 1$$

and

$$\left\|\left(a, -a, 2\sqrt{1-a^2}\right)\right\|_{D(\beta)} = \max\left\{|a|, \frac{1}{2}\left|2Ma + 2N\sqrt{1-a^2}\right|, 1\right\} = 1,$$

where we have used Lemma 1.4.11, part (b). Therefore, the norms are both equal to 1 *at* $b = -a$.

Secondly, suppose that $b = a$; *then, by using the same arguments as in the case of* $b > a$ *with* $(a, b) \in E$, *we have* $\left\|\left(a, b, 2\sqrt{(1-a)(1-b)}\right)\right\|_{D(\beta)} = 1$.

Finally, suppose that $b < a$, *then* $c(a-b) > 0$. *We have two cases to consider. On one hand, assume that* $\tan(2\beta) \leq \frac{c}{a-b}$. *Then, by using the same arguments as in the case of* $b > a$ *with* $(a, b) \in E$, *we have* $\left\|\left(a, b, 2\sqrt{(1-a)(1-b)}\right)\right\|_{D(\beta)} = 1$. *On the other hand, assume that* $\tan(2\beta) \leq \frac{c}{a-b}$, *then, by Theorem 1.4.10 we have*

$$\left\|\left(a, b, 2\sqrt{(1-a)(1-b)}\right)\right\|_{D(\beta)} = \frac{1}{2}\left(|a+b| + \sqrt{(a+b-2)^2}\right) = \frac{1}{2}\left(|a+b| + |a+b-2|\right).$$

Since it is easy to see that $a+b \leq 2$, we have that $a+b-2 \leq 0$, and also since it is easy to prove that $\gamma_3(a) \leq -a$ we have that $b \geq -a$ i.e. $a+b \geq 0$. Thus,

$$\left\|\left(a, b, 2\sqrt{(1-a)(1-b)}\right)\right\|_{D(\beta)} = \frac{1}{2}(a+b-a-b+2) = 1.$$

Thus far, we have shown that $[-1,1]^2 = C \cup D \cup E \subset \pi_{ab}(\mathsf{S}_{D(\beta)})$. On the other hand, if $(a,b) \notin [-1,1]^2$, then either $|a|>1$, in which case $\|(a,b,c)\|_{D(\beta)} \geq |a| > 1$ for all $c \in \mathbb{R}$, or $|b|>1$ and $|a| \leq 1$. In the latter case, by symmetry, it suffices to consider the case $b<-1$ and $|a| \leq 1$. In this case, notice that $a-b>0$.

First, assume that $c>0$ and $\tan(2\beta) > \frac{c}{a-b}$. Then

$$\begin{aligned}\|(a,b,c)\|_{D(\beta)} &= \frac{1}{2}\left(|a+b| + \sqrt{(a-b)^2+c^2}\right)\\ &> \frac{1}{2}\left(|a+b| + |a-b|\right) = |b| > 1.\end{aligned}$$

If $c<0$, then by Theorem 1.4.10 we have

$$\begin{aligned}\|(a,b,c)\|_{D(\beta)} &\geq \frac{1}{2}\left|a+b+\operatorname{sign}(c)\sqrt{(a-b)^2+c^2}\right|\\ &= \frac{1}{2}\left(|a+b| + \sqrt{(a-b)^2+c^2}\right) \geq |b| > 1.\end{aligned}$$

If $c>0$ and $\tan(2\beta) \leq \frac{c}{a-b}$, then by Theorem 1.4.10 we have

$$\begin{aligned}\|(a,b,c)\|_{D(\beta)} &\geq \frac{1}{2}\left|(1+M)a+(1-M)b+cN\right|\\ &= \frac{1}{2}\left(-(1+M)a-(1-M)b-cN\right),\end{aligned}$$

where in the last inequality we have used that $cN<0$ since $c>0$ and $N<0$, and also $(1+M)a+(1-M)b < M(a-b) < 0$. Since $\frac{N}{M} \leq \frac{c}{a-b}$, we also have that

$$\begin{aligned}\|(a,b,c)\|_{D(\beta)} &\geq \frac{1}{2}\left(-(1+M)a-(1-M)b-cN\right)\\ &\geq \frac{1}{2}\left(-(1+M)a-(1-M)b-\frac{1-M^2}{M}a+\frac{1-M^2}{M}b\right)\\ &= \frac{1}{2M}\left(-(1+M)a+(1-M)b\right).\end{aligned}$$

Notice that $\frac{1}{2M}\left(-(1+M)a+(1-M)b\right) > -b = |b|$ since otherwise we have that $b \geq a$ and this is a contradiction. Thus, $\|(a,b,c)\|_{D(\beta)} > |b| > 1$. If $c=0$, then using the fact that the norm is continuous and $\tan(2\beta)>0$, we obtain

$$\|(a,b,c)\|_{D(\beta)} \geq \frac{1}{2}(|a+b|+|a-b|) = |b| > 1.$$

This completes the proof.

Theorem 1.4.18 (Bernal, Muñoz, Rodríguez, and Seoane [10]) *Let $\frac{\pi}{2} < \beta < \frac{3\pi}{4}$ or $\frac{3\pi}{4} < \beta < \pi$ and let C, D and E be the sets defined at the beginning of this section. Define*

$$G_1(a,b) = \begin{cases} -\frac{1}{N}[2+(1+M)a+(1-M)b] & \text{if } (a,b) \in C,\\ 2\sqrt{(1+a)(1+b)} & \text{if } (a,b) \in D,\\ 2\sqrt{(1-a)(1-b)} & \text{if } (a,b) \in E,\end{cases}$$

and $G_2(a,b) = -G_1(-a,-b)$ *for all* $(a,b) \in \pi_{ab}(\mathsf{S}_{D(\beta)})$. *Define also the set*

$$\begin{aligned}\Omega_G = &\left\{(1,b,c) \in \mathbb{R}^3 : \ (1,b) \in \partial\pi_{ab}(\mathsf{S}_{D(\beta)}),\ 0 \geq c \geq G_2(1,b)\right\}\\ &\cup \left\{(-1,b,c) \in \mathbb{R}^3 : \ (-1,b) \in \partial\pi_{ab}(\mathsf{S}_{D(\beta)}),\ 0 \leq c \leq G_1(-1,b)\right\}.\end{aligned}$$

Then the following holds:

(a) $\mathsf{S}_{D(\beta)} = \mathrm{graph}(G_1) \cup \mathrm{graph}(G_2) \cup \Omega_G$ *(see Figures 1.19 and 1.20).*

(b) *The set* $\mathrm{ext}(\mathsf{B}_{D(\beta)})$ *consists of the elements*

$$\begin{aligned}\pm\Bigg(t,\ &\frac{(1+M)t - 2(2+M) + 4\sqrt{(1+M)(1-t)}}{1-M},\\ &-\frac{1}{N}\left[2(1+M)(t-1) + 4\sqrt{(1+M)(1-t)}\right]\Bigg) \quad \text{for} \quad t \in [-1,-M],\end{aligned}$$

$$\pm\left(s,-s,2\sqrt{1-s^2}\right) \quad \text{for} \quad s \in [-M,1],$$

$$\pm\left(-1,r,2\sqrt{2(1-r)}\right) \quad \text{for} \quad r \in \left[\frac{-5-3M+4\sqrt{2(1+M)}}{1-M},1\right],$$

and

$$\pm(1,1,0).$$

PROOF. *Assume that* $\frac{\pi}{2} < \beta < \frac{3\pi}{4}$ *(if* $\frac{3\pi}{4} < \beta < \pi$*, then the result follows by using similar arguments). We know from Theorem 1.4.17 that* $\mathrm{graph}(G_1) \cup \mathrm{graph}(G_2) \subset \mathsf{S}_{D(\beta)}$. *Take* $\pm(-1,b,c)$ *with* $|b| \leq 1$ *and* $0 \leq c \leq G_1(-1,b)$. *We know that* $-(1+b)c \leq 0$. *Observe first that* $-(1+b)c < 0$ *if and only if* $c > 0$ *and* $b \neq -1$. *Then*

$$\begin{aligned}&\|\pm(-1,b,c)\|_{D(\beta)}\\ &= \max\left\{1, \frac{1}{2}\left|1+M-(1-M)b-Nc\right|, \frac{1}{2}\left|1-b-\sqrt{(1+b)^2+c^2}\right|\right\}.\end{aligned}$$

Notice that $1+M-(1-M)b-Nc \geq 0$ *if and only if* $c \geq -\frac{1}{N}[1+M-(1-M)b]$. *Observe that*

$$\begin{aligned}&\frac{1}{2}\left|1+M-(1-M)b-Nc\right|\\ &= \frac{1}{2}(1+M-(1-M)b-Nc)\\ &\leq \frac{1}{2}(1+M-(1-M)b-NG_1(-1,b))\\ &= \begin{cases} 1 & \text{if } b \leq \gamma_3(a) \text{ and } b \geq \widehat{\delta}(a),\\ \frac{1}{2}\left[(1+M)-(1-M)b-2N\sqrt{2(1-b)}\right] & \text{if } (a,b) \in I.\end{cases}\end{aligned}$$

In this case, we are not interested in the set D *because* $a = -1$. *It can be proved after some calculations that*

$$\frac{1}{2}\left[(1+M)-(1-M)b-2N\sqrt{2(1-b)}\right] \leq 1$$

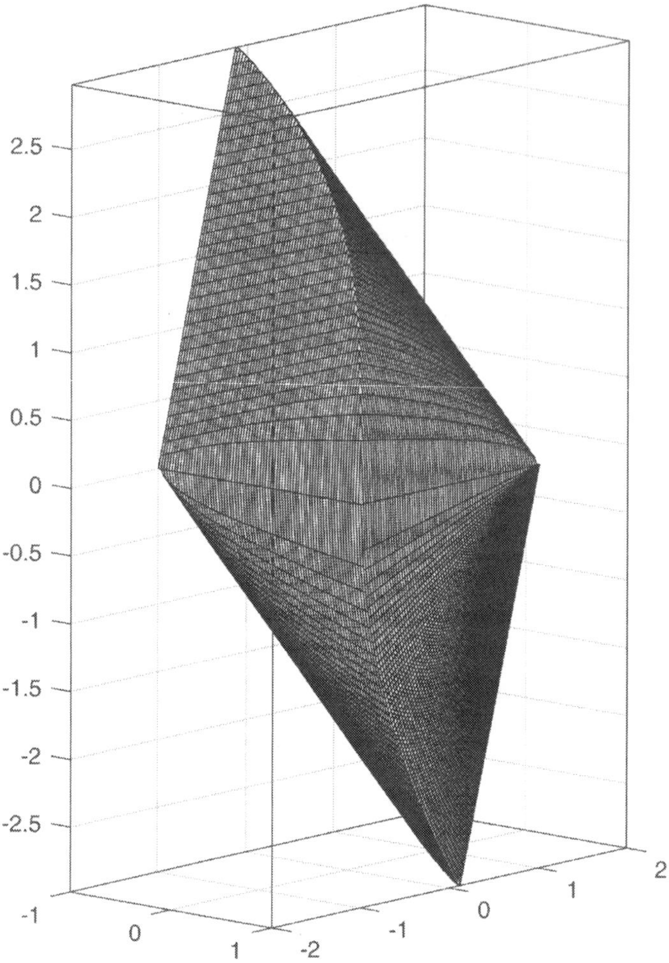

FIGURE 1.19: Representation of $\mathsf{S}_{D(\beta)}$ with $\beta = 3\pi/4 - 0.5$.

for $b \in \left[\frac{-5-3M+4\sqrt{2(1+M)}}{1-M}, 1\right]$ *(of course, here* $\frac{-5-3M+4\sqrt{2(1+M)}}{1-M} \leq 1$*). Therefore, we have proved so far that*

$$\frac{1}{2}\left|1 + M - (1-M)b - Nc\right| \leq 1.$$

It is easy to prove that $\frac{1}{2}\left|1 - b - \sqrt{(1+b)^2 + c^2}\right| \leq 1$ *if and only if* $|c| \leq 2\sqrt{2(1-b)}$ *for* $|b| \leq 1$*. Notice that*

$$\begin{aligned} 0 \leq c &\leq G_1(-1, b) \\ &= \begin{cases} -\frac{1}{N}[(1-M)(1+b)] & \text{if } -1 \leq b \leq \frac{-5-3M+4\sqrt{2(1+M)}}{1-M}, \\ 2\sqrt{2(1-b)} & \text{if } \frac{-5-3M+\sqrt{2(1+M)}}{1-M} \leq b \leq 1. \end{cases} \end{aligned}$$

It is easy to check that

$$-\frac{1}{N}\left[(1-M)(1+b)\right] \leq 2\sqrt{2(1-b)}$$

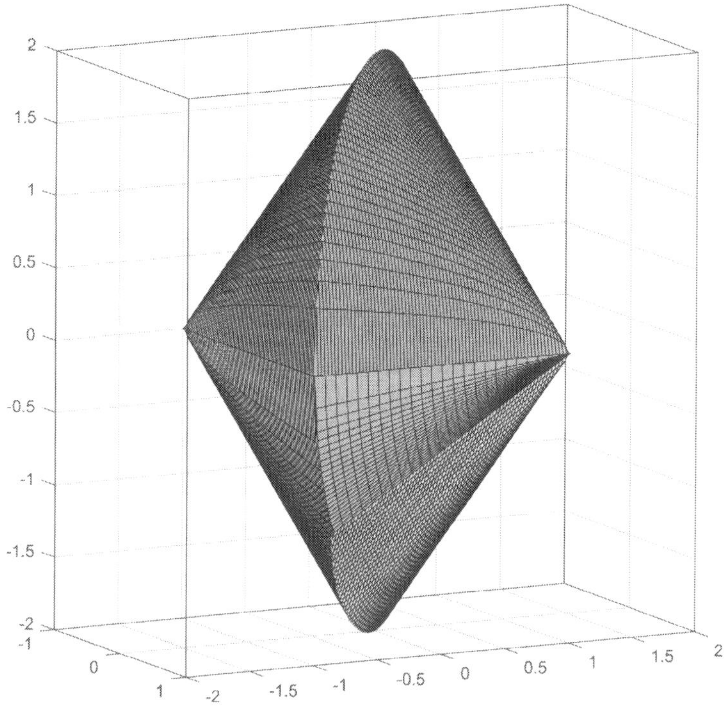

FIGURE 1.20: Representation of $\mathsf{S}_{D(\beta)}$ with $\beta = 3\pi/4 + 0.5$.

for $b \in \left[\frac{-5-3M+4\sqrt{2(1+M)}}{1-M}, 1\right]$. *This proves that*

$$\frac{1}{2}\left|1 - b - \sqrt{(1+b)^2 + c^2}\right| \leq 1.$$

Hence $\| \pm (-1, b, c)\|_{D(\beta)} = 1$. *Notice that whenever* $c = 0$ *or* $b = -1$ *we can use the same formulas as those previously employed but in this case for the third formula of Theorem 1.4.10. Therefore,* $\| \pm (-1, b, c)\|_{D(\beta)} = 1$.

This shows that $\text{graph}(G_1) \cup \text{graph}(G_2) \cup \Omega_G \subset \mathsf{S}_{D(\beta)}$. *In order to prove the reverse inclusion, we proceed as in the proof of Theorem 1.4.14, noting that any straight line passing through the origin meets the set* $\text{graph}(G_1) \cup \text{graph}(G_2) \cup \Omega_G$.

To prove the second part of the statement, notice that the set Ω_G *and the surfaces defined by the formulas* $c = 2\sqrt{(1-a)(1-b)}$, $c = 2\sqrt{(1+a)(1+b)}$, *and* $c = -\frac{1}{N}[2 + (1+M)a + (1-M)b]$, *are ruled sets of* $\mathbb{R}^3$. *This can be seen by applying the same reasoning used in the proof of Theorem 1.4.14. Therefore, the extreme points can only lie in the non-straight intersections of the four surfaces* Ω_G, $c = 2\sqrt{(1-a)(1-b)}$, $c = 2\sqrt{(1+a)(1+b)}$, *and* $c = -\frac{1}{N}[2 + (1+M)a + (1-M)b]$. *The reader can check, by constructing a supporting hyperplane, that all those points are indeed extreme, as listed in the statement.*

Remark 1.4.19 *The set* Ω_G *in Theorem 1.4.18 is nothing but the union of the intersection of* $\mathsf{S}_{D(\beta)}$ *with the planes* $a = -1$ *and* $a = 1$. *More explicitly we have*

$$\begin{aligned}\Omega_G = &\left\{(-1, b, c) \in \mathbb{R}^3 : -1 \leq b \leq 1, 0 \leq c \leq G_1(-1, b)\right\} \\ &\cup \left\{(1, b, c) \in \mathbb{R}^3 : -1 \leq b \leq 1, G_2(1, b) \leq c \leq 0\right\}.\end{aligned}$$

1.5 Some final remarks

As a final remark in this chapter, we would like to point out that there are still several different open questions regarding the geometry of polynomial spaces on non-symmetric (or unbalanced) convex bodies. Some of these are ongoing works by some of the authors, in collaboration with Emma D'Aniello, Manwook Han, Sun Kwang Kim, Jesús Llorente Jorge, Martina Maiuriello, Daniel L. Rodríguez-Vidanes, and Hyung-Joon Tag. A few examples of polynomial spaces on unbalanced convex bodies are the following:

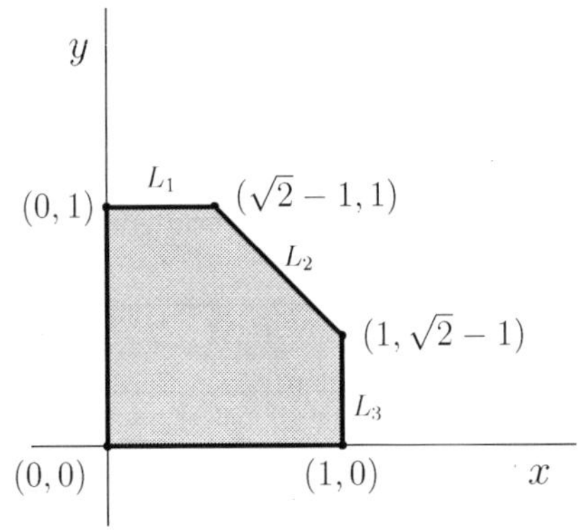

FIGURE 1.21: Sketch of the convex non-symmetric closed region $\mathfrak{H}$.

- The space of 2-homogeneous polynomials on $\mathbb{R}^2$ of the form $ax^2+bxy+cy^2$ endowed with the norm

$$\|(a,b,c)\|_{\mathfrak{H}} = \sup\left\{|ax^2+bxy+cy^2| : (x,y)\in[0,1]^2 \text{ with } x+y\leq\sqrt{2}\right\}.$$

  (See Figure 1.21.)

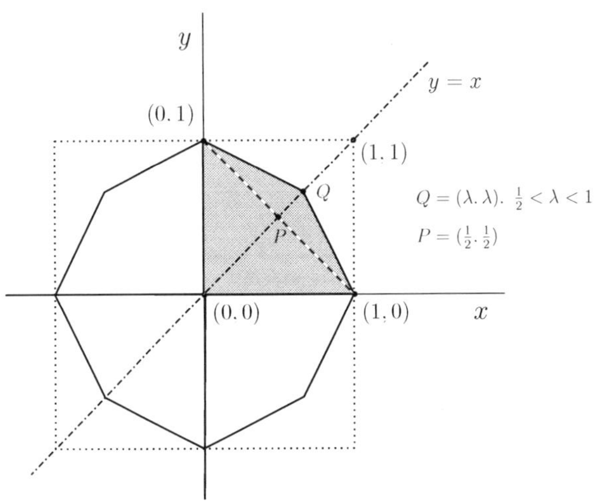

FIGURE 1.22: Sketch (shaded in grey) of the convex non-symmetric closed region $\mathfrak{G}(\lambda)$.

- The space 2-homogeneous polynomials on $\mathbb{R}^2$ of the form $ax^2 + bxy + cy^2$ endowed with the norm

$$\|ax^2 + bxy + cy^2\|_{\mathfrak{G}(\lambda)} = \sup\left\{|ax^2 + bxy + cy^2| : (x, y) \in \mathfrak{G}(\lambda)\right\}$$

where $\mathfrak{G}(\lambda)$ stands for the closed convex region given by

$$\mathfrak{G}(\lambda) = \left\{(x, y) \in [0, 1]^2 \text{ with } y \leq \min\left\{\frac{\lambda - 1}{\lambda}x + 1, \frac{\lambda}{\lambda - 1}(x - 1)\right\}\right\} \tag{1.21}$$

for $\lambda \in \left[\frac{1}{2}, 1\right]$. That is, $\mathfrak{G}(\lambda)$ is exactly the polygon of vertices $(0, 0)$, $(0, 1)$, $(1, 0)$, and (λ, λ). If $\lambda = 1/2$ we have that $\mathfrak{G}(1/2)$ is the simplex (the triangle of vertices $(0, 0)$, $(0, 1)$, and $(1, 0)$) and, if $\lambda = 1$, we get the square $[0, 1]^2$.

- The space of 2-homogeneous polynomials on $\mathbb{R}^2$ of the form $ax^2 + bxy + cy^2$ with the sup norm on the non-symmetric square $\square_h$ given in Figure 1.5. This is part of a work currently in progress by Jesús Llorente Jorge, Daniel L. Rodríguez-Vidanes, Hyung-Joon Tag, and the fourth and fifth authors of this text.

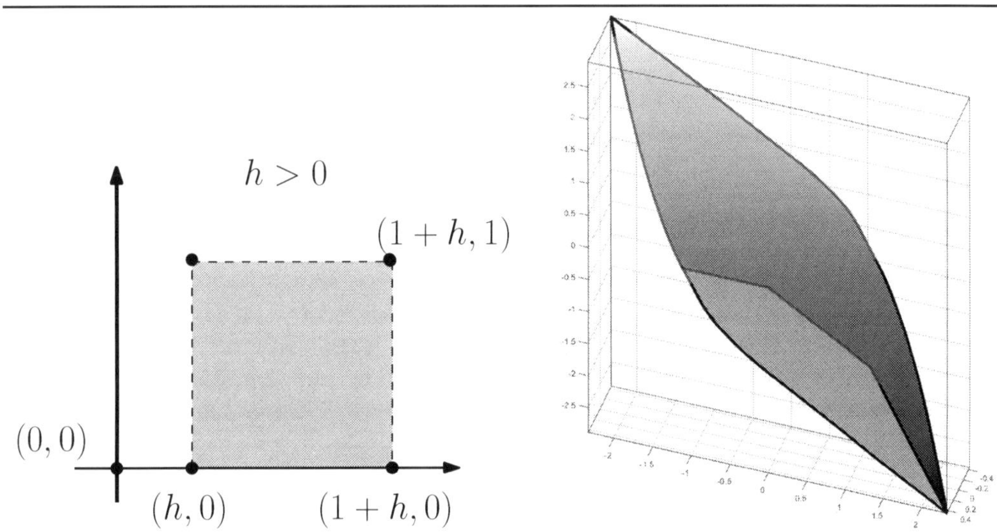

FIGURE 1.23: Sketch (shaded in grey) of the convex non-symmetric square together with the unit sphere of the space of quadratic forms on $\square_h$ for $h = \frac{1}{2}$.

Chapter 2

The Geometry of Homogeneous Trinomials on the Unit Square

2.1 Preliminaries

In addition to the other parts, this text is not only expository in nature; in other words, the entire Chapter 2 is completely original and contains material never published before.

Let us provide a precise definition of the polynomial space that will be examined in this part of the text. Consider the space consisting of the homogeneous trinomials $ax^m + bx^{m-n}y^n + cy^m$ on the plane $\mathbb{R}^2$ where $a, b, c \in \mathbb{R}$ and $m, n \in \mathbb{N}$ are such that $m > n$. That space, endowed with the norm

$$|||ax^m + bx^{m-n}y^n + cy^m|||_{m,n} = \sup\left\{|ax^m + bx^{m-n}y^n + cy^m| : (x, y) \in [-1, 1]^2\right\}$$

will be denoted by $\mathcal{P}^h_{m,n}(\mathbb{R})$. The aim of this chapter is to explore the geometry of the unit ball of $\mathcal{P}^h_{m,n}(\mathbb{R})$, extending and generalizing the results established by Choi, Kim, and Ki in 1998 [23], Aron and Klimek in 2001 [6], and Muñoz and Seoane in 2008 [88]. Furthermore, we complete the investigation of $\mathcal{P}^h_{m,n}(\mathbb{R})$ that was initiated in [63] and [47].

In particular, we will provide a complete description of the unit sphere $\mathsf{S}^h_{m,n}$ of $\mathcal{P}^h_{m,n}(\mathbb{R})$ and characterize the extreme points of the unit ball $\mathsf{B}^h_{m,n}$ of $\mathcal{P}^h_{m,n}(\mathbb{R})$. The solutions to these two questions depend significantly on whether m and n are even or odd. A thorough analysis reveals that the problem must be examined in the following three main cases:

1. Case A: m is odd.
2. Case B: Both m and n are even.
3. Case C: m is even and n is odd.

Cases A and B have been previously studied in [63] and [47], respectively. For the sake of completeness, we will present the known results for each case in the following.

Here, let us take a moment to define and fix some notations that will be used exclusively in the Sections 2.1.1 and 2.1.2:

$$K_{m,n} = \frac{n}{m-n}\left(\frac{m-n}{m}\right)^{\frac{m}{n}}, \quad J_{m,n} = K_{m,m-n}$$

$$L_{m,n} = \frac{m}{m-n}\left(\frac{m-n}{n}\right)^{\frac{n}{m}} \quad \text{and} \quad T_{m,n} = \left(\frac{J_{m,n}}{2}\right)^{\frac{m-n}{m}} = \frac{m-n}{2m}\left(\frac{2n}{m-n}\right)^{\frac{n}{m}}.$$

2.1.1 Case A: m is odd

For Case A, the detailed study of the geometry of the unit sphere $\mathsf{S}^h_{m,n}$ of $\mathcal{P}^h_{m,n}(\mathbb{R})$ needs to be divided into two different subcases, depending on whether $\frac{m}{n} < 2$ or $\frac{m}{n} > 2$

DOI: 10.1201/9781003590637-2

(since $\frac{m}{n} = 2$ is not possible in this case). Although the two subcases within Case A are similar, let us briefly describe below what the geometry of the unit balls "looks like."

As mentioned earlier, the study of the geometry of $\mathsf{B}_{m,n}$ depends strongly on whether m and n are even or odd and each of the four possible choices of the parity of m and n requires a specific treatment (see [88]). As a related space to $\mathcal{P}^h_{m,n}(\mathbb{R})$, consider the space $\mathcal{P}_{m,n}(\mathbb{R})$ consisting of the trinomials in the real line of the form $ax^m + bx^n + c$ endowed with the norm

$$\|ax^m + bx^n + c\|_{m,n} = \sup\{|ax^m + bx^n + c| : x \in [-1, 1]\}.$$

For simplicity, we will frequently use the representation of the polynomials $ax^m + bx^{m-n}y^n + cy^m \in \mathcal{P}^h_{m,n}(\mathbb{R})$ or $ax^m + bx^n + c \in \mathcal{P}_{m,n}(\mathbb{R})$ as the triple $(a, b, c) \in \mathbb{R}^3$, that is,

$$|||(a, b, c)|||_{m,n} = \sup\left\{|ax^m + bx^{m-n}y^n + cy^m| : (x, y) \in [-1, 1]^2\right\}, \text{ and}$$

$$\|(a, b, c)\|_{m,n} = \sup\{|ax^m + bx^n + c| : x \in [-1, 1]\}.$$

The norms $|||\cdot|||_{m,n}$ and $\|\cdot\|_{m,n}$ are tightly related; in fact, using the fact that the elements of $\mathcal{P}^h_{m,n}(\mathbb{R})$ attain their norm on the set $\{(1, y), (x, 1) : x, y \in [-1, 1]\}$ by homogeneity and symmetry, it is not difficult to see that

$$|||(a, b, c)|||_{m,n} = \max\{\|(a, b, c)\|_{m,m-n}, \|(c, b, a)\|_{m,n}\}, \tag{2.22}$$

for every $(a, b, c) \in \mathbb{R}^3$. The previous identity reveals that

$$|||(a, b, c)|||_{m,n} = |||(c, b, a)|||_{m,m-n}, \tag{2.23}$$

for every $(a, b, c) \in \mathbb{R}^3$. The identity (2.23) allows us to simplify the study of the geometry of $\mathsf{B}^h_{m,n}$, at least when m is odd. The case where both m and n are odd can be reduced to the case where m is odd and n is even by swapping a and c on the one hand, and n and $m - n$ on the other. Thus, in order to describe the Case A, we shall focus our attention on the case m odd and n even.

First, let us present the following auxiliary result, that was already stated in page 3 (Lemma 1.1.4) but we state it here again for the sake of completeness, followed by the explicit formulas for the norm $|||\cdot|||_{m,n}$.

Lemma 2.1.1 ([88, Lemma 2.1]) *If $m, n \in \mathbb{N}$ are such that $m > n$ then the equation*

$$|n + mx| = (m - n)|x|^{\frac{m}{m-n}}$$

has only three roots, one at $x = -1$, another one at a point $\lambda_0 \in (-\frac{n}{m}, 0)$ and a third one at a point $\lambda_1 > 0$. In addition to that we have

$$|n + mx| < (m - n)|x|^{\frac{m}{m-n}}, \tag{2.24}$$

if and only if $x < \lambda_0$ or $x > \lambda_1$.

Remark 2.1.2 *The dependence of λ_0, in Proposition 2.1.1, on m and n justifies the notation $\lambda_0(m, n)$ to represent λ_0. The value of $\lambda_0(m, m - n)$ for every odd number m and every even number n with $m > n$ will play an important role in the results for Case A. For short, we put $\mu_0(m, n) = \lambda_0(m, m - n)$, or simply $\mu_0 = \mu_0(m, n)$.*

Proposition 2.1.3 ([63, Theorem 3.4]) *Let $m, n \in \mathbb{N}$ with $m > n$, m odd and n even. Consider the interval $I_{m,n} = [\eta_1, \eta_2]$, where $\eta_1 = -\frac{m}{m-n}$, $\eta_2 = \frac{m}{m-n}\mu_0$ and $\mu_0 = \mu_0(m, n)$*

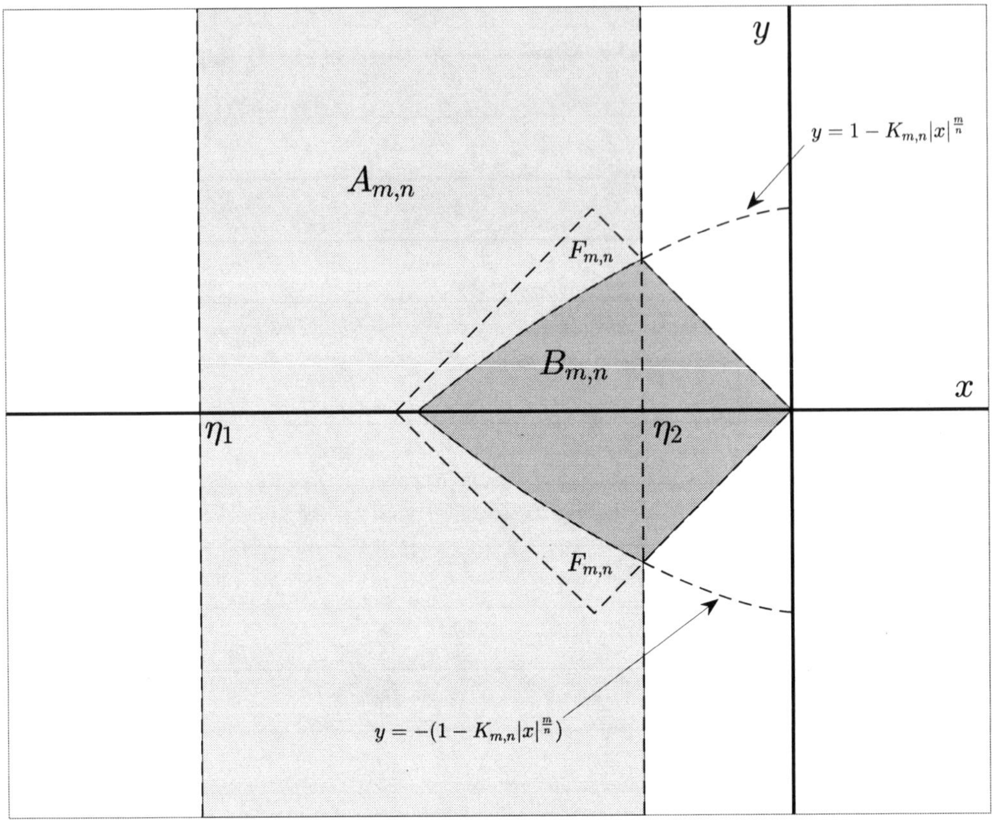

FIGURE 2.1: Regions appearing in the definition of $|||\cdot|||_{m,n}$ in Case A, where m is odd, n is even and $\frac{m}{n} < 2$. The case considered in the picture corresponds to the choice $m = 3$ and $n = 2$.

is the number in $(-\frac{m-n}{m}, 0)$ *introduced in Proposition 2.1.1, and the sets* $A_{m,n}$, $F_{m,n}$, $B_{m,n}$ *and* $\mathcal{B}$ *(see Figures 2.1 and 2.2) given by*

$$\begin{aligned} A_{m,n} &= \left\{(x, y) \in \mathbb{R}^2 : x \in I_{m,n} \text{ and } |y| \geq 1 - K_{m,n}|x|^{\frac{m}{n}}\right\}, \\ F_{m,n} &= \{(x, y) \in \mathbb{R}^2 : x \in I_{m,n} \text{ and } 1 - K_{m,n}|x|^{\frac{m}{n}} < |y| < 1 - |1 + x|\}, \\ \mathcal{B} &= \{(x, y) \in \mathbb{R}^2 : |x + 1| + |y| < 1\}, \\ B_{m,n} &= \mathcal{B} \setminus F_{m,n}. \end{aligned}$$

Then,

$$|||(a, b, c)|||_{m,n} = \begin{cases} \frac{n|a|}{m-n}\left|\frac{(m-n)b}{ma}\right|^{\frac{m}{n}} + |c| & \text{if } a \neq 0 \text{ and } \left(\frac{b}{a}, \frac{c}{a}\right) \in A_{m,n}, \\ |a| & \text{if } a \neq 0 \text{ and } \left(\frac{b}{a}, \frac{c}{a}\right) \in B_{m,n}, \\ |a + b| + |c| & \text{otherwise.} \end{cases} \tag{2.25}$$

Once the formulas for the norm are provided, a different matter is to obtain the set of extreme points for the unit ball. In this Case A, the description of extreme points of $\mathsf{B}^h_{m,n}$ are given in [63], namely:

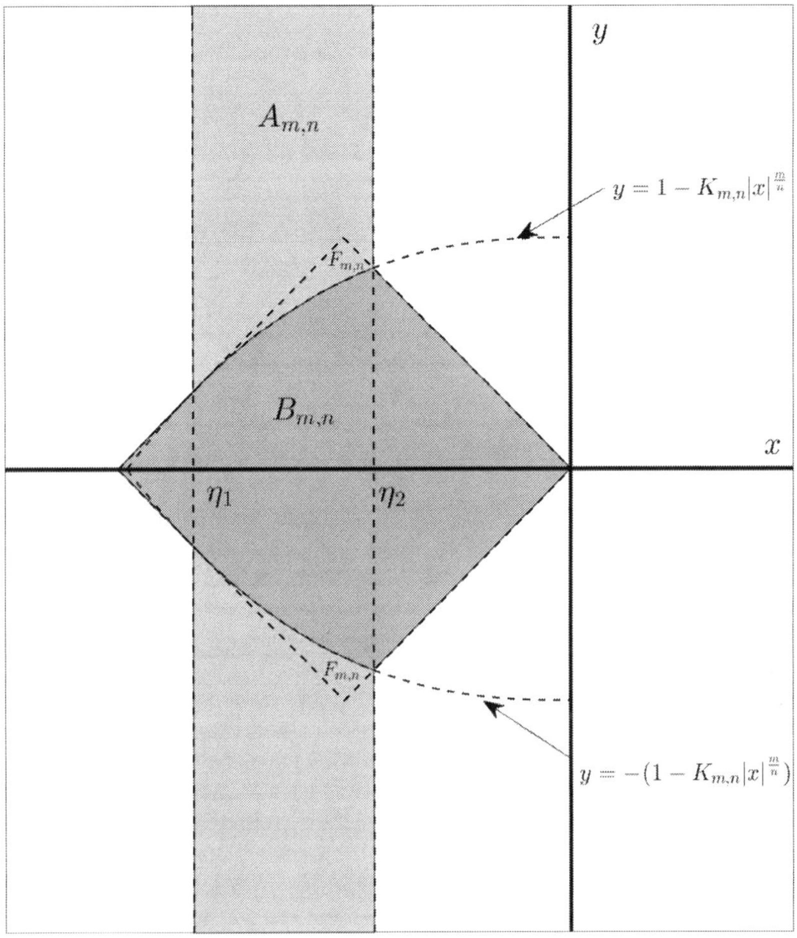

FIGURE 2.2: Regions appearing in the definition of $\|\!|\cdot|\!\|_{m,n}$ in Case A, where m is odd, n is even and $\frac{m}{n} > 2$. The case considered in the picture corresponds to the choice $m = 5$ and $n = 2$.

Proposition 2.1.4 ([63, Theorem 3.6]) *Let $m, n \in \mathbb{N}$ be such that $m > n$, m is odd and n is even and suppose that*

$$a_0 = \frac{m-n}{n}, \quad \eta_1 = -\frac{m}{m-n}, \quad \text{and} \quad \eta_2 = \frac{m}{m-n}\mu_0.$$

We have:

- *If $\frac{m}{n} < 2$, then*

$$\begin{aligned} \operatorname{ext}(\mathsf{B}^h_{m,n}) &= \left\{\pm\left(-1, t, \pm(1 - K_{m,n}|t|^{\frac{m}{n}})\right) : t \in [-\eta_2, L_{m,n}]\right\} \\ &\cup \left\{\pm(0, s, L_{m,n}|s|^{\frac{m}{n}}) : s \in [-1, -a_0]\right\} \\ &\cup \{(\pm 1, 0, 0), (0, 0 \pm 1)\}. \end{aligned}$$

- *If $\frac{m}{n} > 2$, then*

$$\begin{aligned} \operatorname{ext}(\mathsf{B}^h_{m,n}) &= \left\{\pm\left(-1, t, \pm(1 - K_{m,n}|t|^{\frac{m}{n}})\right) : t \in [-\eta_2, -\eta_1]\right\} \\ &\cup \{(\pm 1, 0, 0), (0, 0 \pm 1), \pm(1, -2, 0)\}. \end{aligned}$$

These extreme points are spotted in Figures 2.3 and 2.4 (which are represented for some choices of m and n).

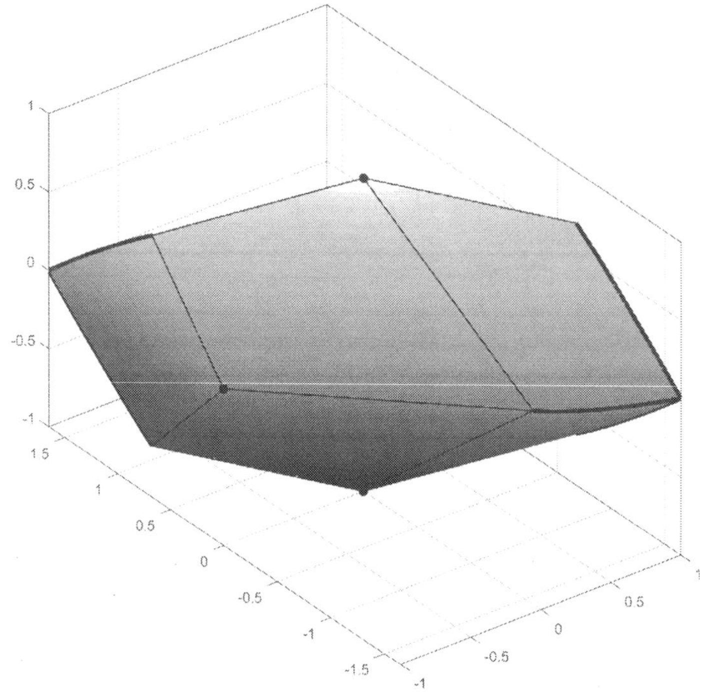

FIGURE 2.3: Sketch of $\mathsf{S}^h_{m,n}$ with $\frac{m}{n} < 2$ in Case A. The picture corresponds with the choice $m = 5$, $n = 4$. The extreme points appear with a thicker line or big dots. The surfaces that form $\mathsf{S}^h_{m,n}$ are delimited by thin lines.

2.1.2 Case B: m and n are both even

Regarding Case B, where both m and n are even, the study of the norm and geometry of the unit sphere is highly technical and complex. In this case, we need to consider three different situations: $\frac{n}{m} \in (0, 1/3)$, $\frac{n}{m} \in [1/3, 2/3]$, and $\frac{n}{m} \in (2/3, 1)$. The outcomes differ quite a bit depending on each situation. To make this part self-contained, we will present these results as well, which can be found in [47].

First, regarding the explicit formula of $|||\cdot|||_{m,n}$ in this case, we introduce the sets $A_{m,n}$, $B_{m,n}$, $C_{m,n}$, and $C_{m,n}$ which will play a crucial role in our analysis. These sets are defined as follows (*note that their definitions differ from those in Section 2.1.1*):

$$A_{m,n} = \left\{(a, c) \in \mathbb{R}^2 : \ a \leq -1 + \frac{n}{m} \text{ and } c \geq -K_{m,n}|a|^{\frac{n-m}{n}} - a\right\}$$

for every $\frac{n}{m} \in \left(0, \frac{2}{3}\right]$. The definition of $B_{m,n}$ depends of whether $\frac{n}{m} \in \left(0, \frac{1}{3}\right)$ or $\frac{n}{m} \in \left[\frac{1}{3}, \frac{2}{3}\right]$. If $\frac{n}{m} \in \left(0, \frac{1}{3}\right)$ then

$$B_{m,n} = \left\{(a, c) \in \mathbb{R}^2 : \ c \in \left[-T_{m,n}, -\frac{n}{m}\right] \text{ and } a \geq \frac{-J_{m,n}}{2}|c|^{-\frac{n}{m-n}}\right\}$$
$$\bigcup \left\{(a, c) \in \mathbb{R}^2 : \ c \leq -T_{m,n} \text{ and } a \geq -J_{m,n}|c|^{-\frac{n}{m-n}} - c\right\}.$$

If $\frac{n}{m} \in \left[\frac{1}{3}, \frac{2}{3}\right]$ then

$$B_{m,n} = \left\{(a, c) \in \mathbb{R}^2 : \ c \leq -\frac{n}{m} \text{ and } a \geq -J_{m,n}|c|^{-\frac{n}{m-n}} - c\right\}.$$

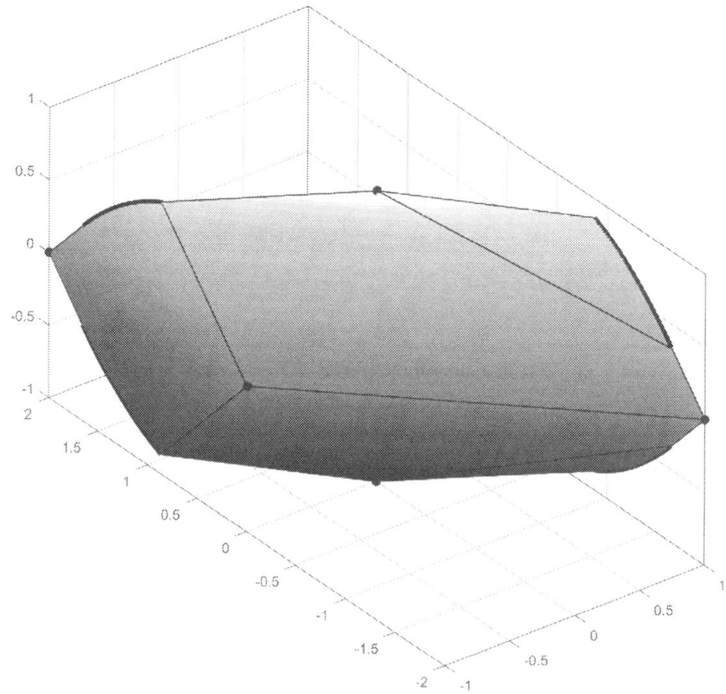

FIGURE 2.4: Sketch of $\mathsf{S}^h_{m,n}$ with $\frac{m}{n} > 2$ in Case A. The picture corresponds with the choice $m = 5$, $n = 2$. The extreme points appear with a thicker line or big dots. The surfaces that form ${\mathsf{S}^h}_{m,n}$ are delimited by thin lines.

Finally

$$\begin{aligned} C_{m,n} &= \{(a,c) \in \mathbb{R}^2 : c \leq a \text{ and } (a,c) \notin \operatorname{Int}(B_{m,n})\}, \\ D_{m,n} &= \{(a,c) \in \mathbb{R}^2 : c \geq a \text{ and } (a,c) \notin \operatorname{Int}(A_{m,n}) \cup \operatorname{Int}(B_{m,n})\}, \end{aligned}$$

where here, and from now on, $\operatorname{Int}(X)$ represents the interior of any set $X \subseteq \mathbb{R}^2$.

Proposition 2.1.5 ([47, Theorem 2.6]) *Let $m, n \in \mathbb{N}$ be even with $m > n$. Consider the sets $A_{m,n}$, $B_{m,n}$, $C_{m,n}$, and $D_{m,n}$, (see Figures 2.5 and 2.6). Letting*

$$\begin{aligned} \Gamma_{m,n}(a,b,c) &= K_{m,n}|a| \left|\frac{a}{b}\right|^{-\frac{m}{n}} + \operatorname{sign}(b)c, \\ \Upsilon(a,b,c) &= \left|\frac{a+b}{2} + c\right| + \left|\frac{a+b}{2}\right|, \end{aligned}$$

for all $(a,b,c) \in \mathbb{R}^3$, then the explicit value for $\|\|(a,b,c)\|\|_{m,n}$ is given by

$$\begin{aligned} \Gamma_{m,n}(a,b,c) \text{ if } b \neq 0 \text{ and } \left(\frac{a}{b}, \frac{c}{b}\right) &\in A_{m,n}, \\ \Gamma_{m,m-n}(c,b,a) \text{ if } b \neq 0 \text{ and } \left(\frac{a}{b}, \frac{c}{b}\right) &\in B_{m,n}, \\ \Upsilon(a,b,c) \text{ if } b \neq 0 \text{ and } \left(\frac{a}{b}, \frac{c}{b}\right) &\in C_{m,n} \text{ or } b = 0 \text{ and } |c| \geq |a|, \\ \Upsilon(c,b,a) \text{ if } b \neq 0 \text{ and } \left(\frac{a}{b}, \frac{c}{b}\right) &\in D_{m,n} \text{ or } b = 0 \text{ and } |c| \leq |a|. \end{aligned}$$

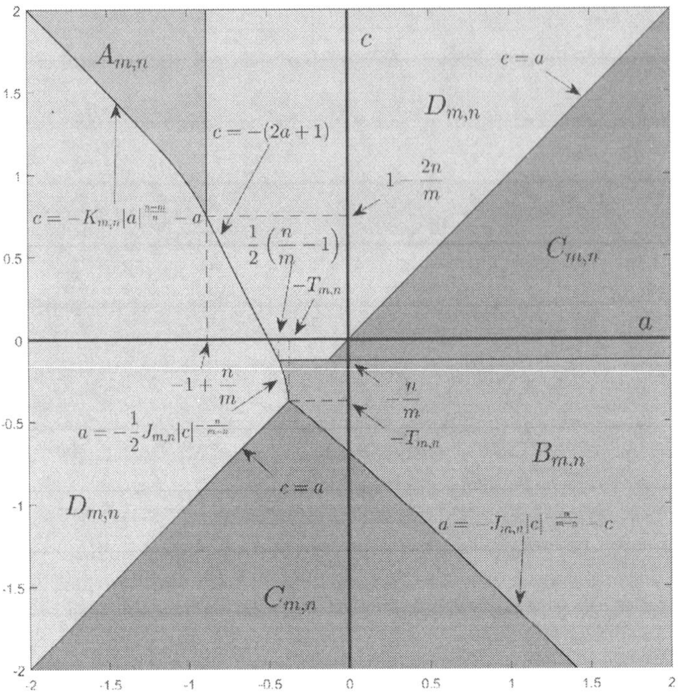

FIGURE 2.5: Representation of the sets $A_{m,n}$, $B_{m,n}$, $C_{m,n}$, and $D_{m,n}$ in Case B for $\frac{n}{m} < \frac{1}{3}$. Here we have considered the case $m = 16$ and $n = 2$.

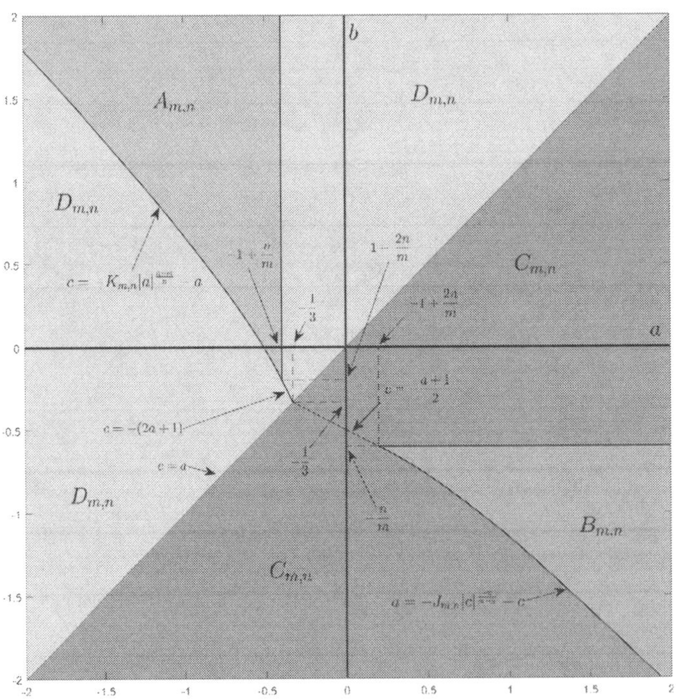

FIGURE 2.6: Representation of the sets $A_{m,n}$, $B_{m,n}$, $C_{m,n}$, and $D_{m,n}$ in Case B for $\frac{1}{3} \leq \frac{n}{m} \leq \frac{2}{3}$. Here we have considered the case $m = 30$ and $n = 18$.

Of course, the description of the extreme points of the corresponding unit sphere is also hard to tackle; however, this was successfully achieved in [47].

Proposition 2.1.6 ([47, Theorem 4.2]) *Let $m, n \in \mathbb{N}$ be even numbers with $m > n$. Also, as earlier, we let $L_{m,n} = \frac{m}{m-n}\left(\frac{m-n}{n}\right)^{\frac{n}{m}}$, $\lambda_0 = -\frac{n}{m-n}$ and $R_{m,n} = 2^{\frac{m-n}{m}} L_{m,n}$. The set of extreme points $\mathrm{ext}(\mathsf{B}^h_{m,n})$ is given by*

1. *If $\frac{n}{m} \in \left(0, \frac{1}{3}\right)$ then*

$$\begin{aligned}\mathrm{ext}(\mathsf{B}^h_{m,n}) = &\left\{\pm\left(-1, L_{m,n}(1-c)^{\frac{n}{m}}, c\right) : c \in [1+\lambda_0, 1]\right\}\\ &\bigcup\left\{\pm\left(-1, R_{m,n}|c|^{\frac{n}{m}}, c\right) : c \in [-1, 2\lambda_0]\right\}\\ &\bigcup\left\{\pm\left(a, L_{m,n}(1-a)^{\frac{m-n}{m}}, -1\right) : a \in [-1, 1]\right\}\\ &\bigcup\{\pm(0,0,1), \pm(1,0,0), \pm(1,-1,1)\}.\end{aligned}$$

2. *If $\frac{n}{m} \in \left[\frac{1}{3}, \frac{2}{3}\right]$ then*

$$\begin{aligned}\mathrm{ext}(\mathsf{B}^h_{m,n}) = &\left\{\pm\left(-1, L_{m,n}(1-c)^{\frac{n}{m}}, c\right) : c \in [1+\lambda_0, 1]\right\}\\ &\bigcup\left\{\pm\left(a, L_{m,n}(1-a)^{\frac{m-n}{m}}, -1\right) : a \in \left[1+\frac{1}{\lambda_0}, 1\right]\right\}\\ &\bigcup\{\pm(0,0,1), \pm(1,0,0), \pm(1,-1,1), \pm(1,-3,1)\}.\end{aligned}$$

3. *If $\frac{n}{m} \in \left(\frac{2}{3}, 1\right)$ then*

$$\begin{aligned}\mathrm{ext}(\mathsf{B}^h_{m,n}) = &\left\{\pm\left(-1, L_{m,n}(1-c)^{\frac{n}{m}}, c\right) : c \in [-1, 1]\right\}\\ &\bigcup\left\{\pm\left(a, R_{m,m-n}|a|^{\frac{m-n}{m}}, -1\right) : a \in \left[-1, \frac{2}{\lambda_0}\right]\right\}\\ &\bigcup\left\{\pm\left(a, L_{m,n}(1-a)^{\frac{m-n}{m}}, -1\right) : a \in \left[1+\frac{1}{\lambda_0}, 1\right]\right\}\\ &\bigcup\{\pm(0,0,1), \pm(1,0,0), \pm(1,-1,1)\}.\end{aligned}$$

Some sketches of the unit spheres together with the corresponding sets of extreme points are shown in Figures 2.7 and 2.8.

Certainly, as the reader can see, Case B is much more complex than Case A, at least at first sight. In fact, when looking into the details and technicalities of the constructions in [63] or [47] (see also the recent monograph [44]), one realizes that these types of problems not easy to tackle at all.

2.2 What we know now: The remaining case

So far, we have briefly discussed the formulas for $|||\cdot|||_{m,n}$ and the complete descriptions of the extreme points of $\mathsf{B}^h_{m,n}$ in Case A and Case B. One of the most challenging aspects of solving Case C is that many of the equations that naturally arise in the computation are implicit, unlike the explicit equations encountered in Case A or Case B. One way to address this

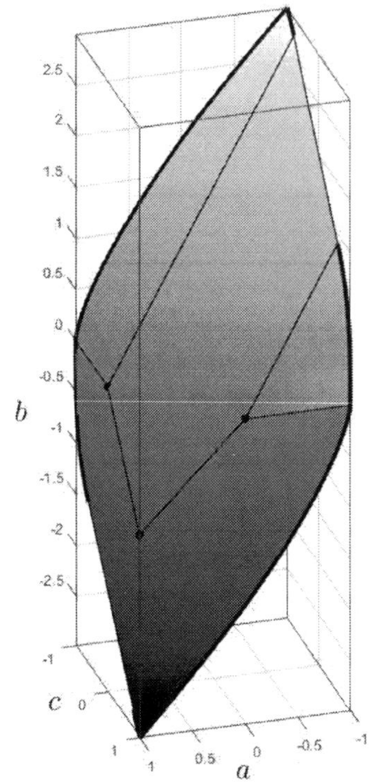

FIGURE 2.7: Representation of $\mathsf{S}^h_{m,n}$ in Case B for $\frac{n}{m} < \frac{1}{3}$. Here we have considered the case $m = 28$ and $n = 8$. The extreme points have been drawn with a thicker line or isolated dots. The different surfaces that form $\mathsf{S}^h_{m,n}$ are delimited by thin lines.

difficulty is by implementing an appropriate change of variables, which requires considering two different coordinate systems simultaneously. Despite these difficulties, the rest of the chapter will be devoted to solving Case C. From now on, unless otherwise specified, m and n will be positive integers such that $m > n$, with m being even and n being odd. Solving this final case will complete the general analysis, covering all possibilities for m and n.

The outline of the rest of this chapter is as follows: Section 2.2.1 is devoted to obtain an explicit formula for $\|\|\cdot\|\|_{m,n}$ for all positive integers $m > n$ with m even and n odd. The formula obtained in Section 2.2.1 will be used in Section 2.2.2 to calculate the projection of $\mathsf{B}^h_{m,n}$ over the plane ac together with a parametrization of $\mathsf{S}^h_{m,n}$. Thanks to the results in Sections 2.2.1 and 2.2.2, we will be able to complete this thorough study by providing the extreme points of $\mathsf{B}^h_{m,n}$ in Section 2.2.3; thus resolving Case C. Additionally, the sphere $\mathsf{S}^h_{m,n}$ will be sketched for several choices of m and n.

The following notations will be useful to understand the rest of the chapter. If C is a convex body, ext(C) will denote the set of extreme points of C. Also, The notation π_{ac} will denote the linear projection given by $\pi_{ac}(a, b, c) = (a, c)$, for every $(a, b, c) \in \mathbb{R}^3$. The plots of $\mathsf{S}^h_{m,n}$ and the projection $\pi_{ac}(\mathsf{S}^h_{m,n})$, together with some other figures were generated using *MATLAB*. All graphs presented here are scaled.

We fix some notations that will be valid only within Sections 2.2.1, 2.2.2, and 2.2.3:

$$K_{m,n} = \frac{n}{m-n}\left(\frac{m-n}{m}\right)^{\frac{m}{n}}, \quad J_{m,n} = \frac{m}{n}\left(\frac{n}{m-n}\right)^{\frac{m-n}{m}}$$

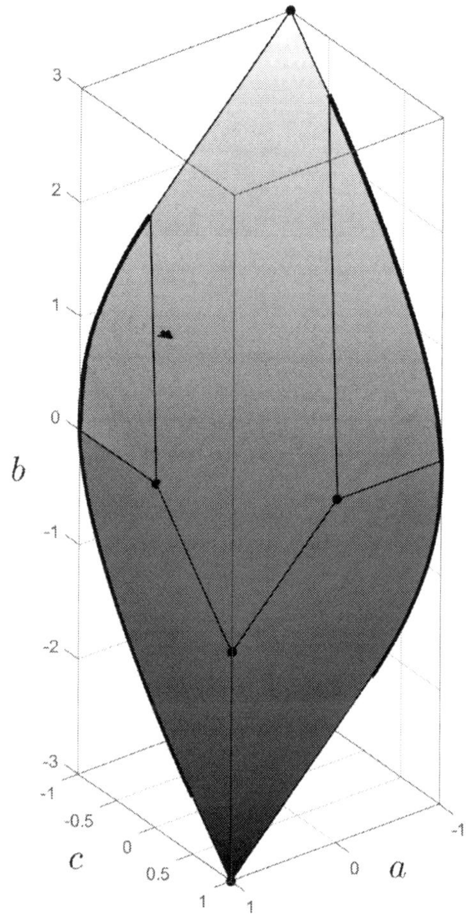

FIGURE 2.8: Representation of $\mathsf{S}^h_{m,n}$ in Case B for $\frac{1}{3} \leq \frac{n}{m} \leq \frac{2}{3}$. Here we have considered the case $m = 20$ and $n = 12$. The extreme points have been drawn with a thicker line or isolated dots. The different surfaces that form $\mathsf{S}^h_{m,n}$ are delimited by thin lines.

Let us emphasize that the number $J_{m,n}$ above is different from the one in the previous Sections 2.1.1 and 2.1.2.

2.2.1 A formula for $|||\cdot|||_{m,n}$ with m even and n odd

Given $m, n \in \mathbb{N}$ with $m > n$ and $(a, b, c) \in \mathbb{R}^3$, recall from (2.22) and (2.23) that

$$|||(a, b, c)|||_{m,n} = \max\{\|(a, b, c)\|_{m,m-n}, \|(c, b, a)\|_{m,n}\},$$

and

$$|||(a, b, c)|||_{m,n} = |||(c, b, a)|||_{m,m-n}.$$

The second identity allows us to simplify some of the forthcoming proofs in the sense that it will be enough to consider the case where $\frac{n}{m} \leq \frac{1}{2}$.

In [88], the authors derived a formula to calculate $\|(a, b, c)\|_{m,m-n}$ and $\|(c, b, a)\|_{m,n}$, which we will state for completeness in the following result:

Theorem 2.2.1 (Muñoz and Seoane [88]) *For every $m, n \in \mathbb{N}$ with m even, n odd and $m > n$, let us define $\mathcal{I}_{m,n}$ as the set of triples $(a, b, c) \in \mathbb{R}^3$ such that*

$$a \neq 0, \quad \left|\frac{nb}{ma}\right| < 1 \text{ and } 1 + \frac{c}{a} < \frac{1}{2}\left[\frac{m-n}{n}\left(\left|\frac{nb}{ma}\right|\right)^{\frac{m}{m-n}} - \left|\frac{b}{a}\right| + 1\right].$$

Then we have

$$\|(a, b, c)\|_{m,n} = \begin{cases} \left|\frac{(m-n)a}{n}\left(\left|\frac{nb}{ma}\right|\right)^{\frac{m}{m-n}} - c\right| & \text{if } (a, b, c) \in \mathcal{I}_{m,n}, \\ |a + c| + |b| & \text{otherwise.} \end{cases} \tag{2.26}$$

In the first main result of this part, we will derive a formula to calculate $|||(a, b, c)|||_{m,n}$ for the case where m is even, n is odd, and $m \geq 2n$. We will introduce necessary definitions and auxiliary results.

Lemma 2.2.2 *Let m and n be positive integers such that m is even, n is odd and $m \geq 2n$. Consider the regions $\mathcal{I}_{m,n}$ and $\mathcal{I}_{m,m-n}$ from Theorem 2.2.1. Define*

$$\mathbb{A}_{m,n} = \{(b, c) \in \mathbb{R}^2 : (1, b, c) \in \mathcal{I}_{m,m-n}\},$$
$$\mathbb{B}_{m,n} = \{(b, c) \in \mathbb{R}^2 : (c, b, 1) \in \mathcal{I}_{m,n}\}.$$

Then

1. $\mathbb{A}_{m,n} = \left\{(\pm b, c) : c < -1/2,\ 0 \leq b < \frac{m}{m-n} \text{ and } f(b) \leq \frac{nb}{mc}\right\}$ *and*

2. $\mathbb{B}_{m,n} = \left\{(\pm b, c) : c \neq 0,\ -1 < \frac{nb}{mc} \leq 0 \text{ and } 0 \leq b \leq g\left(\frac{nb}{mc}\right)\right\}$,

where

$$f(b) = \frac{2nb}{mK_{m,n}b^{\frac{m}{n}} - mb - m},$$
$$g(t) = \frac{2mt}{(m-n)t^{\frac{m}{m-n}} + mt - n}.$$

PROOF. *Let us prove first (1). Using the definition of $\mathcal{I}_{m,m-n}$, we have that $(b, c) \in \mathbb{A}_{m,n}$ if and only if*

$$|b| < \frac{m}{m-n}$$

and

$$c < \frac{1}{2}\left[K_{m,n}|b|^{\frac{m}{n}} - |b| - 1\right].$$

Let $h_1(b) = \frac{1}{2}\left[K_{m,n}|b|^{\frac{m}{n}} - |b| - 1\right]$ for $b \in [-\frac{m}{m-n}, \frac{m}{m-n}]$. Since h_1 is even, we just need to investigate h_1 in $[0, \frac{m}{m-n}]$, where f is given by $h_1(b) = \frac{1}{2}\left[K_{m,n}b^{\frac{m}{n}} - b - 1\right]$. It is elementary to show that $h_1'(b) = \frac{m}{n}K_{m,n}b^{\frac{m-n}{n}} - 1$ vanishes only at $b = \frac{m}{m-n}$ and that $h_1''(\frac{m}{m-n}) > 0$. Hence h_1 has its absolute minimum at $b = \frac{m}{m-n}$ and therefore h_1 is strictly decreasing in $[0, \frac{m}{m-n}]$. Consequently, h_1 attains it maximum at $b = 0$ and so $h_1(b) \leq h(0) = -1/2$ for all $b \in [-\frac{m}{m-n}, \frac{m}{m-n}]$. Now observe that $(b, c) \in \mathbb{A}_{m,n}$ if and only if $(-b, c) \in \mathbb{A}_{m,n}$. This allows us to focus on the pairs (b, c) with non-negative b. Let us take $b \geq 0$. Assume first that $b > 0$. Then $(b, c) \in \mathbb{A}_{m,n}$ is equivalent to

$$0 < b < \frac{m}{m-n} \quad \text{and} \quad c < \frac{1}{2}\left[K_{m,n}|b|^{\frac{m}{n}} - |b| - 1\right] \leq -\frac{1}{2}.$$

On the other hand, the inequality $c < \frac{1}{2}\left[K_{m,n}|b|^{\frac{m}{n}} - |b| - 1\right]$ *is equivalent to*

$$\frac{2}{K_{m,n}|b|^{\frac{m}{n}} - |b| - 1} < \frac{1}{c}.$$

Then multiplying the previous inequality by $\frac{nb}{m}$ *(which is positive) we arrive at the equivalent condition*

$$f(b) = \frac{2nb}{mK_{m,n}|b|^{\frac{m}{n}} - m|b| - m} < \frac{nb}{mc}$$

If $b = 0$ *it is straightforward that* $(0, c) \in \mathbb{A}_{m,n}$ *if and only if* $c < h(0) = -1/2$*. Also* $f(0) = 0$*. Then we have shown that, for* b *with* $0 \leq b < \frac{m}{m-n}$*,*

$$(b, c) \in \mathbb{A}_{m,n} \Longleftrightarrow c < -\frac{1}{2} \text{ and } f(b) \leq \frac{nb}{mc},$$

finishing the proof of (1)*.*

Now we prove (2)*. From the definition of* $\mathcal{I}_{m,n}$ *(see Theorem 2.2.1),* $(a, c) \in \mathbb{B}_{m,n}$ *is equivalent to*

$$c \neq 0, \qquad \left|\frac{b}{c}\right| < \frac{m}{n} \tag{2.27}$$

and

$$\frac{1}{c} < \frac{1}{2}\left[K_{m,m-n}\left|\frac{b}{c}\right|^{\frac{m}{m-n}} - \left|\frac{b}{c}\right| - 1\right]. \tag{2.28}$$

Clearly, $(b, c) \in \mathbb{B}_{m,n}$ *if and only if* $(-b, c) \in \mathbb{B}_{m,n}$*. In the rest of the proof we may assume then that* $b \geq 0$*. Assume first that* $b > 0$*. Writing* (2.27) *and* (2.28) *in terms of* $t = \frac{nb}{mc}$ *we obtain*

$$c \neq 0, \qquad \left|\frac{nb}{mc}\right| < 1$$

and

$$\begin{aligned}\frac{m}{nb}\frac{nb}{mc} &< \frac{1}{2}\left[K_{m,m-n}\left(\frac{m}{n}\right)^{\frac{m}{m-n}}\left|\frac{nb}{mc}\right|^{\frac{m}{m-n}} - \frac{m}{n}\left|\frac{nb}{mc}\right| - 1\right] \\ &= \frac{1}{2}\left[\frac{m-n}{n}\left|\frac{nb}{mc}\right|^{\frac{m}{m-n}} - \frac{m}{n}\left|\frac{nb}{mc}\right| - 1\right]\end{aligned}$$

or equivalently

$$c \neq 0, \quad |t| < 1$$

and

$$\frac{mt}{b} < \frac{1}{2}\left[(m-n)|t|^{\frac{m}{m-n}} - m|t| - n\right]. \tag{2.29}$$

If $h_2(t) = \frac{1}{2}\left[(m-n)|t|^{\frac{m}{m-n}} - m|t| - n\right]$ *for all* t *with* $|t| \leq 1$*, then* h_2 *attains it absolute maximum at* $t = 0$*. Indeed, we just need to notice that* h_2 *is even and that* $h_2'(t) = \frac{m}{2}\left[t^{\frac{n}{m-n}} - 1\right] < 0$ *for all* $t \in [0, 1)$*. Then* $h_2(t) \leq h_2(0) = -\frac{n}{2} < 0$ *for all* t *with* $|t| \leq 1$*. In particular, from* (2.29) *it follows that* $\frac{mt}{b} < 0$*, and since* $b > 0$*, then* $t \in (-1, 0)$ *Consequently,* (2.29) *would be equivalent to*

$$b < \frac{2mt}{(m-n)|t|^{\frac{m}{m-n}} - m|t| - n} = g(t).$$

If $b = 0$ then $t = 0$ and $g(0) = 0$, proving that $(b, c) \in \mathbb{B}_{m,n}$ with $b \geq 0$ is equivalent to

$$c \neq 0, \quad -1 < t \leq 0$$

and

$$b \leq g(t).$$

This concludes the proof of (2).

In the following, on many occasions it will be useful to replace the variable c by the variable t with $t = \frac{nb}{mc}$. Doing so it is straightforward that $(b, c) \in \mathbb{A}_{m,n}$ (respectively $(b, c) \in \mathbb{B}_{m,n}$) if and only if $(b, t) \in A_{m,n}$ (respectively $(b, t) \in B_{m,n}$) where $A_{m,n} = A^1_{m,n} \cup A^2_{m,n}$ and $B_{m,n} = B^1_{m,n} \cup B^2_{m,n}$ with $A^2_{m,n} = -A^1_{m,n}$, $B^2_{m,n} = -B^1_{m,n}$ and

$$A^1_{m,n} = \left\{(b, t) : \ 0 \leq b < \frac{m}{m-n} \text{ and } f(b) \leq t \leq 0\right\},$$

$$B^1_{m,n} = \{(b, t) : \ -1 < t \leq 0 \text{ and } 0 \leq b \leq g(t)\}.$$

It is a good moment to take a look at the representation of $A^1_{m,n}$ and $B^1_{m,n}$ in Figure 2.10 *(we emphasize that the sets $A_{m,n}$ and $B_{m,n}$ in this section are defined differently from those in Sections 2.1.1 and 2.1.2).*

Lemma 2.2.3 *Let m and n be positive integers such that m is even, n is odd and $m \geq 2n$. The equation*

$$mK_{m,n}tb^{\frac{m}{n}} - nb - mt + (m-n)bt^{\frac{m}{m-n}} = 0, \tag{2.30}$$

defines a strictly decreasing function $t = \Lambda_{m,n}(b)$ defined for $b \in \left[0, \frac{m}{m-n}\right]$ such that $\Lambda_{m,n}(0) = 0$. We define $\tau_0 = \Lambda_{m,n}\left(\frac{m}{m-n}\right)$.

PROOF. *If $\Lambda_{m,n}$ exists then it is straightforward that $\Lambda_{m,n}(0) = 0$. Now, differentiating (2.30) with respect to b we arrive at*

$$t' = F(b, t) = \frac{n^2 - m^2 K_{m,n}tb^{\frac{m-n}{n}} - n(m-n)t^{\frac{m}{m-n}}}{mn\left[K_{m,n}b^{\frac{m}{n}} + bt^{\frac{n}{m-n}} - 1\right]}.$$

Hence, if $\Lambda_{m,n}$ exists, it must be a solution to the initial value problem

$$\begin{cases} t' = F(b, t), \\ t(0) = 0. \end{cases} \tag{2.31}$$

Now consider the open subset of $\mathbb{R}^2$ given by

$$\mathcal{U} = \{(b, t) \in \mathbb{R}^2 : K_{m,n}b^{\frac{m}{n}} + bt^{\frac{n}{m-n}} - 1 < 0\}.$$

The set $\mathcal{U}$ has been represented in Figure 2.9. Observe that $(0, 0) \in \mathcal{U}$. Then, since $\frac{\partial F}{\partial b}$ and $\frac{\partial F}{\partial t}$ are both continuous in $\mathcal{U}$, the initial value problem (2.31) has a unique solution that can be extended to $\partial\mathcal{U}$. On the other hand, dividing (2.30) by bt (assuming $bt \neq 0$),

$$mK_{m,n}b^{\frac{m-n}{n}} - \frac{n}{t} - \frac{m}{b} + (m-n)t^{\frac{n}{m-n}} = 0.$$

Differentiating again with respect to b,

$$t' = \frac{-mn - m(m-n)K_{m,n}b^{\frac{m}{n}}}{n^2 t^{\frac{m}{m-n}} + n^2}\frac{t^2}{b^2},$$

which is clearly negative for every plausible value of b and t. Hence $\Lambda_{m,n}$ is strictly decreasing and its domain contains $(-L_{m,n}, L_{m,n})$ and therefore $\left[-\frac{m}{m-n}, \frac{m}{m-n}\right]$ too.

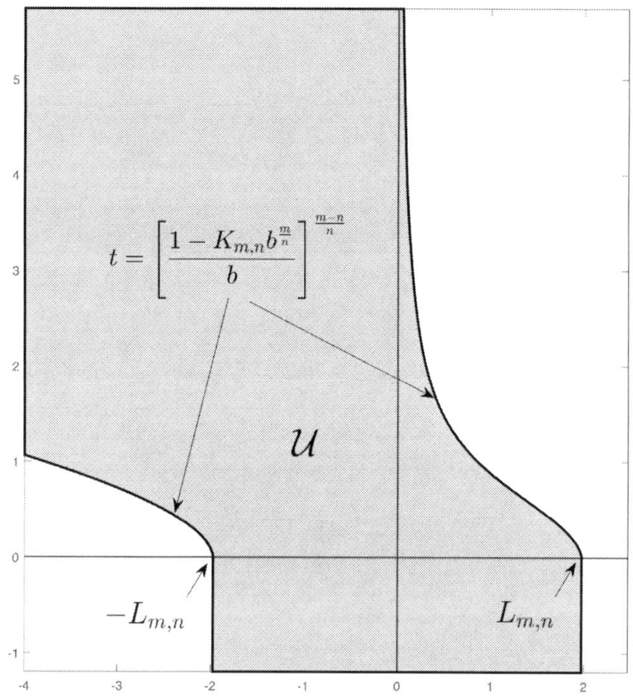

FIGURE 2.9: Representation of $\mathcal{U}$. Here we have considered the case $m = 12$ and $n = 5$.

Remark 2.2.4 *The value of τ_0 for every plausible choice of m and n is the unique solution between -1 and 0 of the equation*

$$(m-n)t^{\frac{m}{m-n}} + (2n-m)t - n = 0. \tag{2.32}$$

The equation (2.32) cannot generally be solved explicitly. However, using specialized software τ_0 can be approximated with precision. We provide below a table with some approximations to τ_0 obtained with Mathematica:

m	n	τ_0
4	1	-0.2560771804
6	1	-0.1359670417
8	1	-0.0911451357
8	3	-0.5472162244
10	1	-0.0681314528
10	3	-0.3536273979
12	1	-0.0542309739
12	3	-0.2560771804
12	5	-0.6823509843

For $m = 4$ and $n = 1$ (or $\frac{m}{n} = 4$ in any case) it can be seen that τ_0 is exactly equal to

$$\tau_0 = -\frac{1}{81}\left(\sqrt[3]{729\sqrt{17}+541} - \frac{206}{\sqrt[3]{729\sqrt{17}+541}} + 19\right).$$

Theorem 2.2.5 *Let m and n be positive integers such that m is even, n is odd and $m > n$. If $m \geq 2n$, then*

$$|||(a,b,c)|||_{m,n} = \begin{cases} \left|K_{m,n}a\left|\frac{b}{a}\right|^{\frac{m}{n}} - c\right| & \text{if } abc \neq 0 \text{ and } \left(\frac{b}{a}, \frac{nb}{mc}\right) \in \mathcal{A}_{m,n}, \\ \left|K_{m,m-n}c\left|\frac{b}{c}\right|^{\frac{m}{m-n}} - a\right| & \text{if } abc \neq 0 \text{ and } \left(\frac{b}{a}, \frac{nb}{mc}\right) \in \mathcal{B}_{m,n}, \\ \max\{|a|,|c|\} & \text{if } b = 0 \text{ and } ac \leq 0, \\ |a+c| + |b| & \text{otherwise,} \end{cases} \tag{2.33}$$

where $\mathcal{A}_{m,n} = \mathcal{A}^1_{m,n} \cup \mathcal{A}^2_{m,n}$, $\mathcal{B}_{m,n} = \mathcal{B}^1_{m,n} \cup \mathcal{B}^2_{m,n}$,

$$\mathcal{A}^1_{m,n} = \left\{(b,t) \in \mathbb{R}^2 : b \in \left(0, \frac{m}{m-n}\right] \text{ and } \Lambda_{m,n}(b) \leq t < 0\right\},$$

$$\mathcal{B}^1_{m,n} = \left\{(b,t) \in \mathbb{R}^2 : b \in \left(0, \frac{m}{m-n}\right] \text{ and } \tau_0 \leq t \leq \Lambda_{m,n}(b)\right\}$$
$$\cup \{(b,t) \in \mathbb{R}^2 : t \in [-1, \tau_0] \text{ and } 0 < b \leq g(t)\}.$$

and $\mathcal{A}^2_{m,n} = -\mathcal{A}^1_{m,n}$, $\mathcal{B}^2_{m,n} = -\mathcal{B}^1_{m,n}$. The regions $\mathcal{A}^1_{m,n}$ and $\mathcal{B}^1_{m,n}$ have been represented in Figure 2.11. On the other hand, if $m < 2n$ then, according to (2.23), $|||(a,b,c)|||_{m,n} = |||(c,b,a)|||_{m,m-n}$.

PROOF. *It is elementary to prove that*

$$|||(0,b,c)|||_{m,n} = |b| + |c|,$$
$$|||(a,b,0)|||_{m,n} = |a| + |b|,$$

for all $a, b, c \in \mathbb{R}$. Also,

$$|||(a,0,c)|||_{m,n} = \max\{|ax^m + cy^m| : (x,y) \in [-1,1]^2\}$$
$$= \begin{cases} \max\{|a|,|c|\} & \text{if } ac \leq 0, \\ |a+c| & \text{if } ac > 0. \end{cases} \tag{2.34}$$

For the rest of the proof we will assume that $a, b, c \neq 0$. In particular we have

$$|||(a,b,c)|||_{m,n} = |a| \cdot \left|\left|\left|\left(1, \frac{b}{a}, \frac{c}{a}\right)\right|\right|\right|_{m,n}$$

and therefore it will be enough to obtain a formula to calculate $|||(1,b,c)|||_{m,n}$ for all $b, c \in \mathbb{R}$ with $b, c \neq 0$.

According to (2.22) we have

$$|||(1,b,c)|||_{m,n} = \max\{\|(1,b,c)\|_{m,m-n}, \|(c,b,1)\|_{m,n}\}.$$

Now using (2.26) we obtain

$$\|(1,b,c)\|_{m,m-n} = \begin{cases} \left|K_{m,n}|b|^{\frac{m}{n}} - c\right| & \text{if } (1,b,c) \in \mathcal{I}_{m,m-n}, \\ |1+c| + |b| & \text{otherwise,} \end{cases}$$

$$\|(c,b,1)\|_{m,n} = \begin{cases} \left|K_{m,m-n}c\left|\frac{b}{c}\right|^{\frac{m}{m-n}} - 1\right| & \text{if } (c,b,1) \in \mathcal{I}_{m,n}, \\ |1+c| + |b| & \text{otherwise.} \end{cases}$$

Since $\|(1,b,c)\|_{m,m-n} = \|(1,-b,c)\|_{m,m-n}$ and $\|(c,b,1)\|_{m,n} = \|(c,-b,1)\|_{m,n}$ we can assume that $b > 0$. From Lemma 2.2.2 we have:

- $(1,b,c)\in\mathcal{I}_{m,m-n}$ *is equivalent to* $(b,t)\in A_{m,n}$.
- $(c,b,1)\in\mathcal{I}_{m,m}$ *is equivalent to* $(b,t)\in B_{m,n}$.

As a matter of fact, since $b>0$ *in fact we have*

- $(1,b,c)\in\mathcal{I}_{m,m-n}$ *is equivalent to* $(b,t)\in A^1_{m,n}$.
- $(c,b,1)\in\mathcal{I}_{m,m}$ *is equivalent to* $(b,t)\in B^1_{m,n}$.

Therefore, if $b>0$ *we have that*

$$|||(1,b,c)|||_{m,n}=\begin{cases}\left|K_{m,n}|b|^{\frac{m}{n}}-c\right| & \text{if } (b,t)\in A^1_{m,n}\setminus B^1_{m,n},\\ \left|K_{m,m-n}c\left|\frac{b}{c}\right|^{\frac{m}{m-n}}-1\right| & \text{if } (b,t)\in B^1_{m,n}\setminus A^1_{m,n},\\ |1+c|+|b| & \text{if } (b,t)\notin A^1_{m,n}\cup B^1_{m,n},\end{cases}$$

and

$$|||(1,b,c)|||_{m,n}=\max\left\{\left|K_{m,n}|b|^{\frac{m}{n}}-c\right|,\left|K_{m,m-n}c\left|\frac{b}{c}\right|^{\frac{m}{m-n}}-1\right|\right\}$$

whenever $(b,t)\in A^1_{m,n}\cap B^1_{m,n}$. *To finish the proof we just need to compare* $\left|K_{m,n}|b|^{\frac{m}{n}}-c\right|$ *and* $\left|K_{m,m-n}c\left|\frac{b}{c}\right|^{\frac{m}{m-n}}-1\right|$ *within* $A^1_{m,n}\cap B^1_{m,n}$.

It is left to the reader to check that the inequality

$$\left|K_{m,n}|b|^{\frac{m}{n}}-c\right|\geq\left|K_{m,m-n}c\left|\frac{b}{c}\right|^{\frac{m}{m-n}}-1\right|$$

can be written using the variables b *and* t *as*

$$\left|K_{m,n}b^{\frac{m}{n}}-\frac{nb}{mt}\right|\geq\left|\frac{m-n}{m}bt^{\frac{n}{m-n}}-1\right|. \tag{2.35}$$

Now, let $F(b,t)$ *and* $G(b,t)$ *be, respectively, the right and the left-hand side of* (2.35). *Notice that*

$$F(b,t)=1-\frac{m-n}{m}bt^{\frac{n}{m-n}}$$

for every $(b,t)\in A^1_{m,n}$ *and*

$$G(b,t)=K_{m,n}b^{\frac{m}{n}}-\frac{nb}{mt}$$

for all $(b,t)\in B^1_{m,n}$ *with* $t\neq 0$. *The equality*

$$G(t,b)=F(t,b)$$

is equivalent to

$$mK_{m,n}tb^{\frac{m}{n}}-nb-mt+(m-n)bt^{\frac{m}{m-n}}=0$$

which is the identity (2.30) *studied in Lemma 2.2.3. Hence* F *and* G *coincide only along the curve* $t=\Lambda_{m,n}(b)$. *Interestingly,* $t=\Lambda_{m,n}(b)$ *and* $b=g(t)$ *meet at the point* $\left(\frac{m}{m-n},\tau_0\right)$. *Indeed, from Lemma 2.2.3, keeping in mind that* $K_{m,n}=\frac{n}{m-n}\left(\frac{m-n}{m}\right)^{\frac{m}{n}}$, *and making*

$$t=\tau_0 \text{ and } b=\frac{m}{m-n},$$

which entails that $\Lambda_{m,n}(b)=\tau_0$, *we have the identity*

$$0=\frac{mn}{m-n}\left(\frac{m-n}{m}\right)^{\frac{m}{n}}\tau_0\left(\frac{m}{m-n}\right)^{\frac{m}{n}}-\frac{mn}{m-n}-m\tau_0+\frac{(m-n)m}{m-n}\tau_0^{\frac{m}{m-n}}.$$

After some straightforward calculations we arrive at

$$(m-n)\tau_0^{\frac{m}{m-n}}=n+(m-n)\tau_0-n\tau_0. \tag{2.36}$$

Next, let us see that

$$g(\tau_0)=\frac{m}{m-n}.$$

Indeed, using equation (2.36), *we have*

$$\begin{aligned}g(\tau_0)&=\frac{2m\tau_0}{(m-n)\tau_0^{\frac{m}{m-n}}+m\tau_0-n}\\&=\frac{2m\tau_0}{n+(m-n)\tau_0-n\tau_0+m\tau_0-n}=\frac{m}{m-n}.\end{aligned}$$

Also, let us notice that equation (2.36) *has only one solution in the interval* $(-1,0)$. *Indeed, using a convexity argument, it is clear that the functions* $(m-n)t^{\frac{m}{m-n}}$ *and* $n+(m-n)t-nt$ *meet, at most, twice. Since* $t=1$ *makes* $(m-n)t^{\frac{m}{m-n}}=n+(m-n)t-nt$ *and we know that there is one solution belonging to* $(-1,0)$, *the above* τ_0 *in unique.*

We conclude that $F(t,b)\geq G(t,b)$ *if* $(b,t)\in A^1_{m,n}\cup B^1_{m,n}$ *and* (b,t) *is below the curve* $t=\Lambda_{m,n}(b)$ *or in the curve, or equivalently in* $\mathcal{B}^1_{m,n}$. *On the contrary* $F(t,b)\leq G(t,b)$ *in the rest of* $A^1_{m,n}\cup B^1_{m,n}$ *or along the curve* $t=\Lambda_{m,n}(b)$, *that is, in* $\mathcal{A}^1_{m,n}$. *Then*

$$|||(1,b,c)|||_{m,n}=\begin{cases}\left|K_{m,n}|b|^{\frac{m}{n}}-c\right| & \textit{if } (b,t)\in\mathcal{A}^1_{m,n},\\ \left|K_{m,m-n}c\left|\frac{b}{c}\right|^{\frac{m}{m-n}}-1\right| & \textit{if } (b,t)\in\mathcal{B}^1_{m,n},\\ |1+c|+|b| & \textit{if } (b,t)\notin\mathcal{A}^1_{m,n}\cup\mathcal{B}^1_{m,n}.\end{cases}$$

We arrive at the desired result by combining the previous formula with

$$|||(a,b,c)|||_{m,n}=|a|\cdot\left|\left|\left|\left(1,\frac{b}{a},\frac{c}{a}\right)\right|\right|\right|_{m,n}.$$

2.2.2 A parametrization of the unit sphere $\mathsf{S}^h_{m,n}$

In order to obtain a parametrization of $\mathsf{S}^h_{m,n}$ it will be necessary to know the projection of $\mathsf{S}^h_{m,n}$ over a plane. The most convenient plane is $b=0$ (also called the ac plane) since the unit ball is symmetric with respect to that plane. The latter is justified with the obvious equation $|||(a,b,c)|||_{m,n}=|||(a,-b,c)|||_{m,n}$ whenever m is even and n odd.

In the main result of this section some notations and definitions will be needed. First we consider two curves that will play an important role, namely $c=\Gamma_{m,n}(a)$ and $c=\Upsilon_{m,n}(a)$. The first curve is defined implicitly using the following lemma.

Lemma 2.2.6 *Let* m,n *be positive integers such that* $m\geq 2n$, m *is even and* n *is odd. Recall the number* $J_{m,n}=\frac{m}{n}\left(\frac{n}{m-n}\right)^{\frac{m-n}{m}}$, *and consider* $\lambda_0=\frac{n}{m-n}$ *as well as the equation*

$$J_{m,n}(1-a)^{\frac{m-n}{m}}|c|^{\frac{n}{m}}-1-a-c=0. \tag{2.37}$$

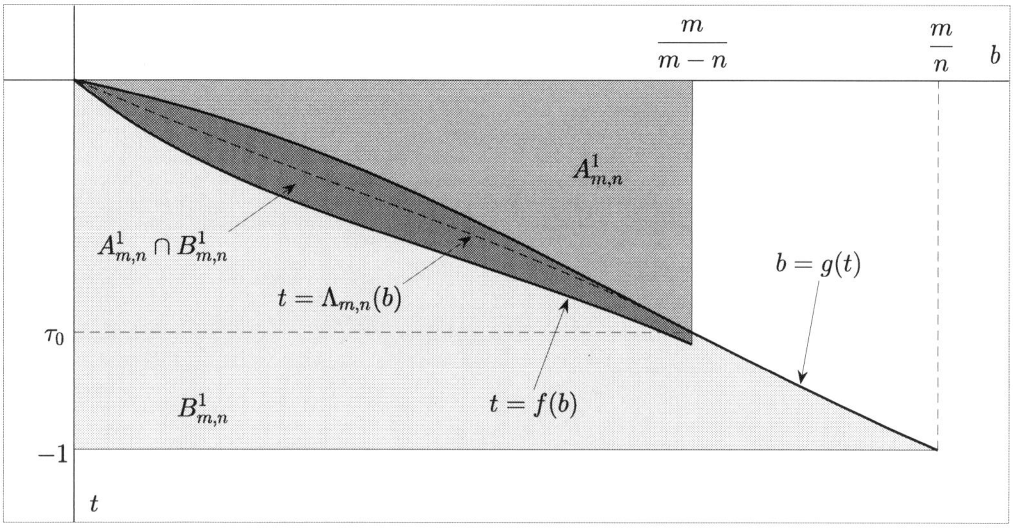

FIGURE 2.10: Representation of $A^1_{m,n}$ and $B^1_{m,n}$. The intersection $A^1_{m,n} \cap B^1_{m,n}$ is the darkest region. We have chosen the values $m = 12$ and $n = 5$.

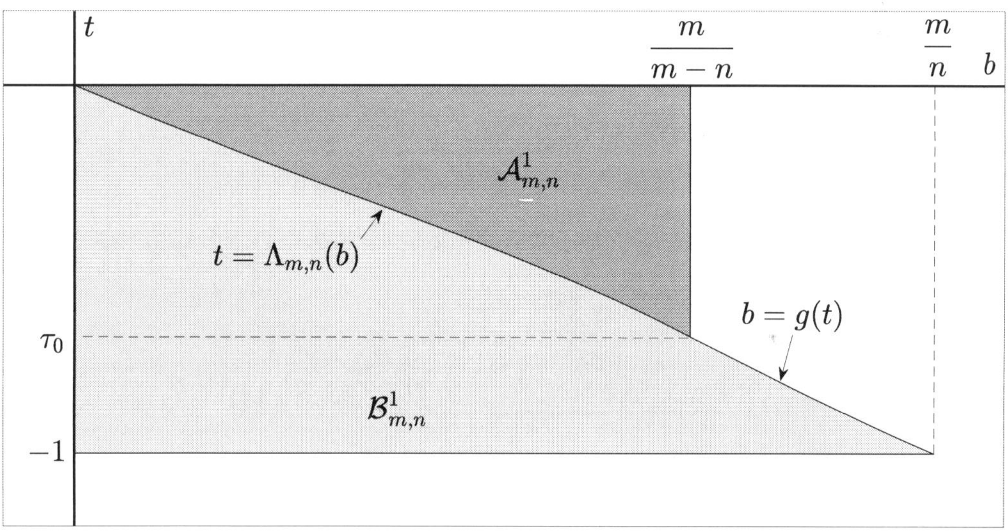

FIGURE 2.11: Representation of $\mathcal{A}^1_{m,n}$ and $\mathcal{B}^1_{m,n}$. Here we have considered the case $m = 12$ and $n = 5$.

Then in (2.37) c can be expressed as a function of a with $a \in (a_0, a_1)$, namely $\Gamma_{m,n}(a)$, where the curve $c = \Gamma_{m,n}(a)$ meets the parallel lines $c = \lambda_0(a-1)$ and $c = \lambda_0 a - 1$ at the points (a_0, c_0) and (a_1, c_1), respectively (see Figure 2.12). Moreover, $\Gamma_{m,n}(a)$ is strictly decreasing and $\Gamma_{m,n}(a) < 0$ for $a \in [a_0, a_1]$. It turns out that

$$a_0 = \frac{n}{m} \quad \text{and} \quad c_0 = -\frac{n}{m}$$

whereas a_1 and c_1 cannot generally be obtained explicitly.

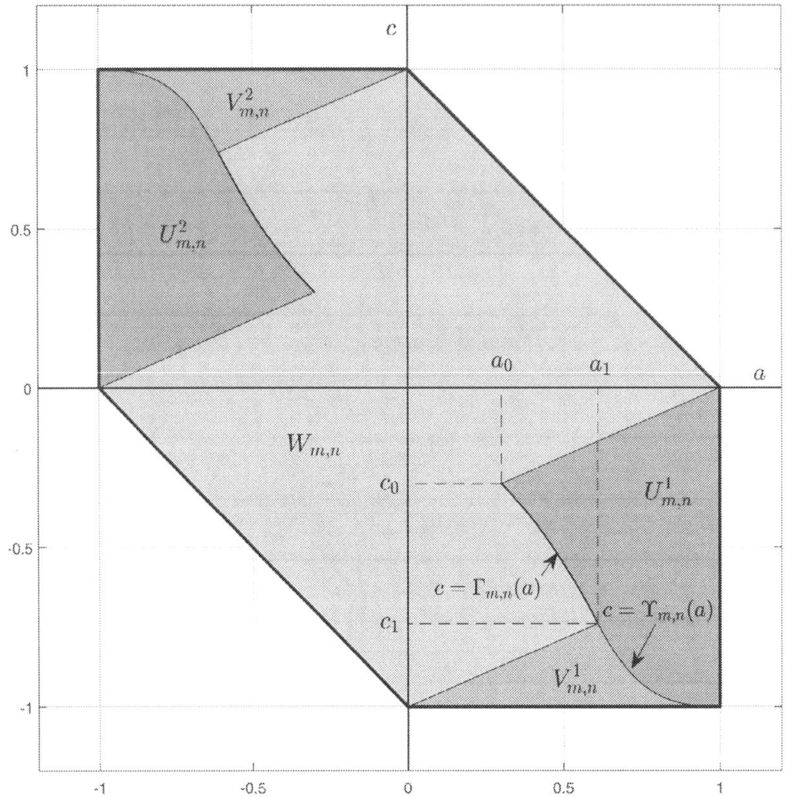

FIGURE 2.12: Representation of $\pi_{a,c}(\mathsf{S}^h_{m,n})$. Here we have considered the case $m = 10$ and $n = 3$.

PROOF. *One can check easily that* (a_0, c_0) *with* $a_0 = \frac{n}{m}$ *and* $c_0 = -\frac{n}{m}$ *is a solution to* (2.37) *and that the line* $c = \lambda_0(a-1)$ *passes though* (a_0, c_0). *Notice that for each fixed* $\overline{a} \in (a_0, a_1)$, *the function*

$$x \in (-1, 0) \mapsto \varphi(x) := J_{m,n}(1-\overline{a})^{\frac{m-n}{m}} |x|^{\frac{n}{m}} - 1 - \overline{a} - x$$

is a strictly decreasing function. It is clear that $\phi(0) = -1 - \overline{a} < 0$. *Observe also that*

$$z \in (0, 1) \mapsto \psi(z) := J_{m,n}(1-z)^{\frac{m-n}{m}} - z$$

is strictly decreasing; so

$$\phi(-1) = J_{m,n}(1-\overline{a})^{\frac{m-n}{m}} - \overline{a} = \psi(\overline{a}) > \psi(a_1) = \frac{1 + a_1 + c_1}{|c_1|^{\frac{n}{m}}} - a_1 > 0,$$

where the last equality holds since the intersection point (a_1, c_1) *satisfies the equation* (2.37). *By Intermediate Value Theorem, this shows that there exists a unique* $\overline{c} \in (c_0, c_1)$ *such that* $\phi(\overline{c}) = 0$, *i.e.,* $(\overline{a}, \overline{c})$ *satisfies the equation* (2.37).

Next, differentiating (2.37) *with respect to* a *we obtain*

$$c' = G(a, c) = -\frac{\frac{m-n}{m} J_{m,n} \left(\frac{|c|}{1-a}\right)^{\frac{n}{m}} + 1}{J_{m,n} \left(\frac{1-a}{|c|}\right)^{\frac{m-n}{m}} + 1} < 0$$

for all $(a, c) \in (a_0, a_1) \times (c_0, c_1)$. *Then if* $\Gamma_{m,n}$ *exists, it must be the solution of the initial value problem for a given point* $\overline{a} \in (a_0, a_1)$,

$$\begin{cases} c'(a) &= G(a, c), \\ c(\overline{a}) &= \overline{c}. \end{cases} \tag{2.38}$$

Since $\frac{\partial G}{\partial a}$ *and* $\frac{\partial G}{\partial c}$ *are both continuous in* $(a_0, a_1) \times (c_0, c_1)$, *the Picard-Lindelöf theorem shows that the initial value problem* (2.38) *has a unique solution (which is* C^1*) around some interval containing* $\overline{a}$. *By uniqueness, this local solution coincides with the one given from Intermediate Value Theorem locally. Since* $\overline{a} \in (a_0, a_1)$ *is given arbitrarily, we conclude that* $\Gamma_{m,n}(a)$ *exists and is well-defined for all* $a \in (a_0, a_1)$. *Also* $\Gamma_{m,n}$ *is strictly decreasing and the curve* $c = \Gamma_{m,n}(a)$ *meets the straight line* $c = \lambda_0 a - 1$ *at the point* (a_1, c_1) *with* $\frac{n}{m} = a_0 < a_1 < 1$ *and* $-\frac{n}{m} = c_0 > c_1 > -1$.

Also, we will consider the curve $c = \Upsilon_{m,n}(a)$ with $a \in [a_1, 1]$ where

$$\Upsilon_{m,n}(a) = -\frac{a^{\frac{m-n}{n}}}{(1-a)^{\frac{m-n}{n}} + a^{\frac{m-n}{n}}} = -\frac{1}{1 + \left(\frac{1-a}{a}\right)^{\frac{m-n}{n}}}.$$

Interestingly, the curves $c = \Gamma_{m,n}(a)$, $c = \Upsilon_{m,n}(a)$ and the straight line $c = \lambda_0 a - 1$ meet at the point (a_1, c_1) (see Figure 2.12). Also, for every pair m, n of positive integers with $m \geq 2n$, m even and n odd, we will use the sets Π, $U_{m,n}$, $U^1_{m,n}$, $U^2_{m,n}$, $V_{m,n}$, $V^1_{m,n}$, $V^2_{m,n}$ and $W_{m,n}$ defined as

$$\begin{aligned} \Pi &= \{(a, c) \in \mathbb{R}^2 : (a, c) \in [-1, 1]^2 \text{ and } |a + c| \leq 1\}, \\ U_{m,n} &= U^1_{m,n} \cup U^2_{m,n}, \\ V_{m,n} &= V^1_{m,n} \cup V^2_{m,n}, \\ W_{m,n} &= \Pi \setminus (U_{m,n} \cup V_{m,n}), \end{aligned}$$

where $U^2_{m,n} = -U^1_{m,n}$, $V^2_{m,n} = -V^1_{m,n}$ and

$$\begin{aligned} U^1_{m,n} &= \big\{(a, c) \in \Pi : a_0 \leq a \leq a_1,\ \lambda_0(a-1) \geq c \geq \Gamma_{m,n}(a)\big\} \\ &\quad \cup \big\{(a, c) \in \Pi : a_1 \leq a \leq 1,\ \lambda_0(a-1) \geq c \geq \Upsilon_{m,n}(a)\big\}, \\ V^1_{m,n} &= \big\{(a, c) \in \Pi : 0 \leq a \leq a_1,\ -1 \leq c \leq \lambda_0 a - 1\big\} \\ &\quad \cup \big\{(a, c) \in \Pi : a_1 \leq a \leq 1,\ -1 \leq c \leq \Upsilon_{m,n}(a)\big\}. \end{aligned}$$

See Figure 2.12 for a representation of Π, $U_{m,n}$, $V_{m,n}$, and $W_{m,n}$.

Theorem 2.2.7 *Let* $m, n \in \mathbb{N}$ *be such that* $m > n$ *with* m *even and* n *odd. Then*

$$\pi_{a,c}(\mathsf{S}^h_{m,n}) = \Pi.$$

PROOF. *Since* $\mathsf{B}^h_{m,n}$ *is symmetric with respect to the plane* $b = 0$, *it is clear that* $\pi_{a,c}(\mathsf{S}^h_{m,n})$ *is the intersection of* $\mathsf{B}^h_{m,n}$ *and the plane* $b = 0$, *that is,*

$$\pi_{a,c}(\mathsf{S}^h_{m,n}) = \mathsf{B}^h_{m,n} \cap \{(a, 0, c) : a, c \in \mathbb{R}\}.$$

In other words, $(a, c) \in \pi_{a,c}(\mathsf{S}^h_{m,n})$ *if and only if* $|||(a, 0, c)|||_{m,n} \leq 1$. *Using* (2.34) *it follows straightforwardly that*

$$\pi_{a,c}(\mathsf{S}^h_{m,n}) = \Pi$$

as desired.

Next we are going to parametrize the unit sphere $\mathsf{S}^h_{m,n}$. Recall that if $m = 2n$ then $\mathcal{P}^h_{m,n}(\mathbb{R})$ is nothing but the space $\mathcal{P}(^2\ell^2_\infty)$, whose unit ball was already studied by Choi and Kim in [20, 21] providing a formula for the norm of any quadratic form $ax^2 + bxy + cy^2$ and a characterization of the extreme polynomials. A parametrization and a representation of the unit sphere of $\mathcal{P}(^2\ell^2_\infty)$ can be found in [63].

Using the parametrization of $\mathsf{S}^h_{m,n}$ mentioned above it will be easy to localize the extreme points of $\mathsf{B}^h_{m,n}$.

Also, we shall consider the functions $F_{m,n}, G_{m,n} : \Pi \to \mathbb{R}$ defined as

$$F_{m,n}(a,c) = \begin{cases} J_{m,n}\,(1-a)^{\frac{m-n}{m}}\,|c|^{\frac{n}{m}} & \text{if } (a,c) \in U^1_{m,n}, \\ J_{m,n}\,(1+a)^{\frac{m-n}{m}}\,c^{\frac{n}{m}} & \text{if } (a,c) \in U^2_{m,n}, \\ J_{m,m-n}\,(1+c)^{\frac{n}{m}}\,a^{\frac{m-n}{m}} & \text{if } (a,c) \in V^1_{m,n}, \\ J_{m,m-n}\,(1-c)^{\frac{n}{m}}\,|a|^{\frac{m-n}{m}} & \text{if } (a,c) \in V^2_{m,n}, \\ 1-|a+c| & \text{if } (a,c) \in W_{m,n}, \end{cases} \tag{2.39}$$

whenever $m \geq 2n$ and

$$G_{m,n}(a,c) = F_{m,m-n}(c,a)$$

if $m \leq 2n$. Observe that for all odd $n \in \mathbb{N}$ we have

$$F_{2n,n}(c,a) = G_{2n,n}(a,c) = \begin{cases} 2\sqrt{c(a-1)} & \text{if } (a,c) \in U^1_{2,1}, \\ 2\sqrt{c(1+a)} & \text{if } (a,c) \in U^2_{2,1}, \\ 2\sqrt{a(1+c)} & \text{if } (a,c) \in V^1_{2,1}, \\ 2\sqrt{a(c-1)} & \text{if } (a,c) \in V^2_{2,1}, \\ 1-|a+c| & \text{if } (a,c) \in W_{2,1}. \end{cases}$$

It was proved in [63, Theorem 2.4] that

$$\mathsf{S}^h_{2n,n} = \operatorname{graph}(F_{2n,n}) \cup \operatorname{graph}(-F_{2n,n}).$$

More generally, in Theorem 2.2.9 we will show that

$$\mathsf{S}^h_{m,n} = \begin{cases} \operatorname{graph}(F_{m,n}) \cup \operatorname{graph}(-F_{m,n}) & \text{if } m \geq 2n, \\ \operatorname{graph}(G_{m,n}) \cup \operatorname{graph}(-G_{m,n}) & \text{if } m \leq 2n, \end{cases}$$

for every $m, n \in \mathbb{N}$ with $m > 2n$, m even and n odd. To prove this the following results will be useful.

Lemma 2.2.8 *If $m, n \in \mathbb{N}$ are such that $m \geq 2n$, m is even and n is odd, we define $\Phi : \Pi \cap \{(a,c) : a \neq 0$ and $c \neq 0\} \to \mathbb{R}^2$ by*

$$\Phi(a,c) = \left(\frac{F_{m,n}(a,c)}{a}, \frac{nF_{m,n}(a,c)}{mc} \right)$$

Then

$$\begin{aligned} \Phi(V^k_{m,n} \setminus \{a=0\}) &= \mathcal{A}^k_{m,n} \cup \{(0,0)\}, \quad \textit{for } k \in \{1,2\}, \\ \Phi(U^k_{m,n} \setminus \{c=0\}) &= \mathcal{B}^k_{m,n} \cup \{(0,0)\}, \quad \textit{for } k \in \{1,2\} \end{aligned}$$

and

$$\Phi(W_{m,n} \setminus \{ac=0\}) = [(\mathbb{R}^2 \setminus \{bt=0\}) \setminus (\mathcal{A}_{m,n} \cup \mathcal{B}_{m,n})] \cup \{(0,0)\}.$$

PROOF. *We present the proof for each region.*

- *The region* $V^k_{m,n}$*: By symmetry, it is suffices to show the case* $k = 1$*. Note that*

$$\bigcup_{0\le\lambda\le\lambda_0}\{(a,\lambda a-1) : 0 < a \le a_\lambda\} = V^1_{m,n}$$

where $\lambda_0 = \frac{n}{m-n}$ *(see Lemma 2.2.6) and* $a_\lambda \in (0, 1]$ *is the value satisfying that* $\lambda a_\lambda - 1 = \Upsilon_{m,n}(a_\lambda)$ *for* $0 \le \lambda \le \lambda_0$*. See Figures 2.12 or 2.13 for a representation of* $V^1_{m,n}$ *and Figure 2.14 for a representation of* $\Phi(V^1_{m,n})$*. We shall check that*

$$\{\Phi(a,\lambda a-1) : 0 < a \le a_\lambda\} = \{(b_\lambda, t) : \Lambda_{m,n}(b_\lambda) \le t < 0\}, \tag{2.40}$$

where

$$b_\lambda := \frac{m}{m-n}\left(\frac{m-n}{n}\lambda\right)^{\frac{n}{m}}.$$

Using (2.39), a direct computation shows that for $0 < a \le a_\lambda$*,*

$$\Phi(a,\lambda a-1) = \left(b_\lambda, \frac{n}{m-n}\left(\frac{m-n}{n}\lambda\right)^{\frac{n}{m}}\frac{a}{\lambda a-1}\right).$$

In particular, in the case when $\lambda = 0$*,*

$$\Phi(a,-1) = (0,0) \quad \text{for every } 0 < a \le 1.$$

Note from (2.30) that $t = \frac{n}{m-n}\left(\frac{m-n}{n}\lambda\right)^{\frac{n}{m}}\frac{a}{\lambda a-1} \ge \Lambda_{m,n}(b_\lambda)$ *if and only if*

$$K_{m,n}(b_\lambda)^{\frac{m}{n}} - \frac{n}{m}\frac{b_\lambda}{t} \ge 1 - \left(\frac{m-n}{m}\right)\frac{b_\lambda}{t}t^{\frac{m}{m-n}} \tag{2.41}$$

provided that $\lambda \ne 0$*. By definition of* b_λ *and* t*, we observe that the left-hand side and right-hand side of (2.41) is* $\frac{1}{a}$ *and* $1 + \left(\frac{\lambda a}{1-\lambda a}\right)^{\frac{n}{m-n}}$*; hence (2.41) is equivalent to that*

$$\frac{1}{a} \ge 1 + \left(\frac{\lambda a}{1-\lambda a}\right)^{\frac{n}{m-n}}. \tag{2.42}$$

Observe that (2.42) is equivalent to saying that $(a, \lambda a - 1)$ *satisfies* $\Upsilon_{m,n}(a) \ge \lambda a - 1$*. Also, the equality in (2.42) holds if and only if* $\Upsilon_{m,n}(a) = \lambda a - 1$*. This proves the claim (2.40); hence* $\Phi(V^1_{m,n} \setminus \{a = 0\}) = \mathcal{A}^1_{m,n} \cup \{(0,0)\}$*.*

- *The region* $U^k_{m,n}$*: Again, by symmetry, we only consider the case* $k = 1$*. For simplicity, put* $\lambda_1 := -\frac{c_1}{1-a_1}$ *and consider*

$$U^1_{m,n} \setminus \{c = 0\} = U^1_{m,n,1} \cup U^1_{m,n,2} \cup \{(1,c) \in \Pi : -1 \le c < 0\},$$

where

$$U^1_{m,n,1} := \{(a,c) \in U_{m,n} : a_1 \le a < 1,\ \Upsilon_{m,n}(a) \le c \le \lambda_1(a-1)\},$$
$$U^1_{m,n,2} := \{(a,c) \in U_{m,n} : a_0 \le a < 1,\ \lambda_1(a-1) < c \le \lambda_0(a-1),\ c \ge \Gamma_{m,n}(a)\}.$$

See Figure 2.13 for a representation of $U^1_{m,n,1}$ *and* $U^1_{m,n,2}$ *and Figure 2.14 for a representation of* $\Phi(U^1_{m,n,1})$ *and* $\Phi(U^1_{m,n,2})$*. It is clear that* $\Phi(1,c) = (0,0)$ *for every* $-1 \le c < 0$*.*

∘ *The region* $U^1_{m,n,1}$*: Our claim is to prove*

$$\Phi(U^1_{m,n,1}) = \mathcal{B}^1_{m,n} \cap \{\tau_0 \le t < 0\}. \tag{2.43}$$

Consider $(a,c) \in U^1_{m,n,1}$ *with* $c = \lambda(a-1)$*. Notice that* λ_1 *is the slope of joining* $(1,0)$ *and* (a_1, c_1)*. Again note from (2.39) that*

$$\Phi(a, \lambda(a-1)) = \left(\frac{1-a}{a}\frac{m}{n}\left(\frac{n}{m-n}\right)^{\frac{m-n}{m}} \lambda^{\frac{n}{m}}, -\left(\frac{n}{m-n}\right)^{\frac{m-n}{m}} \lambda^{\frac{n-m}{m}}\right). \tag{2.44}$$

For simplicity, let us put

$$b = b(a,\lambda) := \left(\frac{1-a}{a}\right)\frac{m}{n}\left(\frac{n}{m-n}\right)^{\frac{m-n}{m}} \lambda^{\frac{n}{m}};$$

$$t = t(\lambda) := -\left(\frac{n}{m-n}\right)^{\frac{m-n}{m}} \lambda^{\frac{n-m}{m}}.$$

Notice that $b(a_1, \lambda_1) = \frac{m}{m-n}$*. Indeed, we can write*

$$b(a_1, \lambda_1) = \frac{1-a_1}{a_1}\left(\frac{n}{m-n}\right)^{-\frac{n}{m}} \lambda_1^{\frac{n}{m}} \left(\frac{m}{m-n}\right). \tag{2.45}$$

On the other hand, since $(a_1, \lambda_1(a_1-1))$ *satisfies (2.37) and* $\frac{n}{m-n}a_1 - 1 = c_1$*, we have that*

$$\begin{aligned}\frac{m}{n}\left(\frac{n}{m-n}\right)^{\frac{m-n}{m}} (1-a_1)^{\frac{m-n}{m}} \lambda_1^{\frac{n}{m}} (1-a_1)^{\frac{n}{m}} &= 1 + a_1 + c_1 \\ &= \frac{m}{m-n}a_1,\end{aligned}$$

which is equivalent to

$$\left(\frac{n}{m-n}\right)^{-\frac{n}{m}} \left(\frac{1-a_1}{a_1}\right) \lambda_1^{\frac{n}{m}} = 1. \tag{2.46}$$

Combining this with (2.45), we conclude that $b(a_1, \lambda_1) = \frac{m}{m-n}$*. We claim that* $t(\lambda_1)$ *coincides with* τ_0*. To this end, observe from (2.30) that* $t(\lambda_1) = \Lambda_{m,n}(b)$ *is equivalent to*

$$\begin{aligned}\left(\frac{n}{m-n}\right)\left(\frac{m-n}{m}\right)^{\frac{m}{n}} b^{\frac{m}{n}} + \frac{n}{m}\left(\frac{n}{m-n}\right)^{\frac{n-m}{m}} \lambda_1^{\frac{m-n}{m}} b \\ = 1 + \left(\frac{m-n}{m}\right) b \left(\frac{n}{m-n}\right)^{\frac{n}{m}} \lambda_1^{-\frac{n}{m}}.\end{aligned} \tag{2.47}$$

Observe that (2.47) is satisfied for $b = \frac{m}{m-n}$ *if and only if*

$$\frac{n}{m-n} + \left(\frac{n}{m-n}\right)^{\frac{n}{m}} \lambda_1^{\frac{m-n}{m}} = 1 + \left(\frac{n}{m-n}\right)^{\frac{n}{m}} \lambda_1^{-\frac{n}{m}}. \tag{2.48}$$

Taking (2.46) into account, we see that (2.48) is equivalent to

$$\frac{n}{m-n} + \left(\frac{1-a_1}{a_1}\right)\lambda_1 = 1 + \left(\frac{1-a_1}{a_1}\right)$$

which is indeed true because of the fact

$$c_1 = \lambda_1(a_1 - 1) = \frac{n}{m-n} a_1 - 1.$$

Consequently, we proved that (2.47) *is satisfied when* $b = \frac{m}{m-n}$, *which implies that* $t(\lambda_1) = \Lambda_{m,n}(\frac{m}{m-n}) = \tau_0$. *Summarizing, we have proved that*

$$\Phi(a, \lambda_1(a-1)) = (b(a,\lambda_1), \tau_0) \quad \text{for all } a_1 \le a < 1 \tag{2.49}$$

and $0 < b(a, \lambda_1) \le b(a_1, \lambda_1) = \frac{m}{m-n}$ *for* $a_1 \le a < 1$. *Next, pick*

$$(a, \lambda(a-1)) \in U^1_{m,n,1}$$

with $\lambda > \lambda_1$. *It is enough to check that the corresponding* $(b(a,\lambda), t(\lambda))$ *satisfies that* $t(\lambda) \le \Lambda_{m,n}(b(a,\lambda))$ *which is equivalent to*

$$K_{m,n} b(a,\lambda)^{\frac{m}{n}} - \frac{n}{m}\frac{b(a,\lambda)}{t(\lambda)} - 1 - \left(\frac{m-n}{m}\right) b(a,\lambda) t(\lambda)^{\frac{n}{m-n}} \le 0. \tag{2.50}$$

By definition of $b(a,\lambda)$ *and* $t(\lambda)$, (2.50) *is equivalent to*

$$\left(\frac{1-a}{a}\right)^{\frac{m}{n}} \lambda + \left(\frac{1-a}{a}\right)\lambda - 1 - \left(\frac{1-a}{a}\right) \le 0$$

which is indeed true because $\lambda(a-1) \ge \Upsilon_{m,n}(a)$. *Also, the equality in* (2.50) *holds if and only if* $\lambda(a-1) = \Upsilon_{m,n}(a)$. *It follows that* (2.43) *holds and the claim is proved.*

- *The region* $U^1_{m,n,2}$: *We claim that*

$$\left\{\Phi(a,c) : (a,c) \in U^1_{m,n,2}\right\} = \mathcal{B}^1_{m,n} \cap \{-1 \le t < \tau_0\}. \tag{2.51}$$

Note first that $a_0 = \frac{n}{m}$ *and* $c_0 = -\frac{n}{m}$; *hence* $-\frac{c_0}{1-a_0} = \frac{n}{m-n}$. *Consider* $(a,c) \in U^1_{m,n,2}$ *with* $c = \lambda(a-1)$. *Recall from* (2.44) *that if* $\lambda = \frac{n}{m-n}$, *then*

$$0 < b(a,\lambda) = \frac{1-a}{a}\frac{m}{m-n} \le b(a_0,\lambda) = \frac{m-n}{n}\frac{m}{m-n} = \frac{m}{n}$$

and $t(\lambda) = -1$. *Now, let* $\frac{n}{m-n} < \lambda < \lambda_1$. *We claim that the corresponding* $b(a,\lambda)$ *and* $t(\lambda)$ *satisfies that*

$$b(a,\lambda) \le 2m \frac{t(\lambda)}{(m-n)|t(\lambda)|^{\frac{m}{m-n}} - m|t(\lambda)| - n} = g(t(\lambda)). \tag{2.52}$$

Note that (2.52) *is equivalent to*

$$\frac{-(m-n)|t(\lambda)|^{\frac{m}{m-n}} + m|t(\lambda)| + n}{m} \le -\frac{2t(\lambda)}{b(a,\lambda)}. \tag{2.53}$$

As a matter of fact, (2.52) *indeed holds since* $c \ge \Gamma_{m,n}(a)$, *that is,*

$$\frac{m}{n}\left(\frac{n}{m-n}\right)^{\frac{m-n}{m}} (1-a)\lambda^{\frac{n}{m}} \le 1 + a + \lambda(a-1).$$

Thus,

$$\begin{aligned}
&\frac{-(m-n)|t(\lambda)|^{\frac{m}{m-n}}+m|t(\lambda)|+n}{m}\\
&\qquad=-\frac{1}{\lambda}\frac{n}{m}+\left(\frac{n}{m-n}\right)^{\frac{m-n}{m}}\lambda^{\frac{n-m}{m}}+\frac{n}{m}\\
&\qquad\leq-\frac{n}{\lambda m}+\frac{n}{\lambda m}\frac{1}{1-a}(1+a+\lambda(a-1))+\frac{n}{m}\\
&\qquad=\frac{n}{\lambda m}\left(-1+\frac{1}{1-a}+\frac{a}{1-a}\right)\\
&\qquad=\frac{2an}{\lambda(1-a)m}=-\frac{2t(\lambda)}{b(a,\lambda)}.
\end{aligned}$$

Combining (2.43), (2.51) with the fact $\Phi(1,c)=(0,0)$ *for every* $-1\leq c<0$, *we conclude* $\Phi(U^1_{m,n}\setminus\{c=0\})=\mathcal{B}^1_{m,n}\cup\{(0,0)\}$.

- *The region* $W_{m,n}$*: By symmetry it will be enough to consider the region* $W_{m,n}\cap\{(a,c)\in\Pi:a>0\}$.

◦ *The region* $P_1:=W_{m,n}\cap\{(a,c)\in\Pi:a,c>0\}$ *(see Figure 2.13): Consider* $(a,c)\in P_1$ *with* $a+c=k$ *and* $0<k\leq 1$. *Then*

$$\Phi(a,c)=\left(\frac{1-k}{a},\frac{n}{m}\frac{1-k}{k-a}\right).$$

Then for each $0<k\leq 1$

$$\{\Phi(a,c):a,c>0 \text{ and } a+c=k\}=\left\{\left(\frac{1-k}{a},\frac{n}{m}\frac{1-k}{k-a}\right):0<a<k\right\}.$$

It is an easy exercise to prove that

$$\bigcup_{0<k<1}\left\{\left(\frac{1-k}{a},\frac{n}{m}\frac{1-k}{k-a}\right):0<a<k\right\}=\{(b,t)\in\mathbb{R}^2:b,t>0\}$$

which implies that

$$\{\Phi(a,c):(a,c)\in P_1\}=\{(b,t)\in\mathbb{R}^2:b,t>0\}\cup\{(0,0)\}.$$

◦ *The region* $P_2:=W_{m,n}\cap\{(a,c)\in\Pi:a>0,\ c<0\}$*: We divide* P_2 *into three regions:* $P_2=P_{2,1}\cup P_{2,2}\cup P_{2,3}$, *where*

$$\begin{aligned}
P_{2,1}&=\left\{(a,c)\in\Pi:0<a<1,\ c=\lambda a-1,\ \frac{n}{m-n}<\lambda\leq\frac{m-n}{n},\ c<\Gamma_{m,n}(a)\right\};\\
P_{2,2}&=\left\{(a,c)\in\Pi:0<a<1,\ c=\lambda a-1,\ \lambda\geq\frac{m-n}{n},\ a+c\leq 0\right\};\\
P_{2,3}&=\left\{(a,c)\in\Pi:0<a<1,\ c=\lambda(a-1),0<\lambda<\frac{n}{m-n},\ a+c\geq 0\right\}.
\end{aligned}$$

See Figure 2.13 for a representation of $P_{2,1}$, $P_{2,2}$, *and* $P_{2,3}$ *and Figure 2.14 for a representation of* $\Phi(P_{2,1})$, $\Phi(P_{2,2})$, *and* $\Phi(P_{2,3})$.

– *The region* $P_{2,1}$: *Pick a point* $(a,c) \in P_{2,1}$ *with* $c = \lambda a - 1$. *Observe that*

$$\Phi(a, \lambda a - 1) = \left(1 + \lambda, \frac{n}{m}\frac{a(1+\lambda)}{\lambda a - 1}\right).$$

It is clear that $\frac{m}{m-n} < 1 + \lambda \leq \frac{m}{n}$. *Observe that* $(b,t) = \left(1+\lambda, \frac{n}{m}\frac{a(1+\lambda)}{\lambda a - 1}\right)$ *satisfies that* $b > g(t)$ *if and only if*

$$(\dagger) := (m-n)\left(\frac{n}{m}\frac{a(1+\lambda)}{|\lambda a - 1|}\right)^{\frac{m}{m-n}} - n\frac{a(1+\lambda)}{|\lambda a - 1|} - n < \frac{2an}{\lambda a - 1}. \tag{2.54}$$

Indeed, (2.54) *holds since* $c < \Gamma_{m,n}(a)$, *that is,*

$$\frac{m}{n}\left(\frac{n}{m-n}\right)^{\frac{m-n}{m}} (1-a)^{\frac{m-n}{m}} |\lambda a - 1|^{\frac{n}{m}} > a(1+\lambda) \tag{2.55}$$

More precisely, applying (2.55) *to* $(\dagger)$, *we have*

$$(\dagger) < \frac{n}{|\lambda a - 1|}(1-a) - n\frac{a(1+\lambda)}{|\lambda a - 1|} - n = \frac{2an}{\lambda a - 1}.$$

Moreover, if $\lambda = \frac{m-n}{n}$, *then*

$$\left\{\Phi\left(a, \frac{m-n}{n}a - 1\right) : 0 < a \leq \frac{n}{m}\right\} = \left\{\left(\frac{m}{n}, t\right) : -1 \leq t < 0\right\};$$

hence

$$\Phi(P_{2,1}) = \left\{(b,t) \in \mathbb{R}^2 : \frac{m}{m-n} < b \leq \frac{m}{n},\, t < 0,\, b > g(t)\right\}. \tag{2.56}$$

– *The region* $P_{2,2}$: *Pick a point* $(a,c) \in P_{2,2}$ *with* $c = \lambda a - 1$. *Recall that* $\Phi(a, \lambda a - 1) = \left(1+\lambda, \frac{n}{m}\frac{a(1+\lambda)}{\lambda a - 1}\right) =: (b,t)$. *Note that* $b = 1 + \lambda \geq \frac{m}{n}$ *and* $a + c = a(1+\lambda) \leq 1$; *hence for* $0 < a < 1$ *we have*

$$0 > t = \frac{n}{m}\frac{a(1+\lambda)}{\lambda a - 1} \geq \frac{n}{m}\left(-\frac{1}{a}\right)a(1+\lambda) = -\frac{n}{m}b.$$

This proves that

$$\Phi(P_{2,2}) = \left\{(b,t) : b \geq \frac{m}{n},\, -\frac{n}{m}b \leq t < 0\right\}. \tag{2.57}$$

– *The region* $P_{2,3}$: *Let* $(a,c) \in P_{2,3}$ *with* $c = \lambda(a-1)$ *and notice that*

$$\Phi(a, \lambda(a-1)) = \left(\frac{(1+\lambda)(1-a)}{a}, -\frac{n}{m}\left(\frac{1+\lambda}{\lambda}\right)\right) =: (b,t)$$

and $\lambda \neq 0$. *It is clear that* $t = -\frac{n}{m}\left(\frac{1+\lambda}{\lambda}\right) < -1$ *since* $\lambda < \frac{n}{m-n}$. *Moreover, as* $a + c = a + \lambda(a-1) \geq 0$, *we have*

$$b = (1+\lambda)\left(\frac{1+\lambda}{\lambda}\right) \leq \frac{1+\lambda}{\lambda} = -\frac{m}{n}t,$$

that is, $t \leq -\frac{n}{m}b$. *This satisfies the equality*

$$\Phi(P_{2,3}) = \left\{(b,t) \in \mathbb{R}^2 : b > 0,\, t < -1,\, t \leq -\frac{n}{m}b\right\}. \tag{2.58}$$

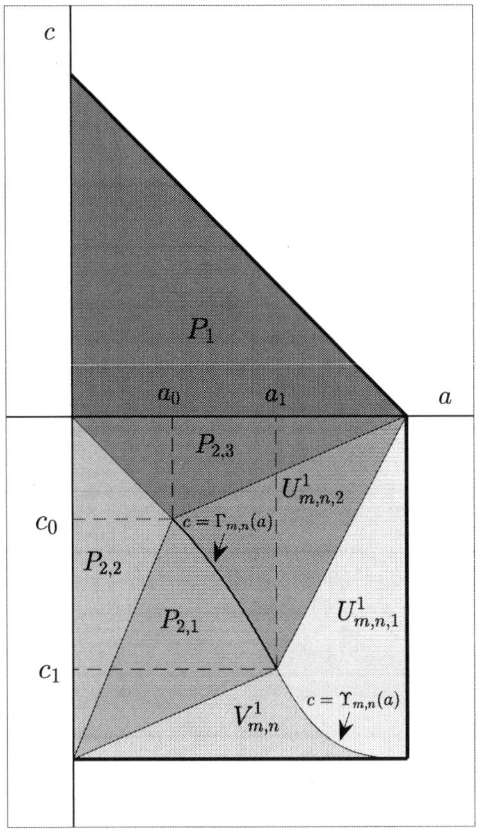

FIGURE 2.13: Representation of the regions P_1, P_2, $P_{2,1}$, $P_{2,2}$, $P_{2,3}$, $U^1_{m,n,1}$, $U^1_{m,n,2}$, and $V^1_{m,n}$ employed in Lemma 2.2.8 and its proof. Here we have considered the case $m = 10$ and $n = 3$.

Consequently, by symmetry again, we complete the proof of the assertion

$$\Phi(W_{m,n} \setminus \{ac = 0\}) = [(\mathbb{R}^2 \setminus \{bt = 0\}) \setminus (\mathcal{A}_{m,n} \cup \mathcal{B}_{m,n})] \cup \{(0, 0)\}.$$

Theorem 2.2.9 *For every $m, n \in \mathbb{N}$ with $m > n$, m even, n odd we have*

$$\mathsf{S}^h_{m,n} = \begin{cases} \text{graph}(F_{m,n}) \cup \text{graph}(-F_{m,n}) & \text{if } m \geq 2n, \\ \text{graph}(G_{m,n}) \cup \text{graph}(-G_{m,n}) & \text{if } m \leq 2n. \end{cases}$$

The reader can find a representation of $\mathsf{S}^h_{m,n}$ in Figure 2.15 for the case $m = 10$ and $n = 3$.

PROOF. *According to (2.23) we have that*

$$\|\!|(a, b, c)|\!\|_{m,n} = \|\!|(c, b, a)|\!\|_{m,m-n},$$

which allows us to focus our efforts on the case where $m \geq 2n$. As a matter of fact we can assume that $m > 2n$.

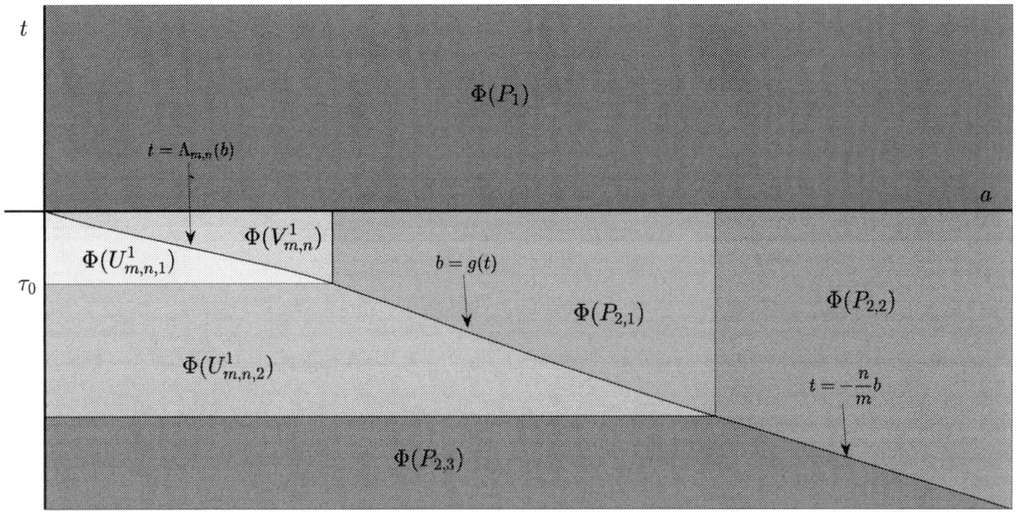

FIGURE 2.14: Representation of the image by Φ of the regions P_1, P_2, $P_{2,1}$, $P_{2,2}$, $P_{2,3}$, $U^1_{m,n,1}$, $U^1_{m,n,2}$, and $V^1_{m,n}$ employed in Lemma 2.2.8 and its proof. Here we have considered the case $m = 10$ and $n = 3$.

Since the mappings $F_{m,n}$ can be easily proved to be concave using elementary differential calculus, the centrally symmetric set

$$\mathsf{S}^{h,*}_{m,n} = \mathrm{graph}(F_{m,n}) \cup \mathrm{graph}(-F_{m,n})$$

must be the unit sphere of a norm in $\mathbb{R}^3$. Hence, in order to prove that $\mathsf{S}^h_{m,n} = \mathsf{S}^{h,*}_{m,n}$ we just need to show that $\mathsf{S}^{h,*}_{m,n} \subset \mathsf{S}^h_{m,n}$. Actually it suffices to show that $\mathrm{graph}(F_{m,n}) \subset \mathsf{S}^h_{m,n}$ due to the symmetry of $\mathsf{S}^h_{m,n}$ with respect to the plane $b = 0$. Let us take $(a, b, c) \in \mathrm{graph}(F_{m,n})$, that is

$$b = F_{m,n}(a, c).$$

We will study separately the five cases $(a, c) \in U^k_{m,n}$, $(a, c) \in V^j_{m,n}$, $1 \leq k, j \leq 2$ and $(a, c) \in W_{m,n}$.

Case 1: $(a, c) \in U^1_{m,n}$

From Lemma 2.2.8 we have that

$$\Phi(a, c) = \left(\frac{F_{m,n}(a, c)}{a}, \frac{nF_{m,n}(a, c)}{mc} \right) \in \mathcal{B}^1_{m,n}.$$

Then according to (2.33) and the fact that

$$F_{m,n}(a, c) = J_{m,n}(1 - a)^{\frac{m-n}{m}} |c|^{\frac{n}{m}},$$

it follows that

$$\begin{aligned}
|||(a, \pm b, c)|||_{m,n} &= \left| K_{m,m-n} c \left| \frac{F_{m,n}(a,c)}{c} \right|^{\frac{m}{m-n}} - a \right| \\
&= \left| K_{m,m-n} \left(J_{m,n} \right)^{\frac{m}{m-n}} \frac{c}{|c|}(1-a) - a \right| \\
&= \left| \frac{c}{|c|}(1-a) - a \right| = |a - 1 - a| = 1.
\end{aligned}$$

In the last two steps we took into consideration that $K_{m,m-n} \left(J_{m,n} \right)^{\frac{m}{m-n}} = 1$ *and that* $c < 0$ *whenever* $(a,c) \in U^1_{m,n}$.

Case 2: $(a,c) \in U^2_{m,n}$

By symmetry, $(-a,-c) \in U^1_{m,n}$ *and* $F_{m,n}(-a,-c) = F_{m,n}(a,c)$. *Then, using the previous case*

$$|||(a,b,c)|||_{m,n} = |||(-a,-b,-c)|||_{m,n} = |||(-a,b,-c)|||_{m,n} = 1.$$

Case 3: $(a,c) \in V^1_{m,n}$

From Lemma 2.2.8 we have that

$$\Phi(a,c) = \left(\frac{F_{m,n}(a,c)}{a}, \frac{nF_{m,n}(a,c)}{mc} \right) \in \mathcal{A}^1_{m,n}.$$

Then according to (2.33) and the fact that

$$F_{m,n}(a,c) = J_{m,m-n} \left(1 + c\right)^{\frac{n}{m}} a^{\frac{m-n}{m}},$$

it follows that

$$\begin{aligned}
|||(a, \pm b, c)|||_{m,n} &= \left| K_{m,n} a \left| \frac{F_{m,n}(a,c)}{a} \right|^{\frac{m}{n}} - c \right| \\
&= \left| K_{m,n} \left(J_{m,m-n} \right)^{\frac{m}{n}} (1+c) - c \right| = 1.
\end{aligned}$$

In the last two steps we took into consideration that $K_{m,n} \left(J_{m,m-n} \right)^{\frac{m}{n}} = 1$ *and that* $a \geq 0$ *whenever* $(a,c) \in V^1_{m,n}$.

Case 4: $(a,c) \in V^2_{m,n}$

By symmetry, $(-a,-c) \in V^1_{m,n}$ *and* $F_{m,n}(-a,-c) = F_{m,n}(a,c)$. *Then, using the previous case*

$$|||(a,b,c)|||_{m,n} = |||(-a,-b,-c)|||_{m,n} = |||(-a,b,-c)|||_{m,n} = 1.$$

Case 5: $(a,c) \in W_{m,n}$

From Lemma 2.2.8 we have that

$$\Phi(a,c) = \left(\frac{F_{m,n}(a,c)}{a}, \frac{nF_{m,n}(a,c)}{mc} \right) \notin \mathcal{A}_{m,n} \cup \mathcal{B}_{m,n}.$$

Assume that $b = F_{m,n}(a,c) = 1 - |a+c| = 0$ *and* $ac \leq 0$. *Observe that* a *and* c *have opposite (or null) signs and that* $|a+c| \leq 1$. *Moreover* $|a+c| = 1$ *only if* $a = \pm 1$ *and* $c = 0$ *or* $a = 0$ *and* $c = \pm 1$. *The using (2.33) we easily have*

$$|||(\pm 1, 0, 0)|||_{m,n} = |||(0, 0, \pm 1)|||_{m,n} = 1.$$

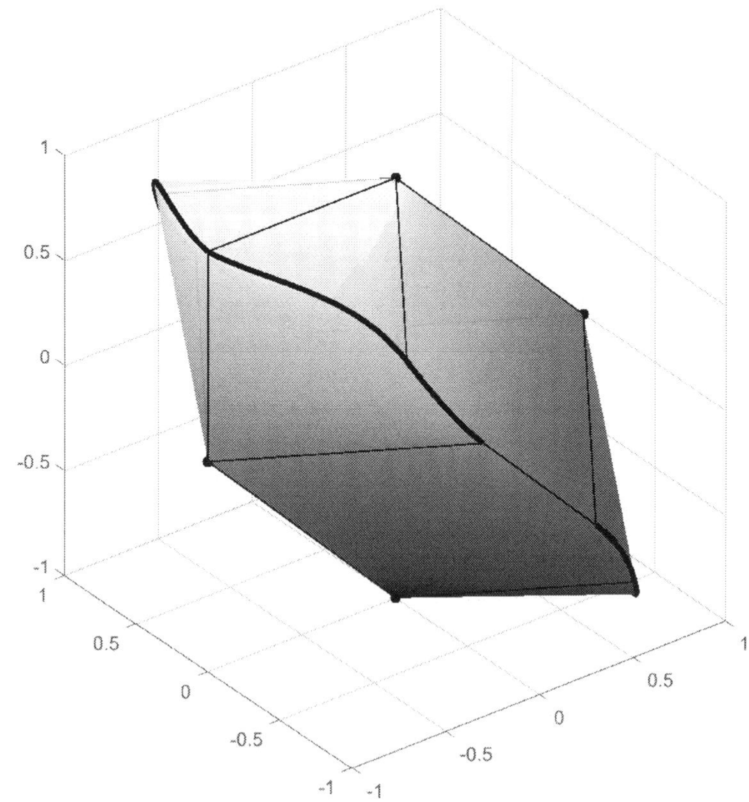

FIGURE 2.15: Representation of $\mathsf{S}_{m,n}^h$. Here we have considered the case $m = 10$ and $n = 3$.

Finally, if $b = F_{m,n}(a, c) = 1 - |a + c| \neq 0$ *or* $ac > 0$ *we have in both cases that* $b = 1 - |a + c| \geq 0$. *Then, according to* (2.33) *it follows that*

$$|||(a, b, c)|||_{m,n} = |a + c| + |b| = |a + c| + 1 - |a + c| = 1.$$

2.2.3 The extreme points of $\mathsf{B}_{m,n}^h$

The parametrization of $\mathsf{S}_{m,n}^h$ found in Theorem 2.2.9 together with the fact that most of the surfaces that take part in the graph of $F_{m,n}$ are ruled surfaces help tremendously to localize the extreme points of $\mathsf{B}_{m,n}^h$. This is addressed in the main result of this section.

Theorem 2.2.10 *Let* m, n *be positive integers with* $m \geq 2n$, m *even and* n *odd. Then* $\text{ext}(\mathsf{B}_{m,n}^h)$ *consists of the points*

$$\pm(1, 0, 0), \quad \pm(0, 0, 1),$$

$$\pm\left(a, \pm J_{m,n} \frac{(1-a)^{\frac{m-n}{m}} a^{\frac{m-n}{m}}}{(1-a)^{\frac{m-n}{n}} + a^{\frac{m-n}{n}}}, -\frac{a^{\frac{m-n}{n}}}{(1-a)^{\frac{m-n}{n}} + a^{\frac{m-n}{n}}}\right), \quad \textit{with } a \in [a_1, 1]$$

and

$$\pm\left(a, 1 - |a + \Gamma_{m,n}(a)|, \Gamma_{m,n}(a)\right), \quad \textit{with } a \in \left[\frac{n}{m}, a_1\right],$$

where $J_{m,n} = \frac{m}{n}\left(\frac{n}{m-n}\right)^{\frac{m-n}{m}}$, a_1 *is the positive number and* $c = \Gamma_{m,n}(a)$ *is the curve from Lemma 2.2.6.*

PROOF. *First we show that the graph of* $\pm F_{m,n}$ *restricted to* $V_{m,n}$ *is a ruled surface (to recall, the function* $F_{m,n}$ *is defined in (2.39), and the region* $V_{m,n}$ *is illustrated in Figure 2.12). Due to the symmetry of* $F_{m,n}$ *and* $V_{m,n}$*, we just need to prove this by restricting* $\pm F_{m,n}$ *to* $V^1_{m,n}$. *Consider the lines in the plane ac that pass though* $(0,-1)$ *and have slope* λ *with* $\lambda \in [0, \lambda_0]$, *where* $\lambda_0 = \frac{n}{m-n}$ *as in Lemma 2.2.6. The equation of each of these lines is given by* $c = \lambda a - 1$. *Let* $(\eta(\lambda), \xi(\lambda))$ *be the intersection point of* $c = \lambda a - 1$ *and the curve* $c = \Upsilon_{m,n}(a)$. *Obviously, the segments* L_λ *that join* $(0,-1)$ *and* $(\eta(\lambda), \xi(\lambda))$, *namely*

$$L_\lambda = \{(a, \lambda a - 1) : a \in [0, \eta(\lambda)]\},$$

cover the whole set $V^1_{m,n}$ *as* λ *ranges over the interval* $[0, \lambda_0]$. *Also,* $\pm F_{m,n}$ *restricted to* L_λ *is linear since, by (2.40)*

$$\pm F_{m,n}(a, \lambda a - 1) = \pm \frac{m}{m-n}\left(\frac{m-n}{n}\lambda\right)^{\frac{n}{m}} a$$

for $a \in [0, \eta(\lambda)]$.

A similar argument reveals that the graph of $\pm F_{m,n}$ *restricted to* $U_{m,n}$ *is a ruled surface as well. In this case we have to consider the straight lines passing through* $(1,0)$ *in the ac plane with slope* $\lambda \in [\lambda_0, \infty)$. *The equation of each of those lines is* $c = \lambda(a-1)$. *Let* $(\eta(\lambda), \xi(\lambda))$ *be the intersection point of* $c = \lambda(a-1)$ *and the curve* $c = \Upsilon_{m,n}(a)$ *or* $c = \Gamma_{m,n}(a)$. *Again, the segments* L_λ *that join* $(1,0)$ *and* $(\eta(\lambda), \xi(\lambda))$ *cover the whole* $U^1_{m,n}$ *as* $\lambda \geq \lambda_0$. *Now it is easy to see that the restriction of* $\pm F_{m,n}$ *to* L_λ *is affine since, by (2.44)*

$$\pm F_{m,n}(a, \lambda(a-1)) = (1-a)\frac{m}{n}\left(\frac{n}{m-n}\right)^{\frac{m-n}{m}} \lambda^{\frac{n}{m}}.$$

Finally, the graph of $\pm F_{m,n}$ *when restricted to* $W_{m,n}$ *is formed by two flat surfaces. We conclude that all the extreme points of* $\mathsf{B}^h_{m,n}$ *must lie in the intersections of the graphs of* $\pm F_{m,n}$ *when restricted to* $V^k_{m,n}$, $U^k_{m,n}$ *with* $k = 1, 2$ *and* $W_{m,n}$. *Some of these intersections produce, in their turn, segments. That is the case of the following intersections:*

- $\operatorname{graph}(F_{m,n}|_{U^1_{m,n}}) \cap \operatorname{graph}(-F_{m,n}|_{U^1_{m,n}}) = [P_1, P_3]$,
- $\operatorname{graph}(F_{m,n}|_{U^2_{m,n}}) \cap \operatorname{graph}(-F_{m,n}|_{U^2_{m,n}}) = [-P_1, -P_3]$,
- $\operatorname{graph}(F_{m,n}|_{V^1_{m,n}}) \cap \operatorname{graph}(-F_{m,n}|_{V^1_{m,n}}) = [P_2, P_3]$,
- $\operatorname{graph}(F_{m,n}|_{V^2_{m,n}}) \cap \operatorname{graph}(-F_{m,n}|_{V^2_{m,n}}) = [-P_2, -P_3]$,
- $\operatorname{graph}(F_{m,n}|_{W_{m,n}}) \cap \operatorname{graph}(-F_{m,n}|_{W_{m,n}}) = [P_2, -P_1] \cup [-P_2, P_1]$,
- $\operatorname{graph}(F_{m,n}|_{V^1_{m,n}}) \cap \operatorname{graph}(F_{m,n}|_{W_{m,n}}) = [P_2, Q_1]$,
- $\operatorname{graph}(-F_{m,n}|_{V^1_{m,n}}) \cap \operatorname{graph}(-F_{m,n}|_{W_{m,n}}) = [P_2, Q_2]$,
- $\operatorname{graph}(F_{m,n}|_{V^2_{m,n}}) \cap \operatorname{graph}(F_{m,n}|_{W_{m,n}}) = [-P_2, -Q_1]$,
- $\operatorname{graph}(-F_{m,n}|_{V^2_{m,n}}) \cap \operatorname{graph}(-F_{m,n}|_{W_{m,n}}) = [-P_2, -Q_2]$,

where $P_1 = (1,0,0)$, $P_2 = (0,0,-1)$, $P_3 = (1,0,-1)$, $Q_1 = (a_0, 1, c_0)$, *and* $Q_2 = (a_0, -1, c_0)$. *Here,* (a_0, c_0) *is the point appeared in Lemma 2.2.6. Those segments can only contain extreme points of* $\mathsf{B}^h_{m,n}$ *at their endpoints, that is, at* $\pm P_1$, $\pm P_2$, $\pm P_3$, $\pm Q_1$,$\pm Q_2$.

There are other intersections that provide both segments and curved lines. That happens in the following cases:

- $\operatorname{graph}(\pm F_{m,n}|_{U^k_{m,n}}) \cap \operatorname{graph}(\pm F_{m,n}|_{W_{m,n}})$ *with* $k = 1, 2$. *These intersections include the curved lines*

$$\pm\left(a, \pm F_{m,n}(a, \Gamma_{m,n}(a)), \Gamma_{m,n}(a)\right) \quad \text{with } a \in [a_0, a_1] \tag{2.59}$$

together with the four segments

$$\pm[P_1, Q_1] \quad \text{and} \quad \pm[P_1, Q_2],$$

Observe that $\pm Q_1$ *and* $\pm Q_2$ *are already in the curves* (2.59). *To see this we just need to put* $a = a_0$ *in* (2.59). *Using the definition of* $F_{m,n}$, (2.59) *reduces to*

$$\pm\left(a, \pm(1 - |a + \Gamma_{m,n}(a)|), \Gamma_{m,n}(a)\right) \quad \text{with } a \in \left[\frac{n}{m}, a_1\right]$$

Finally there are intersections that provide only curved lines:

- $\operatorname{graph}(\pm F_{m,n}|_{U^k_{m,n}}) \cap \operatorname{graph}(\pm F_{m,n}|_{V_{m,n}})$ *with* $k = 1, 2$.
These intersections provide the curves

$$\pm\left(a, \pm F_{m,n}(a, \Upsilon_{m,n}(a)), \Upsilon_{m,n}(a)\right) \quad \text{with } a \in [a_1, 1] \tag{2.60}$$

or equivalently

$$\pm\left(a, \pm J_{m,n}\frac{(1-a)^{\frac{m-n}{m}} a^{\frac{m-n}{m}}}{(1-a)^{\frac{m-n}{n}} + a^{\frac{m-n}{n}}}, -\frac{a^{\frac{m-n}{n}}}{(1-a)^{\frac{m-n}{n}} + a^{\frac{m-n}{n}}}\right), \quad \text{with } a \in [a_1, 1].$$

Observe that putting $a = 1$ *in* (2.60) *we obtain* $\pm P_3$ *so the points* $\pm P_3$ *are already included in the curves* (2.60).

Therefore, the extreme points of $\mathsf{B}^h_{m,n}$ *must lie among the points* $\pm P_1$ *and* $\pm P_2$ *or in the curves described in* (2.59) *and* (2.60).

In the rest of the proof, we will say that a plane H *in* $\mathbb{R}^3$ *is a supporting hyperplane to* $\mathsf{B}^h_{m,n}$ *at* $P \in \mathsf{B}^h_{m,n}$ *if* $H \cap \mathsf{B}^h_{m,n} = \{P\}$. *To finish the proof we just have to construct supporting planes to* $\mathsf{B}^h_{m,n}$ *at each point of the 8 curves represented in* (2.59) *and* (2.60) *and the four points* $\pm P_1$ *and* $\pm P_2$. *The existence of such a plane would guarantee that* P *is an extreme point of* $\mathsf{B}^h_{m,n}$.

Starting with the points $\pm P_1$ *and* $\pm P_2$, *it can be easily seen that the planes*

$$\begin{aligned}
&2(a-1) + c = 0,\\
&2(a+1) + c = 0,\\
&a + 2(c+1) = 0,\\
&a + 2(c-1) = 0,
\end{aligned}$$

are supporting planes to $\mathsf{B}^h_{m,n}$ *at* P_1, $-P_1$, P_2 *and* $-P_2$, *respectively. Moreover, in fact there are infinitely many supporting hyperplanes at* $\pm P_1$ *and* $\pm P_2$. *To prove this we will focus on the point* $P_1 = (1,0,0)$ *since the other cases are similar. We just need to consider the*

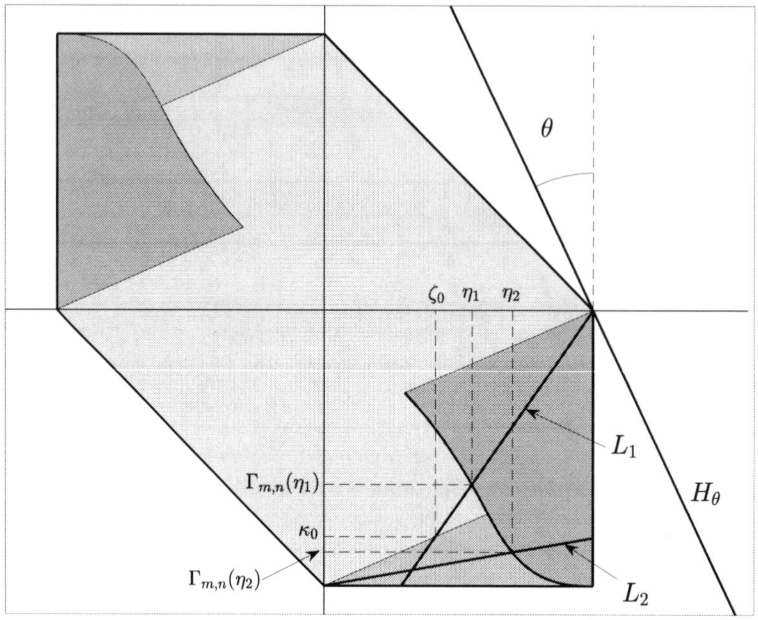

FIGURE 2.16: Representation of the projection onto the ac plane of a supporting hyperplane H_θ to $\mathsf{B}^h_{m,n}$ at $(1,0,0)$. Here we have considered the case $m = 10$, $n = 3$, and $\theta = \pi/7$. Also, we have represented the segments L_1 and L_2 corresponding to two choices of η, namely $\eta_1 = 0.55$ and $\eta_2 = 0.7$, in the proof of Theorem 2.2.10. The point (ζ_0, κ_0) is the intersection of L_1 and the line $c = \lambda_0 a - 1$.

family of hyperplanes H_θ passing through $(1,0,0)$ and parallel to the vectors $(0,1,0)$ and $(\sin\theta, 0, \cos\theta - 1)$ with $\theta \in (0, \pi/2)$ (see Figure 2.16), that is,

$$H_\theta : \cos\theta(a-1) + (1-\sin\theta)c = \begin{vmatrix} a-1 & b & c \\ \sin\theta - 1 & 0 & \cos\theta \\ 0 & 1 & 0 \end{vmatrix} = 0.$$

Obviously $\mathsf{B}^h_{m,n} \cap H_\theta = \{(1,0,0)\}$ for all $\theta \in (0, \pi/2)$.

As for the 8 curves in (2.59) and (2.60), it is sufficient to consider only two cases due to the numerous existing symmetries. For instance take

(a) $(a, F_{m,n}(a, \Gamma_{m,n}(a)), \Gamma_{m,n}(a))$ *with* $a \in [a_0, a_1]$.

(b) $(a, F_{m,n}(a, \Upsilon_{m,n}(a)), \Upsilon_{m,n}(a))$ *with* $a \in [a_1, 1]$.

Let us study the curve appearing in (a). For simplicity we put

$$f_1(a) = F_{m,n}(a, \Gamma_{m,n}(a)),$$
$$g_1(a) = \Gamma_{m,n}(a),$$

ending up with the curve

$$C_1(a) = (a, f_1(a), g_1(a))$$

for $a \in [a_0, a_1]$. Taking an arbitrary $\eta_1 \in [a_0, a_1]$, a supporting plane to $\mathsf{B}^h_{m,n}$ at $C_1(\eta_1)$ must be tangent to the curve C_1 at $C_1(\eta_1)$, or equivalently, parallel to $C_1'(\eta) = (1, f_1'(\eta_1), g_1'(\eta_1))$.

We need another vector to construct a plane passing through $C_1(\eta_1)$. *Let* $\gamma_1(a) = \lambda(a-1)$ *with*

$$\lambda = \frac{\Gamma_{m,n}(\eta_1)}{\eta_1 - 1}.$$

The slope λ *has been chosen so that the line in the ac-plane given by* $c = \gamma_1(a)$ *with* $a \in \mathbb{R}$ *passes through* $(\eta_1, \Gamma_{m,n}(\eta_1))$ *and* $(1,0)$. *Consider the segment* L_1 *which results in intersecting the line* γ_1 *and the projection of* $\mathsf{B}^h_{m,n}$ *onto the ac plane (see Figure 2.16). Also, consider the plane* Π_1 *that contains* L_1 *and is parallel to the b axis. Obviously* Π_1 *contains* $(1,0,0)$. *Recall we have shown above in this proof that* $F_{m,n}$ *is linear on the segment joining* $(1,0)$ *and* $(\eta_1, \Gamma_{m,n}(\eta_1))$. *Therefore, the segment* L_1^* *joining the points* $(1,0,0)$ *and* $C_1(\eta_1)$ *is contained in the graph of* $F_{m,n}$ *restricted to* L_1. *Since* $F_{m,n}$ *is affine on* $W_{m,n}$, *the restriction of* $F_{m,n}$ *to* $L_1 \cap W_{m,n}$ *is a segment, say* $L_1^{\#}$. *Therefore, the graph of* $F_{m,n}$ *restricted to* L_1 *has two confluent segments at* $(\eta_1, \Gamma_{m,n}\eta_1))$, *namely* L_1^* *and* $L_1^{\#}$. *We have represented this situation in Figure 2.17 where we have depicted the plane* Π_1 *with the graph of the restriction of* $F_{m,n}$ *to* L_1. *The remaining vector needed to construct a supporting hyperplane to* $\mathsf{B}^h_{m,n}$ *at* $C_1(\eta_1)$ *can be chosen in many ways. We call that vector* u_δ *and the resulting supporting hyperplane* H_δ *is:*

$$H_\delta : \begin{vmatrix} (a,b,c) - C_1(\eta_1) \\ C_1'(\eta_1) \\ u_\delta \end{vmatrix} = 0.$$

The nature of u_δ *and the index* δ *are revealed next. First consider two unitary vectors* v_{α_1} *and* v_{β_1} *parallel to* $L_1^{\#}$ *and* L_1^*, *respectively. Here* α_1 *and* β_1 *are, respectively, the angles between* $L_1^{\#}$ *and the plane* $b = 0$ *and between* L_1^* *and the plane* $b = 0$ *(see Figure 2.17). Hence, the desired vector* u_δ *can be expressed in canonical coordinates of the plane* Π_1 *as*

$$u_\delta = (\cos(\delta), \sin(\delta))$$

with $\delta \in (\beta_1, \alpha_1)$. *Observe that we have not included the cases* $\delta = \alpha_1, \beta_1$ *because* $\mathsf{B}^h_{m,n} \cap H_{\alpha_1} = L_1^{\#}$ *whereas* $\mathsf{B}^h_{m,n} \cap H_{\beta_1} = L_1^*$.

Finally we consider the curve in (b). Putting

$$\begin{aligned} f_2(a) &= F_{m,n}(a, \Upsilon_{m,n}(a)), \\ g_2(a) &= \Upsilon_{m,n}(a), \end{aligned}$$

we end up with the curve

$$C_2(a) = (a, f_2(a), g_2(a))$$

for $a \in [a_1, 1]$. *Taking an arbitrary* $\eta_2 \in [a_1, 1]$, *a supporting plane to* $\mathsf{B}^h_{m,n}$ *at* $C_2(\eta_2)$ *must be tangent to the curve* C_2 *at* $C_2(\eta_2)$, *or equivalently, parallel to* $C_2'(\eta_2) = (1, f_2'(\eta_2), g_2'(\eta_2))$. *We need another vector to construct a plane passing through* $C_2(\eta_2)$. *Let* $\gamma_2(a) = \lambda a - 1$ *with*

$$\lambda = \frac{\Upsilon_{m,n}(\eta_2) + 1}{\eta_2}.$$

Notice that the slope λ *has been chosen so that the segment* L_2 *in the ac-plane given by* $c = \gamma_2(a)$ *with* $a \in [0,1]$ *passes through* $(\eta_2, \Upsilon(\eta_2))$, *starting at* $(0,-1)$. *Consider the plane* Π_2 *that contains* L_2 *and is parallel to the b axis. Obviously* Π_2 *contains* $(0,0,-1)$. *Recall we have proved above in this proof that* $F_{m,n}$ *is linear on the segment joining* $(0,-1)$ *and* $(\eta_2, \Upsilon(\eta_2))$, *as can be seen in Figure 2.18 where we have depicted the plane* Π_2 *with the representation of the restriction of* $F_{m,n}$ *to* L_2. *Let us call* L_2^* *the segment joining* $(-1,0,0)$ *and* $(\eta_2, F_{m,n}(\eta_2, \Upsilon_{m,n}(\eta_2)), \Upsilon_{m,n}(\eta_2))$. *The remaining vector needed to construct a supporting*

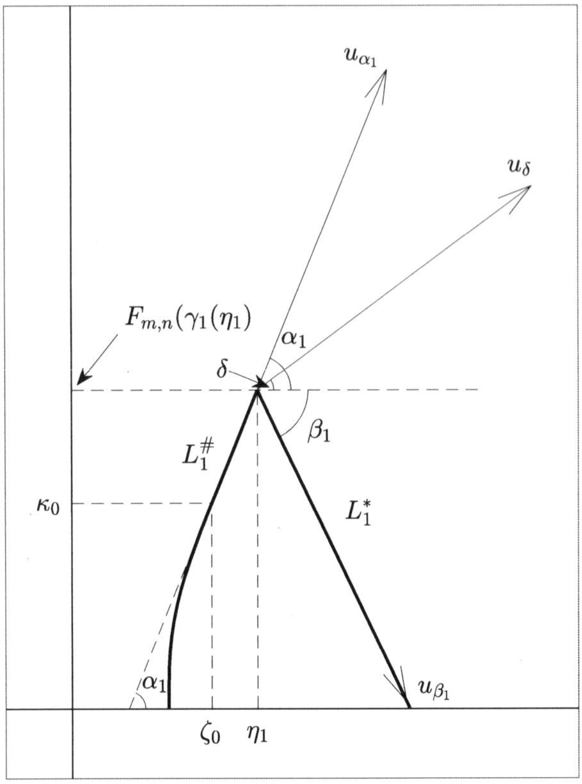

FIGURE 2.17: Representation of the plane Π_1 with the restriction of $F_{m,n}$ to L_1. The mapping $[\zeta_0, \eta_1] \ni a \mapsto F_{m,n}(\gamma_1(a))$ is linear and its graph is the segment $L_1^\#$. Also $[\eta_1, 1] \ni a \mapsto F_{m,n}(\gamma_1(a))$ is linear too and its graph is the segment L_1^*. Here we have considered the case $m = 10$, $n = 3$, $\eta_1 = 0.55$, and $\delta = \pi/5$.

hyperplane to $\mathsf{B}_{m,n}^h$ *at* $C_2(\eta_2)$ *can be chosen in many ways. If we call that vector* v_ω*, then the desired plane would be given by*

$$H_\omega : \begin{vmatrix} (a,b,c) - C_2(\eta_2) \\ C_2'(\eta_2) \\ v_\omega \end{vmatrix} = 0.$$

The meaning of v_ω *and the index* ω *will be explained in the rest of the proof. First consider the angle* α_2 *between the segment* L_2^* *and the plane* $b = 0$*, that is,*

$$\alpha_2 = \arctan(F_{m,n}(C_2(\eta_2))/\eta_2).$$

Also let us choose a vector v_{β_2} *parallel to the tangent line to the graph of* $[0,1] \ni a \mapsto F_{m,n}(\gamma_2(a))$ *at* $a = \eta_2$*, whose slope* s_2 *in the plane* Π_2 *is given by*

$$s_2 = \frac{d}{da}\left[F_{m,n}(\gamma_2(a))\right]|_{a=\eta_2} = -J_{m,n}\left(\frac{1-\lambda\eta_2}{1-\eta_2}\right)^{\frac{n}{m}}\left(1 - \frac{n}{m}\frac{1-\lambda}{1-\lambda\eta_2}\right).$$

Hence, in canonical coordinates of the plane Π_2*,* v_{β_2} *could be expressed as*

$$v_{\beta_2} = (\cos(\beta_2), \sin(\beta_2))$$

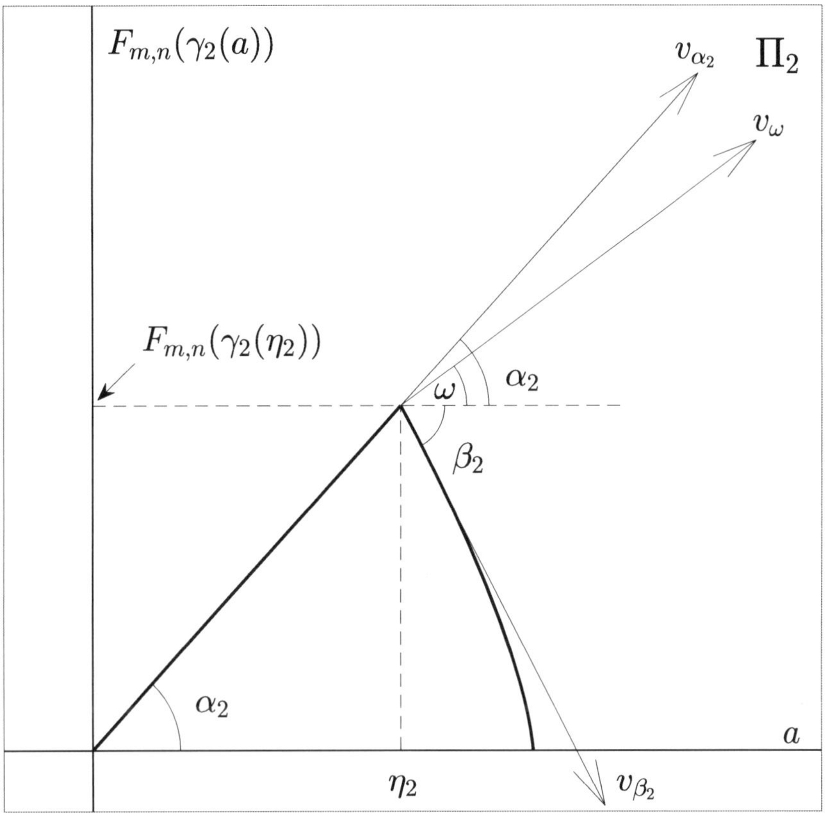

FIGURE 2.18: Representation on the plane Π_2 of the mapping $[0,1] \ni a \mapsto F_{m,n}(\gamma_2(a))$. Here we have considered the case $m = 10$, $n = 3$, $\eta_2 = 0.7$, and $\omega = \pi/5$.

where

$$\beta_2 = \arctan(s_2).$$

Finally, v_ω could be chosen to be any unitary vector between v_{α_2} and v_{β_2} in the plane Π_2, that is

$$v_\omega = (\cos(\omega), \sin(\omega))$$

with $\omega \in [\beta_2, \alpha_2)$ in coordinates of Π_2. Observe that we have not included the case $\omega = \alpha_2$ because $H_{\alpha_2} \cap \mathsf{B}^h_{m,n} = L^*_2$. We recommend to keep an eye on Figure 2.18 to follow the last part of the proof.

Chapter 3

The Krein-Milman Approach, Classical Inequalities and Applications

3.1 Preliminaries

The study of the extreme points of the unit ball of a Banach space plays a crucial role in Functional Analysis. The Krein-Milman theorem (due to Mark Krein and David Milman in 1940, [67]) allows us to describe a convex body (compact and convex with a non-empty interior) entirely through its extreme points. Let us recall the definition of extreme point in a convex set.

Definition 3.1.1 *Given a convex set C in a Banach space, a point $e \in C$ is called extreme if $x, y \in C$ and*

$$\lambda x + (1 - \lambda) y = e,$$

for some $0 < \lambda < 1$, implies that $x = y = e$. Equivalently, $e \in C$ is extreme if and only if $C \setminus \{e\}$ is convex.

The Krein-Milman theorem appears in several contexts. Here we are interested in its version for normed spaces.

Theorem 3.1.1 (Krein-Milman [67]) *Let E be a real normed space and let $K \subseteq E$ be a convex compact set. Then $\operatorname{ext}(K) \neq \varnothing$ and*

$$K = \overline{conv(\operatorname{ext}(K))}$$

where $conv(\operatorname{ext}(K))$ stands for the convex hull of $\operatorname{ext}(K)$.

A more general version of Theorem 3.1.1 can be proved for locally convex spaces. However, we will only use the following finite dimensional version of the Theorem 3.1.1 due to Minkowski and Carathéodory [79, Theorem 916] (see also [103]).

Theorem 3.1.2 (Minkowski and Carathéodory, 1911) *Let K be a compact subset of an n-dimensional Banach space X. Then $conv(K)$ is compact, all extreme points of $conv(K)$ lie in K, and every point $x \in conv(K)$ is the convex hull of at most $(n+1)$ extreme points of $conv(K)$.*

The Krein-Milman theorem is a powerful structural result that reduces complex problems involving convex sets to the study of their extreme points, which are often easier to handle. This is the principle underlying the application of the Krein-Milman theorem in obtaining polynomial inequalities with optimal constants. The principle will be referred to as "Krein-Milman Approach" in this book and its precise formulation is presented next.

Theorem 3.1.3 (*Krein-Milman Approach*) *If $C \subset \mathbb{R}^n$ is a convex compact set and $f : C \to \mathbb{R}$ is a convex function that attains it maximum then there exists an extreme point e of C such that*

$$f(e) = \max\{f(x) : x \in C\}.$$

DOI: 10.1201/9781003590637-3

PROOF. *Assume that $f(e) < M := \max\{f(x) : x \in C\}$ for all $e \in \text{ext}(C)$ and $x_0 \in C$ is such that $f(x_0) = M$. By the Krein-Milman theorem, x_0 must be a convex combination of extreme points of C. That is, there exist $e_1, \ldots, e_p \in \text{ext}(C)$ such that*

$$x_0 = \sum_{k=1}^{p} \lambda_k e_k,$$

with $\lambda_k \in [0,1]$ for all $k \in \{1, \ldots, p\}$ and $\sum_{k=1}^{p} \lambda_k = 1$. By convexity of f, we would then have

$$M = f(x_0) \leq \sum_{k=1}^{p} \lambda_k f(e_k) < \sum_{k=1}^{p} \lambda_k M = M,$$

which is a contradiction.

The Krein-Milman approach, combined with a characterization of the extreme points of the unit ball in a polynomial space, can be used to derive several sharp polynomial inequalities. In this chapter, we present several historically significant polynomial inequalities. Through a variety of examples, we see how the Krein-Milman approach can be applied in concrete settings. We begin with Bernstein- and Markov-type inequalities, which provide bounds on the norm of polynomial derivatives. Next, we present the polynomial Bohnenblust-Hille inequalities. While the Bohnenblust-Hille polynomial constants are notoriously difficult to compute, we show that in certain cases, the Krein-Milman method yields exact values. Polarization constants in several polynomial spaces will also be calculated. Finally, we show how to compute unconditional constants with respect to the canonical basis in certain polynomial spaces using the Krein-Milman approach.

3.2 Bernstein and Markov type inequalities

The first significant problem concerning the estimation of polynomial derivatives was posed in the late 19th century by the renowned Russian chemist D. Mendeleev, best known as the pioneer of the periodic table of elements (see Figure 3.1). His primary interest lay in estimating the derivatives of quadratic polynomials over an interval of the real line. The motivations that led a distinguished chemist like Mendeleev to study a purely mathematical problem are related to the study of the specific gravity of alcoholic solutions. An interesting discussion on this matter can be found in [13]. In this section, we restrict our attention to the problem of estimating polynomial derivatives from a purely mathematical standpoint.

The notion of sup norm will be required to formulate the problem with precision.

> *If $P(x) = ax^2 + bx + c$ is a quadratic polynomial and $\alpha, \beta \in \mathbb{R}$ with $\alpha < \beta$, we recall that the sup norm of P over $[\alpha, \beta]$ is*
>
> $$\|P\|_{[\alpha,\beta]} = \sup\{|P(x)| : x \in [\alpha, \beta]\}.$$

A more precise formulation of Mendeleev's problem is as follows:

> *If $\alpha, \beta \in \mathbb{R}$ with $\alpha < \beta$, what is the smallest constant $\mathcal{M}(\alpha, \beta)$ such that the inequality*
>
> $$|P'(x)| \leq \mathcal{M}(\alpha, \beta)\|P\|_{[\alpha,\beta]}$$
>
> *holds for every quadratic polynomial $P(x) = ax^2 + bx + c$ and every $x \in [\alpha, \beta]$?*

FIGURE 3.1: D. Mendeleev (1834-1907) on the left. On the write we find an early version of D. Mendeleev's periodic table in Russian. The title translates as "Proposal for a system of elements. Based on atomic weights and chemical similarities"

To normalize this problem, we apply the transformation

$$x \to \phi(x) = \frac{\alpha + \beta - (\alpha - \beta)x}{2},$$

which maps $[\alpha, \beta]$ onto $[-1, 1]$. This leads to the relation

$$\mathcal{M}_2(\alpha, \beta) = \frac{2}{\beta - \alpha} \mathcal{M}_2(-1, 1).$$

For brevity, we denote $\mathcal{M}_2(-1, 1)$ simply as $\mathcal{M}_2$, and it is known that $\mathcal{M}_2 = 4$. Mendeleev later communicated this problem to the Markov brothers, who generalized it in various ways to the space $\mathcal{P}_n(\mathbb{R})$ of real polynomials of degree at most n, equipped with the supremum norm on $[-1, 1]$. We will use from now B_n and S_n to denote the unit ball and the unit sphere, respectively, of $\mathcal{P}_n(\mathbb{R})$.

Let $n, k \in \mathbb{N}$ with $n \geq k$, and let $\mathcal{M}_{n,k}$ represent the smallest constant satisfying

$$|P^{(k)}(x)| \leq \mathcal{M}_{n,k} \|P\|_{[-1,1]} \tag{3.61}$$

for all polynomials $P \in \mathcal{P}_n(\mathbb{R})$ and every $x \in [-1, 1]$. Here, $P^{(k)}$ denotes the kth derivative. For simplicity, we write $\mathcal{M}_n$ instead of $\mathcal{M}_{n,1}$. The inequality (3.61) is widely recognized as Markov's inequality for real polynomials. In fact, A. Markov and D. Mendeleev were both corresponding members of the Russian Academy of Sciences in Saint Petersburg, where Mendeleev presumably presented his problem on quadratic polynomials to A. Markov. As mathematicians often do with specific issues like Mendeleev's question on quadratic polynomials, A. Markov generalized Mendeleev's problem to polynomials of arbitrary degree.

Theorem 3.2.1 (A. Markov, 1889) *If $n \in \mathbb{N}$ then $\mathcal{M}_n = n^2$. Equality in* (3.61) *holds for the Chebyshev polynomials of the first kind, defined for $x \in [-1, 1]$ by*

$$T_n(x) = \cos(n \arccos x).$$

FIGURE 3.2: A. A. Markov (1856–1922) on the left and V. A. Markov (1871–1897) on the right.

V. Markov, brother of A. Markov, generalized the previous result for arbitrary derivatives.

Theorem 3.2.2 (V. Markov, 1892) *If $n \in \mathbb{N}$ then*

$$\mathcal{M}_{n,k} = |T_n^{(k)}(\pm 1)| = \frac{n^2(n^2-1^2)\cdots(n^2-(k-1)^2)}{1\cdot 3\cdots(2k-1)}.$$

Equality in (3.61) *holds again for the Chebyshev's polynomials of the first kind.*

The reader can find a picture of the Markov brothers in Figure 3.2.

Beyond these uniform estimates, finer bounds exist within the interior of $[-1,1]$. For any fixed $x \in [-1,1]$, let $\mathcal{B}_{n,k}(x)$ denote the smallest constant satisfying

$$|P^{(k)}(x)| \le \mathcal{B}_{n,k}(x)\|P\|_{[-1,1]} \tag{3.62}$$

for all $P \in \mathcal{P}_n(\mathbb{R})$. We use $\mathcal{B}_n(x)$ instead of $\mathcal{B}_{n,1}(x)$ for brevity. An important upper bound for $\mathcal{B}_n(x)$ was obtained by S. Bernstein.

Theorem 3.2.3 (Bernstein, 1912) *If $n \in \mathbb{R}$ then*

$$\mathcal{B}_n(x) \le \frac{n}{\sqrt{1-x^2}}$$

for all $x \in (-1,1)$.

Although not optimal – particularly near $x = \pm 1$ – the previous bound has important implications, as it appears in one of the simplest proofs of Theorem 3.2.1.

Theorem 3.2.3 can be generalized for arbitrary derivatives.

Theorem 3.2.4 (Duffin and Schaeffer, 1938 [42]**)** *If $k, n \in \mathbb{N}$ and $t \in (-1,1)$, then*

$$|\mathcal{B}_{n,k}(t)| \le \mathcal{F}_{n,k}(t) := \left|\frac{d^k}{dt^k} e^{in\arccos t}\right| = \left\{\left(\frac{d^k}{dt^k}\cos n\theta\right)^2 + \left(\frac{d^k}{dt^k}\sin n\theta\right)^2\right\}^{\frac{1}{2}},$$

where $t = \cos\theta$.

For further insights on Bernstein and Markov polynomial inequalities in the real line, see [13].

The generalization of Theorems 3.2.1 through 3.2.4 has been the subject of intense research, particularly since the 1970s. Although this chapter does not aim to address it in depth, the reader may be interested to know that V. Markov's Theorem was generalized by L. Harris in 2010 (see [60]) for polynomials in a real normed space. Other significant references on Bernstein-Markov inequalities in more general contexts can be found in [86, 97, 101, 102, 107].

The contribution of this section is limited to presenting certain Bernstein–Markov type inequalities obtained through the Krein–Milman approach. To be more precise we will focus on calculating Bernstein functions and Markov constants in the space $\mathcal{P}_n(\mathbb{R})$ of polynomials of degree at most n in Section 3.2.1, in the space of trinomials $\mathcal{P}_{m,n}(\mathbb{R})$ in Section 3.2.2 and in several spaces of quadratic forms in $\mathbb{R}^n$ in Section 3.2.3.

FIGURE 3.3: S. Bernstein (1880–1968).

3.2.1 Calculation of Bernstein functions and Markov constants in $\mathcal{P}_n(\mathbb{R})$

Determining $\mathcal{B}_{n,k}(x)$ exactly is highly complex. Voronovskaja developed a framework addressing this problem and determined $\mathcal{B}_n(x)$ precisely [106]. Gusev later extended these results to $\mathcal{B}_{n,k}(x)$ [57]. These solutions rely heavily on Zolotarev polynomials. Zolotarev polynomials of degree n in x are of the form

$$Z_n(x,\sigma) = x^n - \sigma x^{n-1} + \cdots + a_k x^k + \cdots + a_0 \ ,$$

where σ is a prescribed value for a_{n-1} and the $a_k \in \mathbb{R}$ are otherwise chosen such that the deviation of Z_n from zero is minimum in the interval $[-1,1]$. These are quite difficult to compute explicitly. Thus, an alternative approach utilizes the Krein-Milman theorem. Given fixed $n, k \in \mathbb{N}$ with $n \geq k$ and $x \in [-1,1]$, define the convex function $\Phi_{n,k,x} : \mathsf{B}_n \to \mathbb{R}$ by

$$\Phi_{n,k,x}(P) = |P^{(k)}(x)|.$$

Applying Krein-Milman's theorem, we obtain

$$\mathcal{B}_{n,k}(x) = \max\{\Phi_{n,k,x}(P) : P \in \mathsf{B}_n\} = \max\{\Phi_{n,k,x}(P) : P \in \text{ext}(\mathsf{B}_n)\}. \tag{3.63}$$

This identity facilitates the computation of $\mathcal{B}_{n,k}(x)$ whenever the extreme set $\text{ext}(\mathsf{B}_n)$ is known. Recall that B_n stands for the unit ball in the space of polynomials of a single real variable, of degree at most n, $\mathcal{P}_n(\mathbb{R})$, endowed with the supremum norm on $[-1, 1]$.

A relevant notion when studying the Bernstein function $\mathcal{B}_{n,k}(x)$ is the concept of extremal polynomial.

Definition 3.2.1 *A polynomial $0 \neq P \in \mathcal{P}_n(\mathbb{R})$ is extremal for $\mathcal{B}_{n,k}(x)$ if inequality (3.62) becomes an identity. In that case, if $Q = P/\|P\|$ we would have*

$$\mathcal{B}_{n,k}(x) = \Phi_{n,k,x}(Q) = |Q^{(k)}(x)|.$$

Using the previous ideas, we are going to explicitly compute a simple Bernstein function as an example.

Theorem 3.2.5 *The Bernstein function $\mathcal{B}_2(x)$ is given by*

$$\mathcal{B}_2(x) = \begin{cases} \frac{1}{1-|x|} \text{ if } |x| \le \frac{1}{2}, \\ 4|x| \text{ if } \frac{1}{2} \le |x| \le 1. \end{cases}$$

Consequently, we have $\mathcal{M}_2 = 4$, which completes Mendeleev's original question.

PROOF. *Recall that*

$$\text{ext}\,(\mathsf{B}_2) = \left\{\pm 1, \pm\left(tx^2 \pm 2(\sqrt{2t} - t)x + 1 + t - 2\sqrt{2t}\right) : t \in [1/2, 2]\right\}.$$

Hence,

$$\mathcal{B}_2(x) = 2 \max_{t\in[1/2,2]} \left\{|tx \pm (\sqrt{2t} - t)|\right\} = 2 \max_{t\in[1/2,2]} \left\{t|x| + \sqrt{2t} - t\right\}.$$

Now it all amounts to maximize the one variable function $R(t) = t|x| + \sqrt{2t} - t$ over the interval $[1/2, 1]$. The reader can check easily that R has a unique critical point given by $t_0 = \frac{1}{2(1-|x|)^2}$. The condition $\frac{1}{2} \le t_0 \le 2$ occurs only when $|x| \le \frac{1}{2}$, and since

$$\begin{aligned} R(1/2) &= \frac{|x| + 1}{2}, \\ R(2) &= 2|x|, \\ R(t_0) &= \frac{1}{2(1 - |x|)}, \end{aligned}$$

then

$$\mathcal{B}_2(x) = 2 \max_{t\in[1/2,2]} R(t) = 2 \begin{cases} \max\{R(1/2), R(2), R(t_0)\} \text{ if } |x| \le \frac{1}{2}, \\ \max\{R(1/2), R(2), \} \text{ if } \frac{1}{2} \le |x| \le 1, \end{cases}$$

from which

$$\mathcal{B}_2(x) = \begin{cases} \frac{1}{1-|x|} \text{ if } |x| \le \frac{1}{2}, \\ 4|x| \text{ if } \frac{1}{2} \le |x| \le 1. \end{cases}$$

Now it is easy to see that $\mathcal{M}_2 = \max\{\mathcal{B}_2(x) : x \in [-1, 1]\} = 4$.

On the other hand, using the characterization of $\text{ext}(\mathsf{B}_3)$ given in Theorem 1.1.3 combined with (3.63) it is possible to provide the exact and explicit expression for $\mathcal{B}_3(x)$ (as we will soon see in a forthcoming result, Theorem 3.2.6).

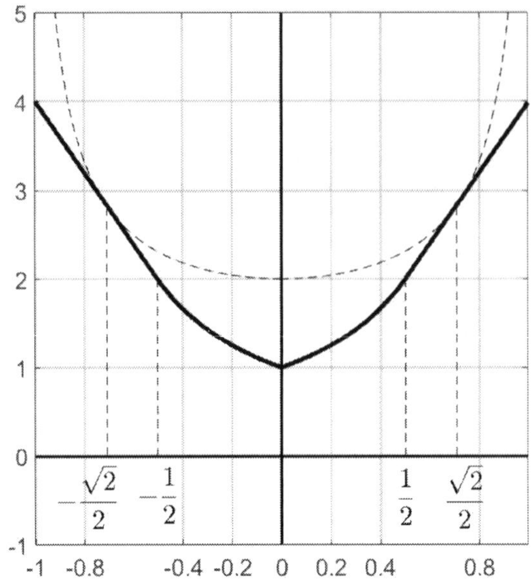

FIGURE 3.4: Representation of $\mathcal{B}_2(x)$ in a solid line. Bernstein's estimate $\frac{2}{\sqrt{1-x^2}}$ appears as a dashed line.

Similarly

$$\mathcal{B}_{3,2}(x) = \begin{cases} \frac{4}{1-9x^2} & \text{if } |x| \leq \frac{1}{9}, \\ \frac{32}{9(|x|-1)^2} & \text{if } \frac{1}{9} \leq |x| \leq \frac{1}{3}, \\ 24|x| & \text{if } |x| \geq \frac{1}{3} \end{cases}$$

The calculation of $\mathcal{B}_3(x)$ and $\mathcal{B}_{3,2}(x)$ is highly non-trivial in any case. Both functions were already known to Voronovskaja and Gusev in the 1960s (see [57, 106]). In fact, much more can be proved using the Krein-Milman approach as shown below. A representation of (3.64) can be found in Figure 3.5. Observe that $\mathcal{M}_3 = 9$.

Theorem 3.2.6 (Araújo, Muñoz, Rodríguez, and Seoane [4]) *The Bernstein function of $\mathcal{P}_3(\mathbb{R})$ is given by*

$$\mathcal{B}_3(x) = \begin{cases} 3(1-4x^2) & \text{if } |x| \leq \frac{\sqrt{7}-2}{6}, \\ \dfrac{7\sqrt{7}+10}{9(|x|+1)} & \text{if } \frac{\sqrt{7}-2}{6} \leq |x| \leq \frac{2\sqrt{7}-1}{9}, \\ \dfrac{-16x^3}{(1-9x^2)(1-x^2)} & \text{if } \frac{2\sqrt{7}-1}{9} \leq |x| \leq \frac{1+2\sqrt{7}}{9}, \\ \dfrac{7\sqrt{7}-10}{9(1-|x|)} & \text{if } \frac{1+2\sqrt{7}}{9} \leq |x| \leq \frac{\sqrt{7}+2}{6}, \\ 3(4x^2-1) & \text{if } |x| \geq \frac{\sqrt{7}+2}{6}. \end{cases} \tag{3.64}$$

moreover, for each $x_0 \geq 0$, the polynomials $P(x)$ given below satisfy $\mathcal{B}_3(x_0) = P'(x_0)$, that is, making equation (3.62) *an equality:*

1. $P(x) = \pm(4x^3 - 3x)$, *for* $x_0 \in \left[0, \frac{\sqrt{7}-2}{6}\right] \cup \left[\frac{\sqrt{7}+2}{6}, \infty\right)$.

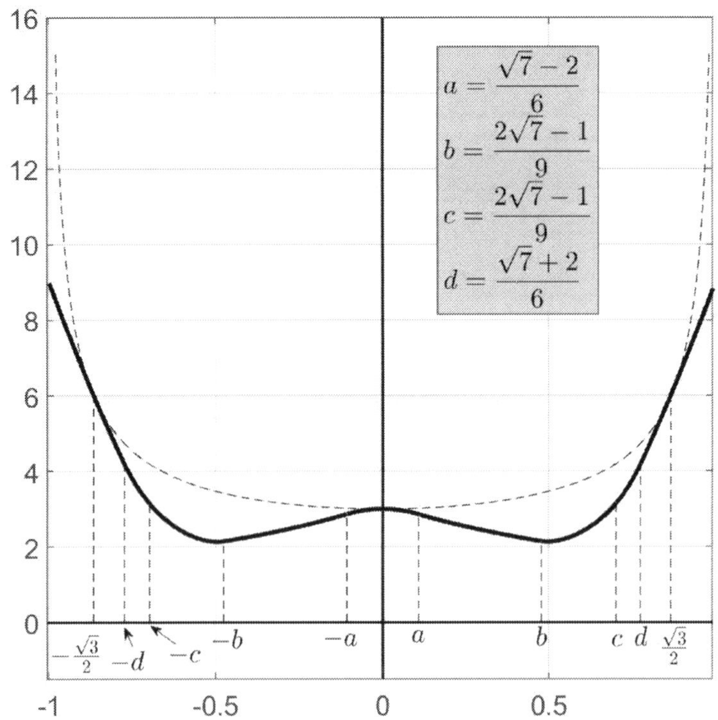

FIGURE 3.5: Representation of $\mathcal{B}_3(x)$ in a solid line. Bernstein's estimate $\frac{3}{\sqrt{1-x^2}}$ appears as a dashed line.

2.

$$P(x) = \pm\left\{1 + \frac{\left[3x - (4-\sqrt{7})x_0 + \sqrt{7} - 1\right]^2 \left[3x + 4\sqrt{7} - 13 - (16 - 4\sqrt{7})x_0\right]}{2(4-\sqrt{7})^3(1+x_0)^3}\right\},$$

for $x_0 \in \left[\frac{\sqrt{7}-2}{6}, \frac{2\sqrt{7}-1}{9}\right]$.

3.

$$P(x) = \pm\left\{1 - \frac{2\left(2x_0x + 1 - 3x_0^2\right)^2 \left[(12x_0^2 - 4)x_0x + 9x_0^4 - 2x_0^2 + 1\right]}{(9x_0^4 - 10x_0^2 + 1)^2}\right\},$$

for $x_0 \in \left[\frac{2\sqrt{7}-1}{9}, \frac{2\sqrt{7}+1}{9}\right]$.

4. $P(x) = \pm\left\{1 + \frac{27\left[x-(4+\sqrt{7})x_0+\sqrt{7}+3\right]^2(x-1)}{2(4+\sqrt{7})^3(1-x_0)^3}\right\}$, *for* $x_0 \in \left[\frac{2\sqrt{7}+1}{9}, \frac{\sqrt{7}+2}{6}\right]$.

Notice that if $x_0 \leq 0$ *and* P *is extremal for* $\mathcal{B}_3(|x_0|)$, *then* $\pm P(-x)$ *are extremal for* $\mathcal{B}_3(x_0)$.

PROOF. *Given that* $\mathcal{B}_3$ *is evidently an even function, it is sufficient to evaluate* $\mathcal{B}_3(x_0)$ *for non-negative values of* x_0. *Through direct examination, it becomes apparent that the*

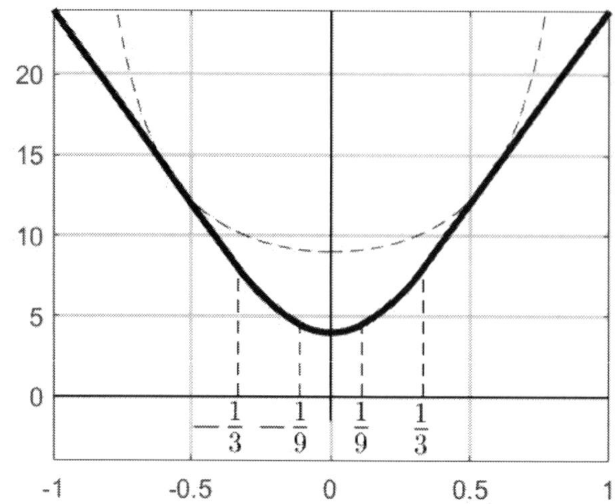

FIGURE 3.6: Sketch of $\mathcal{B}_{3,2}(x)$ (solid line) compared to the Duffin and Schaeffer's estimate $\mathcal{F}_{3,2}(x) = \frac{3\sqrt{-8x^2+9}}{\sqrt{(1-x^2)^3}}$ (dashed line).

maximum of the absolute values of the derivatives at x_0 for the extreme polynomials of types (i), (ii), and (iii) as described in Theorem 1.1.3 is given by

$$\max\left\{\frac{3}{4}(1+x_0)^2, 4x_0\right\} = \begin{cases} \frac{3}{4}(1+x_0)^2 & \text{if } x_0 \in [0, 1/3], \\ 4x_0 & \text{if } x_0 \geq 1/3. \end{cases} \tag{3.65}$$

Now we delve into the polynomials of type (iv) as described in Theorem 1.1.3. The derivatives of these polynomials, up to a sign, are represented by the functions

$$f^+_{x_0}(q) = \frac{4(3q^2x_0 - 3qx_0^2 + q - x_0)}{(1-q^2)^2},$$
$$f^-_{x_0}(q) = \frac{4(3q^2x_0 + 3qx_0^2 + q - x_0)}{(1-q^2)^2}.$$

Our goal is to optimize the absolute value $|f^{\pm}_{x_0}(q)|$ within the interval $q \in (-1/3, 0)$ or, equivalently by continuity, $q \in [-1/3, 0]$. The critical points of $f^{\pm}_{x_0}(q)$ are determined by solving the equation

$$\frac{df^{\pm}_{x_0}(q)}{dq} = \frac{4(3q^2+1)(-3x_0^2 \pm 2qx_0 + 1)}{(1-q^2)^3} = 0,$$

which results in the critical points

$$q_1 = \frac{3x_0^2 - 1}{2x_0} \quad \textit{and} \quad q_2 = -\frac{3x_0^2 - 1}{2x_0},$$

corresponding to the unique critical points of $f^+_{x_0}$ and $f^-_{x_0}$, respectively.

It is important to note that $q_1 \in [-1/3, 0]$ if and only if $x_0 \in \left[\frac{-1+2\sqrt{7}}{9}, \frac{\sqrt{3}}{3}\right]$, whereas $q_2 \in [-1/3, 0]$ if and only if $x_0 \in \left[\frac{\sqrt{3}}{3}, \frac{1+2\sqrt{7}}{9}\right]$. Therefore, if $x_0 \in \left[0, \frac{-1+2\sqrt{7}}{9}\right] \cup \left[\frac{1+2\sqrt{7}}{9}, 1\right]$, *we have*

$$\begin{aligned}\sup_{q\in[-1/3,0]} |f^{\pm}_{x_0}(q)| &= \max\{|f^{\pm}_{x_0}(0)|, |f^{\pm}_{x_0}(-1/3)|\}\\ &= \max\left\{4x_0, \left|-\frac{81}{16}x_0^2 \pm \frac{27}{8}x_0 + \frac{27}{16}\right|\right\}\\ &= \begin{cases} -\frac{81}{16}x_0^2 + \frac{27}{8}x_0 + \frac{27}{16} & \text{if } x_0 \in \left[0, \frac{-1+2\sqrt{7}}{9}\right],\\ \frac{81}{16}x_0^2 - \frac{27}{8}x_0 - \frac{27}{16} & \text{if } x_0 \in \left[\frac{1+2\sqrt{7}}{9}, 1\right].\end{cases}\end{aligned}$$

Also, it can be readily checked that

$$\left|f^{\pm}_{x_0}\left(\pm\frac{3x_0^2-1}{2x_0}\right)\right| = \frac{-16x_0^3}{9x_0^4 - 10x_0^2 + 1} \geq \max\left\{4x_0, \left|-\frac{81}{16}x_0^2 \pm \frac{27}{8}x_0 + \frac{27}{16}\right|\right\}$$

for $x_0 \in \left[\frac{-1+2\sqrt{7}}{9}, \frac{1+2\sqrt{7}}{9}\right]$. *Consequently, the latter implies that*

$$\sup_{q\in[-1/3,0]} |f^{\pm}_{x_0}(q)| = \begin{cases} -\dfrac{81}{16}x_0^2 + \dfrac{27}{8}x_0 + \dfrac{27}{16} & \text{if } x_0 \in \left[0, \frac{-1+2\sqrt{7}}{9}\right],\\ \dfrac{-16x_0^3}{9x_0^4 - 10x_0^2 + 1} & \text{if } x_0 \in \left[\frac{-1+2\sqrt{7}}{9}, \frac{1+2\sqrt{7}}{9}\right],\\ \dfrac{81}{16}x_0^2 + \dfrac{27}{8}x_0 - \dfrac{27}{16} & \text{if } x_0 \in \left[\frac{1+2\sqrt{7}}{9}, 1\right].\end{cases} \tag{3.66}$$

We will now proceed to optimize the first derivatives evaluated at x_0 *for the polynomials classified as type (v) in Theorem 1.1.3. These derivatives are represented by the following functions*

$$\begin{aligned} g^{+}_{x_0}(t) &= \frac{(t-x_0)(t-3x_0+2)}{(1+t)^2},\\ g^{-}_{x_0}(t) &= \frac{(t+x_0)(t+3x_0+2)}{(1+t)^2},\end{aligned}$$

for $t \in (-1/2, 1)$. *Note that, to maximize the absolute value* $|g^{\pm}_{x_0}(t)|$ *over the interval* $t \in (-1/2, 1)$, *it is equally valid to consider the closed interval* $[-1/2, 1]$ *due to continuity. The critical points of the functions* $g^{\pm}_{x_0}(t)$ *can be determined by solving the following equations*

$$\begin{aligned} \frac{dg^{+}_{x_0}(t)}{dt} &= \frac{2(-3x_0^2 + 2tx_0 + 1)}{(1+t)^3} = 0,\\ \frac{dg^{-}_{x_0}(t)}{dt} &= \frac{2(-3x_0^2 - 2tx_0 + 1)}{(1+t)^3} = 0.\end{aligned}$$

Therefore, the values $t_1 = \frac{3x_0^2-1}{2x_0}$ *and* $t_2 = -\frac{3x_0^2-1}{2x_0}$ *represent the unique critical points of the functions* $g^{+}_{x_0}(t)$ *and* $g^{-}_{x_0}(t)$, *respectively. Additionally, the condition* $t_1 \in [-1/2, 1]$ *holds if and only if* $x_0 \in \left[\frac{-1+\sqrt{13}}{6}, 1\right]$, *while the condition* $t_2 \in [-1/2, 1]$ *is satisfied if and only if* $x_0 \in \left[\frac{1}{3}, \frac{1+\sqrt{13}}{6}\right]$. *By considering the absolute values* $|g^{+}_{x_0}(t)|$ *and* $|g^{-}_{x_0}(t)|$ *separately, we obtain the following results:*

$$\sup_{t\in[-1/2,1]} |g^{+}_{x_0}(t)| = \begin{cases} \max\{|g^{+}(-1/2)|, |g^{+}(1)|\} & \text{if } x_0 \in \left[0 \frac{-1+\sqrt{13}}{6}\right],\\ \max\{|g^{+}(-1/2)|, |g^{+}(1)|, |g^{+}(t_1)|\} & \text{if } x_0 \in \left[\frac{-1+\sqrt{13}}{6}, 1\right],\end{cases}$$

$$\sup_{t\in[-1/2,1]} |g^{-}_{x_0}(t)| = \begin{cases} \max\{|g^{-}(-1/2)|, |g^{-}(1)|\} & \text{if } x_0 \in \left[0, \frac{1}{3}\right] \cup \left[\frac{1+\sqrt{13}}{6}, 1\right],\\ \max\{|g^{-}(-1/2)|, |g^{+}(1)|, |g^{-}(t_2)|\} & \text{if } x_0 \in \left[\frac{1}{3}, \frac{1+\sqrt{13}}{6}\right].\end{cases}$$

With the help of Figure 3.7 it follows that

$$\sup_{t\in[-1/2,1]} |g^{\pm}_{x_0}(t)| = \begin{cases} |g^{\pm}_{x_0}(-1/2)| & \textit{if } x_0 \in [0,\lambda_0] \cup \left[\frac{1+\sqrt{13}}{6}, 1\right], \\ |g^{-}_{x_0}(t_2)| & \textit{if } x_0 \in \left[\lambda_0 \frac{1+\sqrt{13}}{6}, 1\right], \end{cases}$$

$$= \begin{cases} -12x_0^2+3 & \textit{if } x_0 \in [0,\lambda_0], \\ \frac{(x_0+1)^2}{-3x_0^2+2x_0+1} & \textit{if } x_0 \in \left[\lambda_0, \frac{1+\sqrt{13}}{6}\right], \\ 12x_0^2-3 & \textit{if } x_0 \in [\frac{1+\sqrt{13}}{6}, 1], \end{cases} \tag{3.67}$$

where $\eta_0 \approx 0.3651$ *is the smallest root of the equation* $|g^{\pm}_{x_0}(-1/2)| = |g^{-}_{x_0}(t_2)|$ *in* $[0,1]$ *(see Figure 3.7). It is important to say that* η_0 *does not significantly influence the ultimate outcome.*

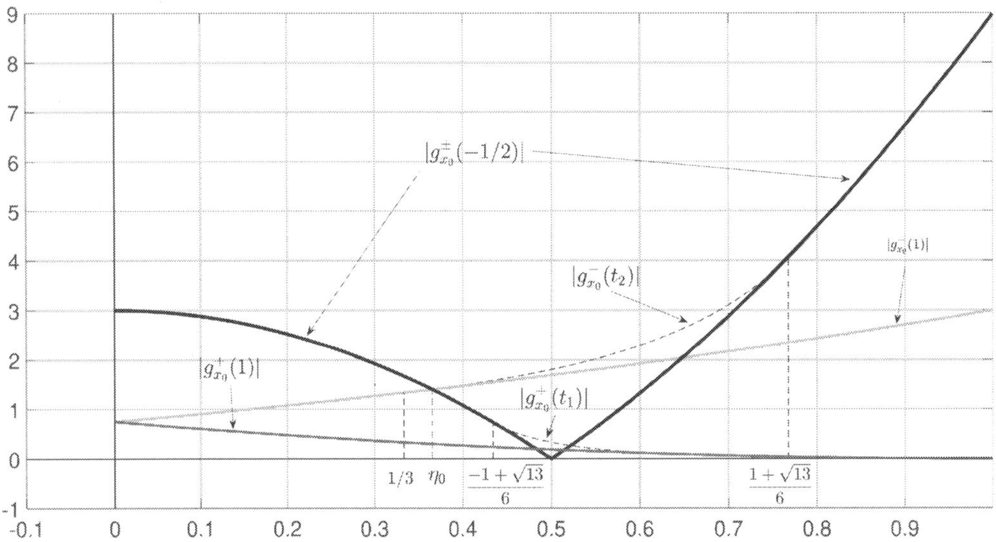

FIGURE 3.7: Illustrative calculation of $\sup_{t\in[-1/2,1]} |g^{\pm}_{x_0}(t)|$.

We finally arrive at the analysis of the polynomials of type (vi), parametrized in the set

$$\mathcal{R} = \left\{(r,s) : -1 \le r < s \le 1,\ s \ge \min\left\{3r+2, \frac{r+2}{3}\right\}\right\}.$$

See Figure 3.8 for a sketch of $\mathcal{R}$. *Differentiating the polynomials of type (vi), up to a sign we have*

$$h^{+}_{x_0}(r,s) = \frac{-12(r-x_0)(s-x_0)}{(s-r)^3},$$

$$h^{-}_{x_0}(r,s) = \frac{12(r+x_0)(s+x_0)}{(s-r)^3},$$

for $(r,s) \in \mathcal{R}$.

Our objective is to optimize the expression $|h^{\pm}_{x_0}(r,s)|$ *within the region* $\mathcal{R}$. *To identify the critical points of the functions* $h^{+}x_0(r,s)$ *and* $h^{-}x_0(r,s)$, *we proceed by calculating their*

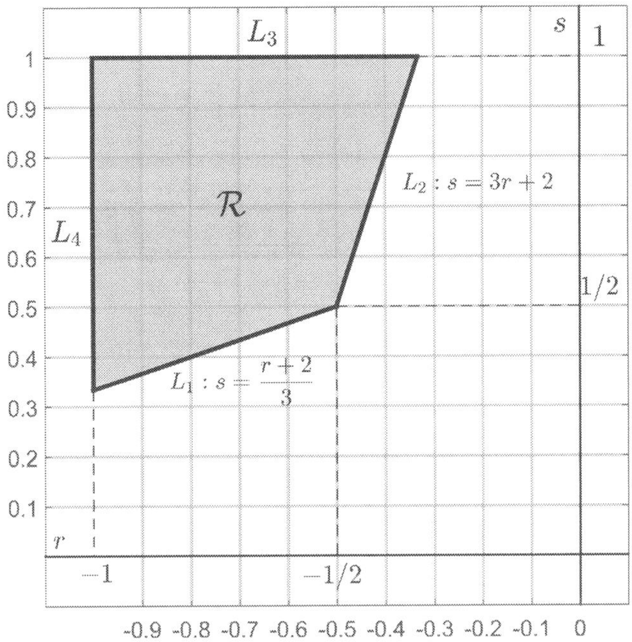

FIGURE 3.8: Sketch of the set of parameters $\mathcal{R}$ in the (r, s) plane used to index the polynomials of type (vi) in Theorem 1.1.3.

respective gradients.

$$\nabla h^{+}_{x_0}(r,s) = \left(\frac{12(s-x_0)(2r+s-3x_0)}{(r-s)^4}, -\frac{(r-x_0)(r+2s-3x_0)}{(r-s)^4}\right),$$
$$\nabla h^{-}_{x_0}(r,s) = \left(-\frac{12(s+x_0)(2r+s+3x_0)}{(r-s)^4}, \frac{(r+x_0)(r+2s+3x_0)}{(r-s)^4}\right).$$

It is evident that $\nabla h^{\pm}_{x_0}(r,s) \neq (0,0)$ *within the region* $\mathcal{R}$. *Consequently, the maximum value of* $|h^{\pm}_{x_0}(r,s)|$ *in* $\mathcal{R}$ *is achieved on the boundary* $\partial\mathcal{R}$. *Given the symmetry of* $\mathcal{R}$, *and the symmetry of* $h^{+}x_0(r,s)$ *and* $h^{-}x_0(r,s)$ *with respect to the line* $r = -s$, *it suffices to maximize* $|h^{+}x_0(r,s)|$ *on* $\partial\mathcal{R}$. *Observe that the restriction of* $|h^{+}x_0(r,s)|$ *to the segments* L_3 *and* L_4 *is a monotonic function. However, the restrictions of* $h^{+}_{x_0}(r,s)$ *to* L_1 *and* L_2, *denoted as* h_1 *and* h_2, *respectively, may exhibit local extrema. We have*

$$h_1(r) = -\frac{27(r-x_0)(r-3x_0+2)}{2(r-1)^3} \quad \text{with } r \in \left[-1, -\frac{1}{2}\right],$$
$$h_2(r) = \frac{3(r-x_0)(3r+2-x_0)}{2(r+1)^3} \quad \text{with } r \in \left[-\frac{1}{2}, -\frac{1}{3}\right],$$

and

$$h_1'(r) = \frac{27\left[r^2 + (-8x_0+6)r + 9x_0^2 - 10x_0 + 2\right]}{2(r-1)^4},$$
$$h_2'(r) = \frac{3\left[-3r^2 + (8x_0+2)r - 3x_0^2 + 2x_0 + 2\right]}{2(r+1)^4}.$$

It is straightforward to check that the function $h_1(r)$ *possesses two critical points, denoted as* r_1^- *and* r_1^+. *Similarly, the function* $h_2(r)$ *also has two critical points, labelled* r_2^- *and* r_2^+.

These critical points are determined by the following expressions

$$r_1^{\pm} = (4 \pm \sqrt{7})x_0 \mp \sqrt{7} - 3,$$
$$r_2^{\pm} = \frac{4 \pm \sqrt{7}}{3}x_0 + \frac{\pm\sqrt{7}+1}{3}.$$

It is evident that r_1^- *never falls within the interval* $\left[-1, -\frac{1}{2}\right]$, *and* r_2^+ *is never within* $\left[-\frac{1}{2}, -\frac{1}{3}\right]$. *Additionally,* r_1^+ *lies in* $\left[-1, -\frac{1}{2}\right]$ *if and only if* x_0 *belongs to the interval* $I_1 := \left[\frac{1+2\sqrt{7}}{9}, \frac{2+\sqrt{7}}{6}\right]$. *Similarly,* r_2^- *is within* $\left[-\frac{1}{2}, -\frac{1}{3}\right]$ *if and only if* x_0 *is in the interval* $I_2 := \left[\frac{-2+\sqrt{7}}{6}, \frac{-1+2\sqrt{7}}{9}\right]$. *Consequently, the supremum of* $|h_{x_0}^{\pm}(r,s)|$ *over the region* $\mathcal{R}$ *is determined by*

$$\begin{cases} \max\{|h_{x_0}^+(-1,\frac{1}{3})|, |h_{x_0}^+(-1,1)|, |h_{x_0}^+(-\frac{1}{3},1)|, |h_{x_0}^+(-\frac{1}{2},\frac{1}{2})|\} & \text{if } x_0 \notin I_1 \cup I_2, \\ \max\{|h_{x_0}^+(-1,\frac{1}{3})|, |h_{x_0}^+(-1,1)|, |h_{x_0}^+(-\frac{1}{3},1)|, |h_{x_0}^+(-\frac{1}{2},\frac{1}{2})|, |h_2(r_2^-)|\} & \text{if } x_0 \in I_2, \\ \max\{|h_{x_0}^+(-1,\frac{1}{3})|, |h_{x_0}^+(-1,1)|, |h_{x_0}^+(-\frac{1}{3},1)|, |h_{x_0}^+(-\frac{1}{2},\frac{1}{2})|, |h_1(r_1^+)|\} & \text{if } x_0 \in I_1. \end{cases}$$

On the other hand

$$\left|h_{x_0}^+\left(-1,\frac{1}{3}\right)\right| = \left|\frac{27(x_0+1)(3x_0-1)}{16}\right|,$$
$$|h_{x_0}^+(-1,1)| = \left|\frac{3(x_0-1)(x_0+1)}{2}\right|,$$
$$\left|h_{x_0}^+\left(-\frac{1}{3},1\right)\right| = \left|\frac{27(x_0-1)(3x_0+1)}{16}\right|,$$
$$\left|h_{x_0}^+\left(-\frac{1}{2},\frac{1}{2}\right)\right| = |12x_0^2 - 3|,$$
$$|h_2(r_2^-)| = \frac{7\sqrt{7}+10}{9(1+x_0)},$$
$$|h_1(r_1^+)| = \frac{7\sqrt{7}-10}{9(1-x_0)}.$$

Comparing these functions, we end up with (see Figure 3.9)

$$\sup_{(r,s)\in\mathcal{R}} |h_{x_0}^{\pm}(r,s)| = \begin{cases} 3 - 12x_0^2 & \text{if } 0 \le x_0 \le \frac{\sqrt{7}-2}{6}, \\ \frac{7\sqrt{7}+10}{9(1+x_0)} & \text{if } \frac{\sqrt{7}-2}{6} \le x_0 \le \frac{-1+2\sqrt{7}}{9}, \\ \frac{27(1-x_0)(3x_0+1)}{16} & \text{if } \frac{-1+2\sqrt{7}}{9} \le x_0 \le \frac{\sqrt{3}}{3}, \\ \frac{27(1+x_0)(3x_0-1)}{16} & \text{if } \frac{\sqrt{3}}{3} \le x_0 \le \frac{1+2\sqrt{7}}{9}, \\ \frac{7\sqrt{7}-10}{9(1-x_0)} & \text{if } \frac{1+2\sqrt{7}}{9} \le x_0 \le \frac{\sqrt{7}+2}{6}, \\ 12x_0^2 - 3 & \text{if } \frac{\sqrt{7}+2}{6} \le x_0 \le 1. \end{cases} \tag{3.68}$$

Finally, the result in (3.64) *is derived by comparing the functions presented in* (3.65), (3.66), (3.67), *and* (3.68). *This comparison is illustrated in Figure 3.10.*

Regarding the extremal polynomials, we may assume without loss of generality that $x_0 \ge 0$. *It is evident that the polynomials* $P(x) = \pm(4x^3 - 3x)$ *are extremal for all* $x_0 \in \left[0, \frac{\sqrt{7}-2}{6}\right] \cup \left[\frac{\sqrt{7}+2}{6}, \infty\right)$, *which corresponds to case* (1).

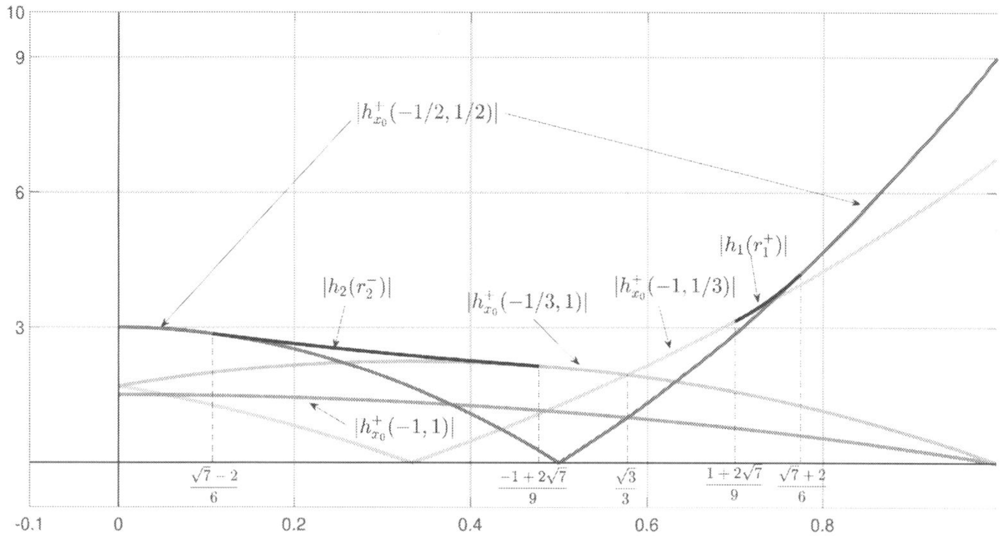

FIGURE 3.9: Visual calculation of $\sup_{(r,s)\in\mathcal{R}} |h^{\pm}_{x_0}(r,s)|$.

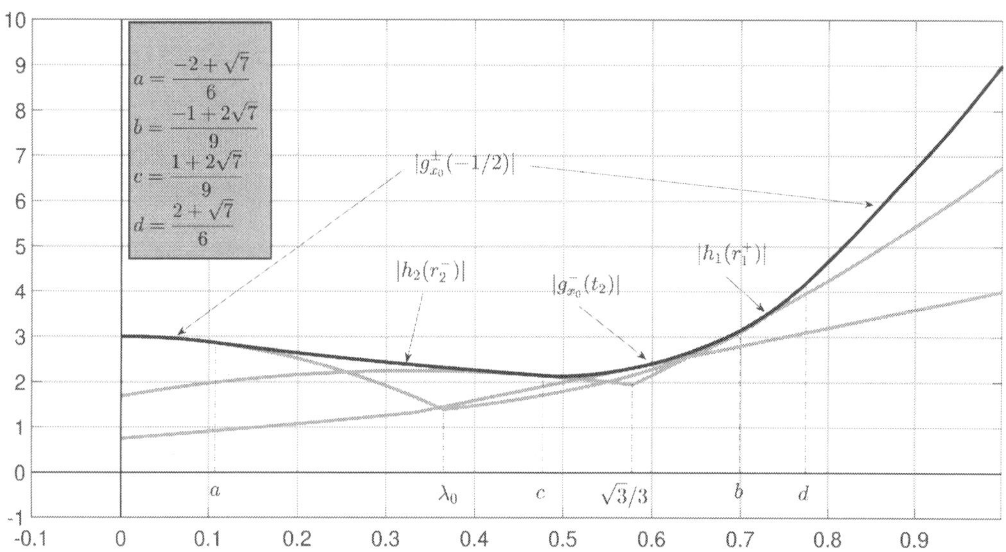

FIGURE 3.10: Comparison of the functions appearing in (3.65), (3.66), (3.67), and (3.68).

For $x_0 \in \left[\frac{\sqrt{7}-2}{6}, \frac{-1+2\sqrt{7}}{9}\right]$, *we have shown that* $\mathcal{B}_3(x_0)$ *is achieved by polynomials of type (vi) in the form* $P(x) = \pm\left[1+\dfrac{4}{(s-r)^3}(x-r)^2\left(x-\dfrac{3s-r}{2}\right)\right]$, *with* $r = r_2^- = \frac{4-\sqrt{7}}{3}x_0 + \frac{-\sqrt{7}+1}{3}$ *and* $s = 3r_2^- + 2 = (4-\sqrt{7})x_0 - \sqrt{7}+3$. *Substituting these values, we obtain the polynomials in* (2).

For $x_0 \in \left[\frac{2\sqrt{7}-1}{9}, \frac{2\sqrt{7}+1}{9}\right]$, *it is already known that* $\mathcal{B}_3(x_0)$ *is attained by polynomials of type (iv) given by* $P(x) = \pm\left[1-\dfrac{1}{(1-q^2)^2}(x-q)^2\left(4qx+2+2q^2\right)\right]$ *with* $q = q_2 = -\frac{3x_0^2-1}{2x_0}$, *which are precisely the polynomials in case* (3) *of the statement.*

Finally, for $x_0 \in \left[\frac{2\sqrt{7}+1}{9}, \frac{\sqrt{7}+2}{6}\right]$*, we have established that* $\mathcal{B}_3(x_0)$ *is attained by polynomials of type (vi) in the form* $P(x) = \pm\left[1 + \frac{4}{(s-r)^3}(x-r)^2\left(x - \frac{3s-r}{2}\right)\right]$ *with* $r = r_1^+ = (4+\sqrt{7})x_0 - \sqrt{7} - 3$ *and* $s = \frac{r_1^+ + 2}{3} = \frac{4+\sqrt{7}}{3}x_0 - \frac{\sqrt{7}+1}{3}$*. Direct substitution leads to the polynomials described in case (4) of the statement.*

Theorem 3.2.7 (Araújo, Muñoz, Rodríguez, and Seoane [4]) *The Bernstein function corresponding to the second derivative of the polynomials in* $\mathcal{P}_3(\mathbb{R})$ *is given by*

$$\mathcal{B}_{3,2}(x) = \begin{cases} \dfrac{4}{1-9x^2} & \text{if } |x| \le \frac{1}{9}, \\ \dfrac{32}{9(|x|-1)^2} & \text{if } \frac{1}{9} \le |x| \le \frac{1}{3}, \\ 24|x| & \text{if } |x| \ge \frac{1}{3}. \end{cases} \tag{3.69}$$

Furthermore, for each $x_0 \in \mathbb{R}$*, the polynomials (of norm equal to 1)* $P(x)$ *given below are extremal for* $\mathcal{B}_{3,2}$*, i.e.,* $\mathcal{B}_{3,2}(x_0) = P''(x_0)$*:*

1.

$$P(x) = \pm\left[1 - \frac{2}{(1-9x_0^2)^2}(x-3x_0)^2\left(6x_0x + 9x_0^2 + 1\right)\right], \quad \text{for } x_0 \in [0, 1/9],$$
$$P(x) = \pm\left[1 + \frac{2}{(1-9x_0^2)^2}(x+3x_0)^2\left(6x_0x - 9x_0^2 - 1\right)\right], \quad \text{for } x_0 \in [-1/9, 0].$$

2.

$$P(x) = \pm\left[1 + \frac{2(4x-9x_0+5)^2(x-1)}{27(1-x_0)^3}\right], \quad \text{for } x_0 \in [1/9, 1/3],$$
$$P(x) = \pm\left[1 - \frac{2(4x+9x_0-5)^2(x+1)}{27(1-x_0)^3}\right], \quad \text{for } x_0 \in [-1/3, -1/9].$$

3. $P(x) = \pm(4x^3 - 3x)$ *for* $|x_0| \ge 1/3$.

PROOF. *Given that* $\mathcal{B}_{3,2}$ *is an even function, it is sufficient to evaluate* $\mathcal{B}_{3,2}(x_0)$ *for* $x_0 \ge 0$. *By direct inspection, the maximum absolute value of the derivatives with respect to* x *at* x_0 *for the extremal polynomials of types (i), (ii), and (iii) in Theorem 1.1.3 is* $\max 4, \frac{3}{2}(x_0+1) = 4$.

Next, we focus on the polynomials of type (iv) in Theorem 1.1.3, where we identify two distinct categories of polynomials. To maximize the absolute value of their second derivatives with respect to x *at* x_0*, we need to optimize the following functions*

$$f_{x_0}^+(q) = \frac{4}{(1-q^2)^2}(3q^2 - 6qx_0 + 1),$$
$$f_{x_0}^-(q) = \frac{4}{(1-q^2)^2}(3q^2 + 6qx_0 + 1),$$

with $q \in \left(-\frac{1}{3}, 0\right)$*. Since*

$$\frac{df_{x_0}^{\pm}}{dq}(q) = -\frac{8(3q^2+1)(q \pm 3x_0)}{(q^2-1)^3},$$

we observe that $3x_0$ *and* $-3x_0$ *are critical points of the functions* $f^+_{x_0}(q)$ *and* $f^-_{x_0}(q)$*, respectively. Clearly,* $3x_0$ *does not fall within the interval* $\left(-\frac{1}{3},0\right)$ *since* $x_0>0$*. Conversely,* $-3x_0$ *lies within* $\left(-\frac{1}{3},0\right)$ *if and only if* $x_0\in\left(0,\frac{1}{9}\right)$*. Therefore,*

$$\sup_{q\in(-1/3,0)}|f^{\pm}_{x_0}(q)|=\begin{cases}\max\{|f^{\pm}(0)|,|f^{\pm}(-1/3)|,|f^{-}(-3x_0)|\} & \textit{if } x_0\in[0,\frac{1}{9}],\\ \max\{|f^{\pm}(0)|,|f^{\pm}(-1/3)|\} & \textit{if } x_0\geq\frac{1}{9}.\end{cases}$$
$$=\begin{cases}\frac{4}{1-9x_0^2} & \textit{if } x_0\in[0,\frac{1}{9}],\\ \frac{81}{8}x_0+\frac{27}{8} & \textit{if } x_0\geq\frac{1}{9}.\end{cases}\tag{3.70}$$

Note that the function (3.70) *exceeds 4, thus the contributions of the polynomials of types (i), (ii), and (iii) to the calculation of* $\mathcal{B}(x_0)$ *can be disregarded.*

Next, we turn our attention to the polynomials of type (v) in Theorem 1.1.3. By evaluating their second derivatives with respect to x*, we need to optimize the following functions*

$$g^+_{x_0}(t)=\frac{2(-2t+3x_0-1)}{(t+1)^2},$$
$$g^-_{x_0}(t)=\frac{2(-2t-3x_0-1)}{(t+1)^2},$$

for $t\in(-1/2,1)$*. Since*

$$\frac{dg^{\pm}_{x_0}}{dt}(t)=\frac{4(t\mp 3x_0)}{(t+1)^3},$$

It is evident that $g^+x_0(t)$ *and* $g^-x_0(t)$ *have critical points at* $t=3x_0$ *and* $t=-3x_0$*, respectively. Considering that* $x_0\geq 0$*, the conditions* $3x_0\in(-1/2,1)$ *and* $-3x_0\in(-1/2,1)$ *are equivalent to* $x_0\in(0,1/3)$ *and* $x_0\in(0,1/6)$*, respectively. Therefore,*

$$\sup_{t\in(-1/2,1)}|g^{\pm}_{x_0}(t)|=\begin{cases}\max\{|g^{\pm}_{x_0}(-1/2)|,|g^{\pm}_{x_0}(1)|,|g^{\pm}_{x_0}(\pm 3x_0)|\} & \textit{if } x_0\in\left[0,\frac{1}{6}\right],\\ \max\{|g^{\pm}_{x_0}(-1/2)|,|g^{\pm}_{x_0}(1)|,|g^{+}_{x_0}(3x_0)|\} & \textit{if } x_0\in\left[\frac{1}{6},\frac{1}{3}\right],\\ \max\{|g^{\pm}_{x_0}(-1/2)|,|g^{\pm}_{x_0}(1)|\} & \textit{if } x_0\geq\frac{1}{3}.\end{cases}$$
$$=\begin{cases}\max\left\{24x_0,\frac{3}{2}(1\pm x_0),\frac{1}{1\pm 3x_0}\right\} & \textit{if } x_0\in\left[0,\frac{1}{6}\right],\\ \max\left\{24x_0,\frac{3}{2}(1\pm x_0),\frac{1}{1+3x_0}\right\} & \textit{if } x_0\in\left[\frac{1}{6},\frac{1}{3}\right],\\ \max\left\{24x_0,\frac{3}{2}(1\pm x_0)\right\} & \textit{if } x_0\geq\frac{1}{3}.\end{cases}$$
$$=\begin{cases}\frac{2}{1-3x_0} & \textit{if } x_0\in\left[0,\frac{1}{6}\right],\\ 24x_0 & \textit{if } x_0\geq\frac{1}{6}.\end{cases}\tag{3.71}$$

Finally, we need to optimize the second derivative of the polynomials of type (vi) as described in Theorem 1.1.3. The second derivatives of these polynomials, disregarding the sign, are expressed as follows

$$h^+_{x_0}(r,s)=\frac{12(r+s-2x_0)}{(r-s)^3},$$
$$h^-_{x_0}(r,s)=\frac{12(r+s+2x_0)}{(r-s)^3}.$$

It is straightforward that

$$\nabla h^+_{x_0}(r,s)=\left(-\frac{24(r+2s-3x_0)}{(r-s)^4},\frac{24(2r+s-3x_0)}{(r-s)^4}\right),$$
$$\nabla h^-_{x_0}(r,s)=\left(\frac{24(2r+s+3x_0)}{(r-s)^4},-\frac{24(r+2s+3x_0)}{(r-s)^4}\right).$$

From this, we determine that the unique critical points of $h^+_{x_0}(r,s)$ *and* $h^-_{x_0}(r,s)$ *are* $\pm(x_0,x_0)$*, neither of which lies within the region* $\mathcal{R}$ *in the* (r,s) *plane (refer to Figure 3.8). This indicates that the maximum values of both* $|h^+_{x_0}(r,s)|$ *and* $|h^-_{x_0}(r,s)|$ *occur on the boundary* $\partial\mathcal{R}$*.*

We only need to analyze $|h^+_{x_0}(r,s)|$ *since the calculations for* $|h^-_{x_0}(r,s)|$ *are nearly identical. Additionally, due to the symmetry of* $|h^+_{x_0}(r,s)|$ *and the region* $\mathcal{R}$ *with respect to the line* $s=-r$*, it suffices to optimize* $|h^+_{x_0}(r,s)|$ *on the segments* L_1 *and* L_4 *shown in Figure 3.8. Let*

$$h_1(r) = h^+_{x_0}(r,(r+2)/3) = \frac{27(4r-6x_0+2)}{2(r-1)^3},$$
$$h_2(r) = h^+_{x_0}(-1,s) = \frac{12(2x_0-s+1)}{(s+1)^3},$$

with $r\in[-1,-1/2]$ *and* $s\in[1/3,1]$*, be the restrictions of* $h^+_{x_0}$ *to* L_1 *and* L_4*, respectively. It is straightforward to check that*

$$h_1'(r) = -\frac{27(4r-9x_0+5)}{(r-1)^4},$$
$$h_2'(r) = -\frac{24(3x_0-s+2)}{(s+1)^4}.$$

From this, we find that h_1 *and* h_2 *each have a unique critical point at* $r_0=\frac{9x_0-5}{4}$ *and* $s_0=3x_0+2$*, respectively. In fact,* r_0 *lies within the interval* $[-1,1/2]$ *if and only if* $x_0\in[1/9,1/3]$*, whereas* s_0 *never falls within the interval* $[1/3,1]$*. Additionally,* $h_1(-1)=h_2(1/3)$*. Therefore,*

$$\sup_{(r,s)\in\partial\mathcal{R}} |h^+_{x_0}(r,s)| = \begin{cases} \max\{|h_1(-1)|,|h_1(\frac{-1}{2})|,|h_2(1)|,|h_1(r_0)|\} & \text{if } x_0\in[\frac{1}{9},\frac{1}{3}],\\ \max\{|h_1(-1)|,|h_1(\frac{-1}{2})|,|h_2(1)|\} & \text{otherwise.}\end{cases}$$

Observe that $|h_1(-1)|=\frac{81}{8}x_0+\frac{27}{8}$*,* $h_1(-1/2)=24x_0$*,* $h_2(1)=3x_0$ *and* $|h_1(r_0)|=\frac{32}{9(1-x_0)^2}$*, and that* $\frac{32}{9(1-x_0)^2}\geq\max\{\frac{81}{8}x_0+\frac{27}{8},24x_0,3x_0\}$ *whenever* $x_0\in[1/9,1/3]$*. To sum it up,*

$$\sup_{(r,s)\in\partial\mathcal{R}} |h^+_{x_0}(r,s)| = \begin{cases} \frac{81}{8}x_0+\frac{27}{8} & \text{if } x_0\in[0,\frac{1}{9}],\\ \frac{32}{9(1-x_0)^2} & \text{if } x_0\in[\frac{1}{9},\frac{1}{3}],\\ 24x_0 & \text{otherwise.}\end{cases} \tag{3.72}$$

Finally, the result follows straightforwardly by comparing the functions (3.70), (3.71), *and* (3.72).

To determine the extremal polynomials for every $x_0\in\mathbb{R}$*, we can assume that* $x_0\geq 0$*. If* $x_0\in[0,1/9]$*, according to the above proof,* $\mathcal{B}(x_0)$ *is attained for the polynomials* $P(x)=\pm\left[1+\dfrac{2}{(1-q^2)^2}(x+q)^2\left(2qx-1-q^2\right)\right]$ *with* $q=-3x_0$*.*

Additionally, if $x_0\in[1/9,1/3]$*, it has been shown in this proof that* $\mathcal{B}_{3,2}(x_0)$ *is attained for the polynomials* $P(x)=\pm\left[1+\dfrac{4}{(s-r)^3}(x-r)^2\left(x-\dfrac{3s-r}{2}\right)\right]$ *with* $r=r_0=\frac{9x_0-5}{4}$ *and* $s=\frac{r_0+2}{3}=\frac{9x_0+3}{4}$*.*

To complete the study of the extremal polynomials, it is straightforward that if $x_0\geq 1/3$*, then* $\mathcal{B}_{3,2}(x_0)$ *is achieved for* $P(x)=\pm(4x^3-3x)$*.*

3.2.2 Markov and Bernstein inequalities in spaces of trinomials in the real line

In this section, we consider the spaces of trinomials with constant term,

$$\mathcal{P}_{m,n}(\mathbb{R}) := \{ax^m + bx^n + c : \ a, b, c \in \mathbb{R}\},$$

endowed with the norm

$$\|ax^m + bx^n + c\|_{m,n} := \sup_{x\in[-1,1]} |ax^m + bx^n + c|$$

and provide explicit formulas for the smallest constant $\mathcal{M}_{m,n}(x)$ in the inequality

$$|p'(x)| \leq \mathcal{M}_{m,n}(x)\|p\|, \tag{3.73}$$

for any $p \in \mathcal{P}_{m,n}(\mathbb{R})$ and $x \in [-1, 1]$ fixed.

We are also interested in obtaining the Markov type constants $M_{m,n}$ which best fit in the inequality

$$\|p'\| \leq M_{m,n}\|p\|$$

for all $p \in \mathcal{P}_{m,n}(\mathbb{R})$.

Let $\mathsf{B}_{m,n}$ be the closed unit ball of $\mathcal{P}_{m,n}(\mathbb{R})$. Since we can assume that $\|p\| \leq 1$ in (3.73), calculating $\mathcal{M}_{m,n}(x)$ for a fixed x reduces to maximizing the convex function $\Phi_{m,n} : \mathsf{B}_{m,n} \to \mathbb{R}$ defined by

$$\Phi_{m,n}(p) = |p'(x)|$$

for all $p \in \mathsf{B}_{m,n}$. Using the Krein-Milman approach, we obtain

$$\mathcal{M}_{m,n}(x) = \sup\{\Phi_{m,n}(p) : p \in \mathsf{B}_{m,n}\} = \sup\{\Phi_{m,n}(p) : p \in \text{ext}(\mathsf{B}_{m,n})\} \tag{3.74}$$

The fact that $\text{ext}(\mathsf{B}_{m,n})$ depends heavily on whether m and n are even or odd will lead us to analyze multiple cases separately.

3.2.2.1 $\mathcal{M}_{m,n}(x)$ and $M_{m,n}(x)$ for $m, n \in \mathbb{N}$ odd

Let us recall, once more, the following extremely simple and useful result (that already appeared as Lemma 1.1.4, page 3 and, later again, in page 55) which provides the definition of (by now) an important number, that we have been denoting by λ_0.

Lemma 3.2.8 (Muñoz and Seoane [88]) *If $m, n \in \mathbb{N}$ are odd numbers with $m > n$ then the equation*

$$|n + mx| = (m - n)|x|^{\frac{m}{m-n}} \tag{3.75}$$

has only three roots, one at $x = -1$, another one at a point $\lambda_0 \in (-\frac{n}{m}, 0)$ and a third one at a point $\lambda_1 > 0$.

Although this previous lemma has appeared before in this text, let us now make a brief remark (regarding the value of it) that has not been commented previously. The value of $\lambda_0 = \lambda_0(m, n)$ for every $m, n \in \mathbb{N}$ odd must be approximated numerically for most choices of m and n, however some values of λ_0 can be obtained using symbolic calculus like $\lambda_0(3, 1) = -\frac{1}{4}$ or $\lambda_0(5, 3) = \dfrac{4 + \sqrt[3]{10} - 2\sqrt[3]{100}}{6}$. The following table provides the approximation to 5 decimal digits of 21 values of λ_0.

λ_0	$m=3$	$m=5$	$m=7$	$m=9$	$m=11$	$m=13$
$n=1$	-0.25	-0.13471	-0.09072	-0.06795	-0.05414	-0.04491
$n=3$	—	-0.52145	-0.34142	-0.25000	-0.19558	-0.15983
$n=5$	—	—	-0.65076	-0.47306	-0.36750	-0.29838
$n=7$	—	—	—	-0.72537	-0.56186	-0.45475
$n=9$	—	—	—	—	-0.77380	-0.62538
$n=11$	—	—	—	—	—	-0.80774

Recall that according to Theorem 1.1.6 we have that

$$\text{ext}(\mathsf{B}_{m,n}) = \left\{ \pm \left(t, -\frac{m}{(m-n)^{\frac{m-n}{m}} n^{\frac{n}{m}}} t^{\frac{n}{m}}, 0 \right) : \frac{n}{m-n} \leq t \leq \frac{n}{n+m\lambda_0} \right\} \cup \{\pm(0,0,1)\},$$

where λ_0 is given in Lemma 3.2.8. Then plugging $\text{ext}(\mathsf{B}_{m,n})$ in (3.74) the following is obtained.

Theorem 3.2.9 (Muñoz, Sarantopoulos, and Seoane [87]**)** *Let $m > n$ be odd natural numbers and*

$$I_{m,n} = \left[\left(\frac{|\lambda_0| n}{m} \right)^{\frac{1}{m-n}}, \left(\frac{n}{m} \right)^{\frac{1}{m-n}} \right],$$

then

$$\mathcal{M}_{m,n}(x) = \begin{cases} \dfrac{mn}{n+m\lambda_0} \cdot x^{n-1} \cdot \left| x^{m-n} + \lambda_0 \right| & \text{if } |x| \in [0,1] \setminus I_{m,n}, \\ n\left(\dfrac{n}{m}\right)^{\frac{n}{m-n}} \cdot \dfrac{1}{|x|} & \text{if } |x| \in I_{m,n}, \end{cases} \tag{3.76}$$

where λ_0 is given in Lemma 3.2.8. See Figure 3.11 for a representation of $\mathcal{M}_{m,n}$.

Moreover

$$M_{m,n} = \mathcal{M}_{m,n}(\pm 1) = \frac{mn(1+\lambda_0)}{n+m\lambda_0}.$$

and equality in (3.73) *is attained for the polynomials*

$$p(x) = \frac{\pm(nx^m + \lambda_0 m x^n)}{n+m\lambda_0}.$$

To establish Theorem 3.2.9, we begin by presenting and proving a series of technical lemmas that will serve as essential tools in the argument.

Lemma 3.2.10 *Let $m > n$ be odd natural numbers, and let λ_0 be the real number from Lemma 3.2.8. Then, the following inequality holds*

$$|\lambda_0| \cdot \frac{n}{m} < |\lambda_0| \cdot \frac{1 - |\lambda_0|^{\frac{n}{m-n}}}{1 - |\lambda_0|^{\frac{m}{m-n}}} < \frac{n}{m}. \tag{3.77}$$

PROOF. *To prove that $|\lambda_0| \cdot \frac{n}{m} < |\lambda_0| \cdot \frac{1-|\lambda_0|^{\frac{n}{m-n}}}{1-|\lambda_0|^{\frac{m}{m-n}}}$, recall that $|\lambda_0| < \frac{n}{m} < 1$. Consider the inequality*

$$\frac{n}{m} < \frac{1-x^n}{1-x^m}. \tag{3.78}$$

We will show that this inequality holds for every x in the interval $\left(0, \left(\frac{n}{m}\right)^{\frac{1}{m-n}}\right)$.

For $0 < x < 1$, inequality (3.78) *is equivalent to*

$$m - n > mx^n - nx^m. \tag{3.79}$$

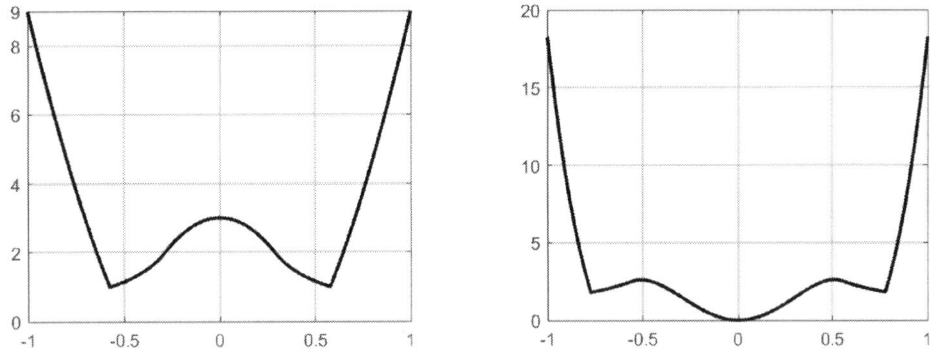

$\mathcal{M}_{3,1}(x)$ $\qquad\qquad$ $\mathcal{M}_{5,3}(x)$

FIGURE 3.11: Some representations of $\mathcal{M}_{m,n}(x)$ with $m, n \in \mathbb{N}$ odd.

Since $mx^n - nx^m$ is strictly increasing on $(0, 1)$ (as can be check by calculating its derivative), it meets the line $y = m - n$ at most at one point, which occurs when $x = 1$. Consequently, $mx^n - nx^m$ remains smaller than $m - n$ over $(0, 1)$, confirming (3.79) for every $x \in \left(0, \left(\frac{n}{m}\right)^{\frac{1}{m-n}}\right)$.

For the second inequality, proving $|\lambda_0| \cdot \frac{1-|\lambda_0|^{\frac{n}{m-n}}}{1-|\lambda_0|^{\frac{m}{m-n}}} < \frac{n}{m}$ requires simple calculations and the use of the equation $n + m\lambda_0 = (m - n)|\lambda_0|^{\frac{m}{m-n}}$ (Equation (3.75)).

Lemma 3.2.11 *Let $m > n$ be odd natural numbers, and let λ_0 be the real number from Lemma 3.2.8. Define the functions*

$$f(x) = \frac{mn}{m-n} \cdot x^{n-1} \cdot |x^{m-n} - 1|,$$
$$g(x) = \frac{mn}{n + m\lambda_0} \cdot x^{n-1} \cdot |x^{m-n} + \lambda_0|.$$

Then, we have $g(x) \geq f(x)$ whenever

$$0 \leq |x| \leq \left(\frac{|\lambda_0|n}{m}\right)^{\frac{1}{m-n}} \quad or \quad \left(\frac{n}{m}\right)^{\frac{1}{m-n}} \leq |x| \leq 1. \tag{3.80}$$

PROOF. *By symmetry, assume $x > 0$. It is straightforward to check that $f(x)$ and $g(x)$ meet at*

$$x_1 = \left(\frac{n}{m}\right)^{\frac{1}{m-n}}, \quad x_2 = \left(|\lambda_0| \cdot \frac{1 - |\lambda_0|^{\frac{n}{m-n}}}{1 - |\lambda_0|^{\frac{m}{m-n}}}\right)^{\frac{1}{m-n}}.$$

Using Lemma 3.2.10, it follows that neither x_1 nor x_2 lie within the interior of the intervals in (3.80). Consequently, one of the functions f or g must be larger (or equal) on each interval. Clearly, $f(1) < g(1)$ and also

$$f\left(\left(\frac{|\lambda_0|n}{m}\right)^{\frac{1}{m-n}}\right) < g\left(\left(\frac{|\lambda_0|n}{m}\right)^{\frac{1}{m-n}}\right).$$

Using equation (3.75), we obtain

$$|\lambda_0|^{\frac{m}{m-n}} < |\lambda_0|,$$

which is clearly true since $-1 < -\frac{n}{m} < \lambda_0 < 0$. This completes the proof.

Lemma 3.2.12 *Let $m > n$ be odd natural numbers, and let λ_0 be the real number from Lemma 3.2.8. Define the functions*

$$f(x) = \frac{mn}{m-n} \cdot x^{n-1} \cdot |x^{m-n} - 1|,$$
$$g(x) = \frac{mn}{n+m\lambda_0} \cdot x^{n-1} \cdot |x^{m-n} + \lambda_0|,$$
$$h(x) = n\left(\frac{n}{m}\right)^{\frac{n}{m-n}} \cdot \frac{1}{|x|}.$$

Then, we have $h(x) \geq \max\{f(x), g(x)\}$ whenever

$$\left(\frac{|\lambda_0|n}{m}\right)^{\frac{1}{m-n}} \leq |x| \leq \left(\frac{n}{m}\right)^{\frac{1}{m-n}}. \tag{3.81}$$

PROOF. *We prove that if (3.81) holds, then $h(x) \geq f(x)$ and $h(x) \geq g(x)$. Assume $x > 0$ by symmetry.*

(i) The function $x^n - x^m$ is strictly increasing over $\left(0, \left(\frac{n}{m}\right)^{\frac{1}{m-n}}\right)$, so its maximum value is attained at $x = \left(\frac{n}{m}\right)^{\frac{1}{m-n}}$. This leads to the bound $f(x) \leq h(x)$.

(ii) Similarly, we show that $g(x) \leq h(x)$ using monotonicity arguments and checking endpoint evaluations. The details follow similar calculations as in (i), completing the proof.

PROOF. ***(of Theorem 3.2.9)*** *From (3.74) we have*

$$\mathcal{M}_{m,n}(x) = \sup_{p \in \text{ext}(\mathsf{B}_{m,n})} |p'(x)|.$$

To compute the previous supremum, we will use the characterization of $\text{ext}(\mathsf{B}_{m,n})$ given just before the statement of Theorem 3.2.9. It suffices to consider the polynomials

$$p_t(x) = \pm\left(tx^m - \frac{m}{(m-n)^{\frac{m-n}{m}} n^{\frac{n}{m}}} t^{\frac{n}{m}} x^n\right),$$

where $\frac{n}{m-n} \leq t \leq \frac{n}{n+m\lambda_0}$. Therefore,

$$\begin{aligned}
\mathcal{M}_{m,n}(x) &= \sup_{\frac{n}{m-n} \leq t \leq \frac{n}{n+m\lambda_0}} \left|\frac{dp_t}{dx}(x)\right| \\
&= \sup_{\frac{n}{m-n} \leq t \leq \frac{n}{n+m\lambda_0}} \left|mtx^{m-1} - \frac{mnt^{\frac{n}{m}}}{(m-n)^{\frac{m-n}{m}} n^{\frac{n}{m}}} x^{n-1}\right| \\
&= \sup_{\frac{n}{m-n} \leq t \leq \frac{n}{n+m\lambda_0}} \left|m \cdot x^{n-1} \cdot \left[tx^{m-n} - \left(\frac{n}{m-n}\right)^{\frac{m-n}{m}} t^{\frac{n}{m}}\right]\right|.
\end{aligned}$$

Now, let $R(t) = m \cdot x^{n-1} \cdot \left[tx^{m-n} - \left(\frac{n}{m-n}\right)^{\frac{m-n}{m}} t^{\frac{n}{m}}\right]$. Clearly, the above supremum will be attained at either $t = \frac{n}{m-n}$, $t = \frac{n}{n+m\lambda_0}$, or at a critical point of $R(t)$ within the interval $\left(\frac{n}{m-n}, \frac{n}{n+m\lambda_0}\right)$. It is easy to see that the only critical point of $R(t)$ corresponds to

$$t_0 = \frac{n}{m-n} \cdot \left(\frac{n}{m}\right)^{\frac{m}{m-n}} \cdot \frac{1}{|x|^m},$$

which leads to

$$R(t_0) = n\left(\frac{n}{m}\right)^{\frac{n}{m-n}}\frac{1}{|x|},$$

provided that $\frac{n}{m-n} \leq t_0 \leq \frac{n}{n+m\lambda_0}$. *The latter is equivalent to*

$$\left(\frac{|\lambda_0|n}{m}\right)^{\frac{1}{m-n}} \leq |x| \leq \left(\frac{n}{m}\right)^{\frac{1}{m-n}}.$$

Thus

$$\begin{aligned}\mathcal{M}_{m,n}(x) &= \sup_{\frac{n}{m-n}\leq t\leq \frac{n}{n+m\lambda_0}} |R(t)| \\ &= \begin{cases}\max\left\{\left|R\left(\frac{n}{m-n}\right)\right|, \left|R\left(\frac{n}{n+m\lambda_0}\right)\right|, n\left(\frac{n}{m}\right)^{\frac{n}{m-n}}\frac{1}{|x|}\right\} & \text{if } x\in I_1,\\ \max\left\{\left|R\left(\frac{n}{m-n}\right)\right|, \left|R\left(\frac{n}{n+m\lambda_0}\right)\right|\right\} & \text{if } x\in I_2,\end{cases}\end{aligned}$$

where

$$I_1 = \left\{x\in\mathbb{R} : \left(\frac{|\lambda_0|n}{m}\right)^{\frac{1}{m-n}} \leq |x| \leq \left(\frac{n}{m}\right)^{\frac{1}{m-n}}\right\},$$

$$I_2 = \left\{x\in\mathbb{R} : 0 \leq |x| \leq \left(\frac{|\lambda_0|n}{m}\right)^{\frac{1}{m-n}} \text{ or } \left(\frac{n}{m}\right)^{\frac{1}{m-n}} \leq |x| \leq 1\right\}.$$

Notice that

$$R\left(\frac{n}{m-n}\right) = \frac{mnx^{n-1}}{m-n}\cdot|x^{m-n}-1|$$

and

$$R\left(\frac{n}{n+m\lambda_0}\right) = \frac{mnx^{n-1}}{n+m\lambda_0}\cdot\left|x^{m-n}+\lambda_0\right|.$$

To finish the proof, we use Lemmas 3.2.11 and 3.2.12.

3.2.2.2 $\mathcal{M}_{m,n}(x)$ for m odd and n even

Recall that according to Theorem 1.1.7, if $m, n \in \mathbb{N}$ are such that m is odd, n is even and $m > n$, then

$$\operatorname{ext}(\mathsf{B}_{m,n}) = \{\pm(0,2,-1), \pm(1,1,-1), \pm(1,-1,1), \pm(0,0,1)\}.$$

Therefore, if $x \in [-1,1]$ is fixed, by (3.74), we have

$$\mathcal{M}_{m,n}(x) = \sup_{p\in\operatorname{ext}(\mathsf{B}_{m,n})} |p'(x)|.$$

In the previous supremum, it suffices to consider the polynomials

$$p_1(x) = \pm(2x^n - 1), \quad p_2(x) = \pm(x^m + x^n - 1), \quad \text{and} \quad p_3(x) = \pm(x^m - x^n + 1).$$

Thus

$$\begin{aligned}\mathcal{M}_{m,n}(x) &= \max\{|p_1'(x)|, |p_2'(x)|, |p_3'(x)|\}\\ &= \max\{2n|x|^{n-1}, |mx^{m-1}+nx^{n-1}|, |mx^{m-1}-nx^{n-1}|\}\\ &= \max\{2n|x|^{n-1}, mx^{m-1}+n|x|^{n-1}\}\\ &= |x|^{n-1}\max\{2n, m|x|^{m-n}+n\},\end{aligned}$$

and since $2n \leq m|x|^{m-n}+n$ provided $\left(\frac{n}{m}\right)^{\frac{1}{m-n}} \leq |x|$, we finally arrive at the sharp estimate:

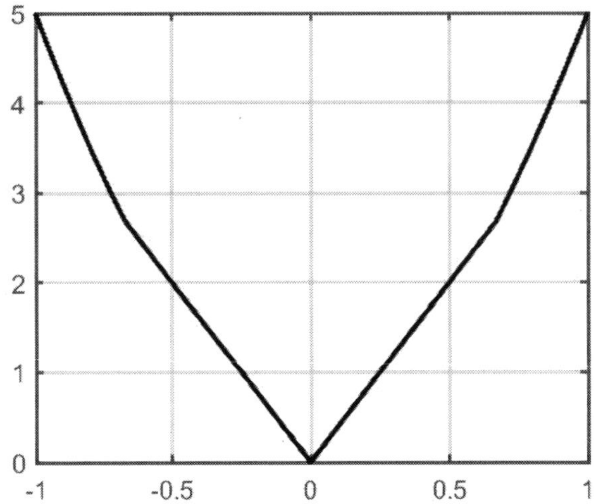

FIGURE 3.12: $\mathcal{M}_{3,2}(x)$. In the general case $\mathcal{M}_{m,n}(x)$, being m odd and n even, is of similar shape. Here, $M_{3,2} = 5$.

Theorem 3.2.13 (Muñoz, Sarantopoulos, and Seoane [87]) *Let $m, n \in \mathbb{N}$ be such that m is odd, n is even and $m > n$. Then*

$$\mathcal{M}_{m,n}(x) = \begin{cases} 2n|x|^{n-1} & \text{if } 0 \leq |x| \leq \left(\frac{n}{m}\right)^{\frac{1}{m-n}}, \\ mx^{m-1} + n|x|^{n-1} & \text{if } \left(\frac{n}{m}\right)^{\frac{1}{m-n}} \leq |x| \leq 1. \end{cases} \tag{3.82}$$

Moreover

$$M_{m,n} = \mathcal{M}_{m,n}(\pm 1) = m + n,$$

and equality is attained in (3.73) *for the polynomials $p(x) = \pm(x^m \pm x^n - 1)$.*

3.2.2.3 $\mathcal{M}_{m,n}(x)$ for $m, n \in \mathbb{N}$ with m even

If m is even n is odd, then according to Theorem 1.1.9, we have

$$\text{ext}(\mathsf{B}_{m,n}) = \left\{\pm\big(t, \pm\Gamma(t), 1 - t - \Gamma(t)\big) : \frac{n}{m} \leq t \leq 2\right\} \cup \{\pm(0,0,1)\},$$

where $\Gamma(t)$ is the implicit function of t defined by the equation

$$\frac{(m-n)t}{n}\left(\frac{n\Gamma(t)}{mt}\right)^{\frac{m}{m-n}} = 2 - t - \Gamma(t). \tag{3.83}$$

The fact that $\text{ext}(\mathsf{B}_{m,n})$ is not completely explicit makes the case m even and n odd particularly difficult to analyze. A similar thing occurs in the case where both m and n are even. Recall that from Theorem 1.1.10 we have

$$\text{ext}(\mathsf{B}_{m,n}) = \{\pm(s, \Lambda(s), -1 - s - \Lambda(s))\} \cup \{\pm(t, -\Upsilon(t), 1)\} \cup \{\pm(0,0,1)\},$$

where $s \in [\gamma_0, -2]$, $t \in [-\gamma_1, -\gamma_0]$, $\gamma_0 = -\frac{2}{m-n} \cdot \left(\frac{m^m}{n^n}\right)^{\frac{1}{m-n}}$, $\gamma_1 = \frac{-2n}{m-n}$ and Λ and Υ are defined by

$$\Lambda(s) = -\Gamma(|s|) \quad \text{and} \quad \Upsilon(t) = \left(\frac{2m}{m-n}\right)^{\frac{m-n}{m}} \cdot \left(\frac{mt}{n}\right)^{\frac{n}{m}}.$$

In the following we will give explicit formulas for $\mathcal{M}_{m,n}(x)$ in the cases $m = 2n$ with n being either odd or even.

$\mathcal{M}_{2n,n}(x)$ for $n \in \mathbb{N}$ odd

In this case (3.83) can be solved explicitly as Aron and Klimek proved in [6] to obtain

$$\text{ext}(\mathsf{B}_{2n,n}) = \{\pm(t, \pm 2(\sqrt{2t} - t), 1 + t - 2\sqrt{2t}) : \ t \in [1/2, 2]\} \cup \{\{\pm(0, 0, 1)\}.$$

Therefore, if $x \in [-1, 1]$ is fixed, by (3.74) we have

$$\mathcal{M}_{2n,n}(x) = \sup_{p \in \text{ext}(\mathsf{B}_{m,n})} |p'(x)|.$$

We now proceed to compute this supremum explicitly.

Theorem 3.2.14 (Muñoz, Sarantopoulos, and Seoane [87]) *Let $n \in \mathbb{N}$ be odd. Then*

$$\mathcal{M}_{2n,n}(x) = \begin{cases} \dfrac{n|x|^{n-1}}{1 - x^n} & \text{if } |x| \le \dfrac{1}{\sqrt[n]{2}}, \\ 4n|x|^{2n-1} & \text{if } \dfrac{1}{\sqrt[n]{2}} \le |x| \le 1. \end{cases}$$

See Figure 3.13 for a representation of $\mathcal{M}_{6,3}(x)$. Moreover, if $n \in \mathbb{N}$ is odd, then

$$M_{2n,n} = \mathcal{M}_{2n,n}(\pm 1) = 4n,$$

and equality is attained in (3.73) for the polynomials $p(x) = \pm(2x^{2n} - 1)$.

PROOF. *If $x \in [-1, 1]$ from (3.74) we have that*

$$\mathcal{M}_{2n,n}(x) = \sup_{p \in \text{ext}(\mathsf{B}_{2n,n})} |p'(x)|.$$

Using the characterization of $\text{ext}(\mathsf{B}_{2n,n})$ *given above one notices fast that it is enough to consider the polynomials*

$$p_t(x) = \pm(tx^{2n} \pm 2(\sqrt{2t} - t)x^n + 1 + t - 2\sqrt{2t}),$$

where $t \in [1/2, 2]$. Therefore,

$$\begin{aligned} \mathcal{M}_{2n,n}(x) &= \max\left\{\left|\frac{dp_t}{dx}(x)\right| : t \in [1/2, 2]\right\} \\ &= 2nx^{n-1} \max\{|tx^n \pm (\sqrt{2t} - t)| : t \in [1/2, 2]\} \\ &= 2nx^{n-1} \max\{t|x|^n + \sqrt{2t} - t : t \in [1/2, 2]\}. \end{aligned}$$

Define $R(t) = t|x|^n + \sqrt{2t} - t$. The supremum will be achieved at either $t = \frac{1}{2}$, $t = 2$, or at a critical point of $R(t)$ within the interval $\left(\frac{1}{2}, 2\right)$. A straightforward calculation reveals that the only critical point of $R(t)$ is $t_0 = \frac{1}{2(1-x^n)^2}$, which results in

$$2nx^{n-1}R(t_0) = \frac{nx^{n-1}}{1 - |x|^n},$$

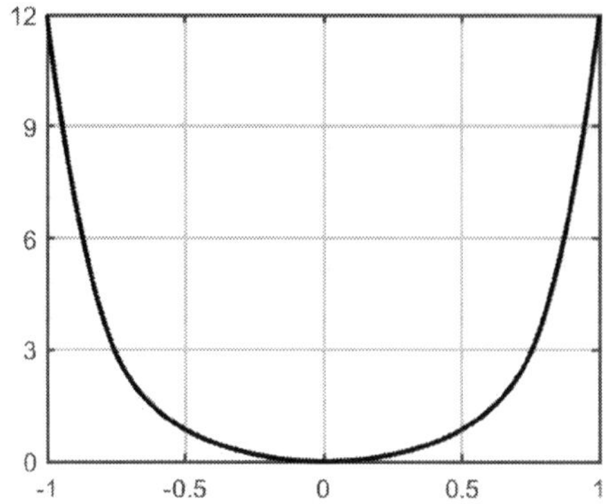

FIGURE 3.13: $\mathcal{M}_{2n,n}(x)$ with $n \in \mathbb{N}$ odd and $n > 3$ is similar to $\mathcal{M}_{6,3}(x)$.

provided that $\frac{1}{2} \le t_0 \le 2$. *This condition is equivalent to* $|x| \le \frac{1}{\sqrt[n]{2}}$. *Consequently, we obtain*

$$\begin{aligned}\mathcal{M}_{2n,n}(x) &= 2nx^{n-1} \sup_{\frac{1}{2}\le t\le 2} |R(t)| \\ &= 2nx^{n-1}\cdot\begin{cases}\max\{|R(1/2)|,|R(2)|,|R(t_0)|\} & \text{if } |x|\le \frac{1}{\sqrt[n]{2}},\\ \max\{|R(1/2)|,|R(2)|\} & \text{if } \frac{1}{\sqrt[n]{2}}\le |x|\le 1\end{cases} \\ &= \begin{cases}\max\left\{n|x|^{n-1}(1-|x|^n), 4n|x|^{2n-1}, \frac{n|x|^{n-1}}{1-|x|^n}\right\} & \text{if } |x|\le \frac{1}{\sqrt[n]{2}},\\ \max\left\{n|x|^{n-1}(1-|x|^n), 4n|x|^{2n-1}\right\} & \text{if } \frac{1}{\sqrt[n]{2}}\le |x|\le 1.\end{cases}\end{aligned}$$

The result now follows straightforwardly.

$\mathcal{M}_{2n,n}(x)$ **for** $n \in \mathbb{N}$ **even**

In this case, $\mathrm{ext}(\mathsf{B}_{2n,n})$ can be explicitly described. In fact, it is proven in [88] that

$$\mathrm{ext}(\mathsf{B}_{2n,n}) = \{\pm(t, 2(\sqrt{2t}-t), 1+t-2\sqrt{2t}) \quad \text{and} \quad \pm(t, -2\sqrt{2t}, 1) : t \in [2,8]\} \cup \{\pm(0,0,1)\}.$$

Thus, for a fixed $x \in [-1,1]$, using (3.74), we obtain

$$\mathcal{M}_{m,n}(x) = \sup_{p\in \mathrm{ext}(\mathsf{B}_{m,n})} |p'(x)|.$$

The previous supremum is calculated in the following result.

Theorem 3.2.15 *Let* $n \in \mathbb{N}$ *be even, then*

$$\mathcal{M}_{2n,n}(x) = \begin{cases} 8n(-2|x|^{2n-1}+|x|^{n-1}) & \text{if } 0\le |x|\le \left(\frac{1}{4}\right)^{\frac{1}{n}},\\ \frac{n}{|x|} & \text{if } \left(\frac{1}{4}\right)^{\frac{1}{n}} \le |x| \le \left(\frac{1}{2}\right)^{\frac{1}{n}},\\ \frac{n|x|^{n-1}}{1-x^n} & \text{if } \left(\frac{1}{2}\right)^{\frac{1}{n}} \le |x| \le \left(\frac{3}{4}\right)^{\frac{1}{n}},\\ 8n(2|x|^{2n-1}-|x|^{n-1}) & \text{if } \left(\frac{3}{4}\right)^{\frac{1}{n}} \le |x| \le 1.\end{cases} \tag{3.84}$$

The reader can find a representation of $\mathcal{M}_{8,4}(x)$ *in Figure 3.14. Moreover*

$$M_{2n,n} = \mathcal{M}_{2n,n}(\pm 1) = 8n$$

and equality is attained in (3.73) *for the polynomials* $p(x) = \pm(8x^{2n} - 8x^n + 1)$.

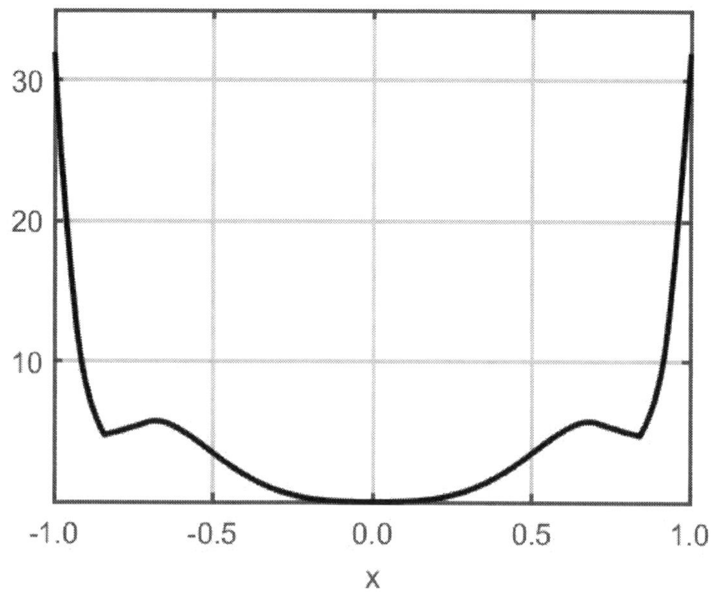

FIGURE 3.14: Representation of $\mathcal{M}_{4,2}(x)$. Observe that $M_{8,4} = 32$.

PROOF. *We will proceed similarly to the proof of Theorem 3.2.14. If $x \in [-1,1]$ recall that*

$$\mathcal{M}_{2n,n}(x) = \sup_{p \in \text{ext}(\mathsf{B}_{2n,n})} |p'(x)|.$$

Now we just need to use the characterization of $\text{ext}(\mathsf{B}_{2n,n})$ *given above. In fact we just have to focus on the polynomials of the form*

$$p_t(x) = \pm(tx^{2n} \pm 2(\sqrt{2t} - t)x^n + 1 + t - 2\sqrt{2t}) \quad \textit{and} \quad q_t(x) = \pm(tx^{2n} - 2\sqrt{2t}x^n + 1),$$

where $t \in [2,8]$. As in the proof of Theorem 3.2.14 we arrive at

$$\max_{t\in[2,8]} \left|\frac{dp_t}{dx}(x)\right| = 2nx^{n-1} \max\{|tx^n + (\sqrt{2t} - t)| : t \in [2,8]\}$$

$$= \begin{cases} \max\left\{|8nx^{n-1}(1-2x^n)|, 4n|x|^{2n-1}, \frac{n|x|^{n-1}}{1-x^n}\right\} & \textit{if } \sqrt[n]{\frac{1}{2}} \leq |x| \leq \sqrt[n]{\frac{3}{4}}, \\ \max\left\{|8nx^{n-1}(1-2x^n)|, 4n|x|^{2n-1}\right\} & \textit{if } |x| \leq \sqrt[n]{\frac{1}{2}} \textit{ or } \sqrt[n]{\frac{3}{4}} \leq |x| \leq 1, \end{cases}$$

and also

$$\max_{t\in[2,8]} \left|\frac{dq_t}{dx}(x)\right| = 2nx^{n-1} \max\{|tx^n - \sqrt{2t}| : t \in [2,8]\}$$

$$= \begin{cases} \max\left\{|8nx^{n-1}(1-2x^n)|, 4n|x|^{n-1}|x^n - 1|, \frac{n}{|x|}\right\} & \textit{if } \sqrt[n]{\frac{1}{4}} \leq |x| \leq \sqrt[n]{\frac{1+\sqrt{2}}{4}}, \\ \max\left\{|8nx^{n-1}(1-2x^n)|, 4n|x|^{n-1}|x^n - 1|\right\} & \textit{if } |x| \leq \sqrt[n]{\frac{1}{4}} \textit{ or } \sqrt[n]{\frac{1+\sqrt{2}}{4}} \leq |x| \leq 1. \end{cases}$$

Clearly, for $n \in \mathbb{N}$ even, we have

$$\mathcal{M}_{2n,n}(x) = \max\left\{\max_{t\in[2,8]} \left|\frac{dp_t}{dx}(x)\right|, \max_{t\in[2,8]} \left|\frac{dq_t}{dx}(x)\right|\right\}.$$

Now, it suffices to consider $x > 0$ and compare the above functions over the intervals

$$\left[0, \sqrt[n]{\frac{1}{4}}\right], \left[\sqrt[n]{\frac{1}{4}}, \sqrt[n]{\frac{2}{5}}\right], \left[\sqrt[n]{\frac{2}{5}}, \sqrt[n]{\frac{1}{2}}\right], \left[\sqrt[n]{\frac{1}{2}}, \sqrt[n]{\frac{1+\sqrt{2}}{4}}\right], \left[\sqrt[n]{\frac{1+\sqrt{2}}{4}}, \sqrt[n]{\frac{3}{4}}\right] \text{ and } \left[\sqrt[n]{\frac{3}{4}}, 1\right].$$

Proceeding in the same manner as in the proof of Theorem 3.2.14, we arrive at (3.84).

3.2.3 Calculation of Bernstein functions and Markov constants in spaces of quadratic forms in $\mathbb{R}^2$

In this section, we will consider the space $\mathcal{Q}(\mathbb{R}^2)$ consisting of all the quadratic forms in $\mathbb{R}^2$, that is, the polynomials of the form $ax^2 + bxy + cy^2$ for $a, b, c \in \mathbb{R}$. We will endow $\mathcal{Q}(\mathbb{R}^2)$ with several norms. To be more precise, if $K \subset \mathbb{R}^2$ is a convex body and we define the sup norm of a quadratic form $ax^2 + bxy + c^2$ as

$$\|ax^2 + bxy + c^2\|_K := \sup\{|ax^2 + bxy + c^2| : (x, y) \in K\},$$

then we endow $\mathcal{Q}(\mathbb{R}^2)$ with the norm $\|\cdot\|_K$. Combining the geometric results of Chapter 1 together with the Krein-Milman approach we will obtain the Bernstein functions and the Markov constants corresponding to the polynomial spaces $(\mathcal{Q}(\mathbb{R}^2), \|\cdot\|_K)$ for several choices of K. Specifically, for each fixed $(x, y) \in K$ we will determine the best estimate $\mathcal{M}_K(x, y)$ in the inequality

$$\|DP(x, y)\|_K \leq \mathcal{M}_K(x, y)\|P\|_K \tag{3.85}$$

for every $P \in \mathcal{Q}(\mathbb{R}^2)$. Notice that $DP(x, y)$ is a linear form on $\mathbb{R}^2$. In this setting, the space $\mathcal{L}(\mathbb{R}^2)$ of linear forms on $\mathbb{R}^2$ is endowed with the norm

$$\|L\|_K = \sup\{|L(x, y)| : (x, y) \in K\}.$$

The Bernstein-type function $\mathcal{M}_K(x, y)$ provides a uniform constant in (3.85) by considering its supremum over K, that is, if

$$M_K = \sup\{|\mathcal{M}_K(x, y)| : (x, y) \in K\}$$

then we obtain the Markov-type inequality

$$\|DP(x, y)\|_K \leq M_K\|P\|_K \tag{3.86}$$

for all $P \in \mathcal{Q}(\mathbb{R}^2)$ and every $(x, y) \in K$.

We will also consider variants of the Bernstein-Markov-type inequalities (3.85) and (3.86) in this section. For a fixed $(x, y) \in K$ the best constant $\mathcal{M}_{K,2}(x, y)$

$$\|\nabla P(x, y)\|_2 \leq \mathcal{M}_{K,2}(x, y)\|P\|_K, \tag{3.87}$$

for every $P \in \mathcal{Q}(\mathbb{R}^2)$ provides an optimal estimate on the length $\|\nabla P(x, y)\|_2$ of the gradient. Also, if

$$M_{K,2} = \sup\{|\mathcal{M}_{K,2}(x, y)| : (x, y) \in K\}$$

then

$$\|\nabla P(x, y)\|_2 \leq M_{K,2}\|P\|_K \tag{3.88}$$

for all $P \in \mathcal{Q}(\mathbb{R}^2)$ and every $(x, y) \in K$.

The Bernstein functions $\mathcal{M}_K(x, y)$ and $\mathcal{M}_{K,2}(x, y)$ together with the Markov constants M_K and $M_{K,2}$ will be calculated in the following sections for several choices of K using the

Krein-Milman approach. Let B_K be the closed unit ball of $(\mathcal{Q}(\mathbb{R}^2), \|\cdot\|_K)$. We can assume that $\|P\|_K \leq 1$ in (3.85) so in order to calculate $\mathcal{M}_K(x,y)$ for a fixed $(x,y) \in K$ we just have to maximize the convex function $\Phi_K : \mathsf{B}_K \to \mathbb{R}$ defined by

$$\Phi_K(P) = \|\nabla P(x,y)\|_K$$

for all $P \in \mathsf{B}_K$. Using the Krein-Milman approach we would have

$$\mathcal{M}_K(x,y) = \sup\{\Phi_K(P) : P \in \mathsf{B}_K\} = \sup\{\Phi_K(P) : P \in \operatorname{ext}(\mathsf{B}_K)\}. \tag{3.89}$$

Similarly, for a fixed $(x,y) \in K$, the maximum of the convex function $\Phi_{K,2} : \mathsf{B}_K \to \mathbb{R}$ defined by

$$\Phi_{K,2}(P) = \|\nabla P(x,y)\|_2$$

for all $P \in \mathsf{B}_K$ is $\mathcal{M}_{K,2}(x,y)$. Using the Krein-Milman approach we would have

$$\mathcal{M}_{K,2}(x,y) = \sup\{\Phi_K(P) : P \in \mathsf{B}_K\} = \sup\{\Phi_{K,2}(P) : P \in \operatorname{ext}(\mathsf{B}_K)\}. \tag{3.90}$$

3.2.3.1 The case where K is the simplex

In this section, we will consider the case where K is the triangle in $\mathbb{R}^2$ of vertices $(0,0)$, $(0,1)$, and $(1,0)$ (or simplex for short). This triangle will be represented by Δ and the space $(\mathcal{Q}(\mathbb{R}^2), \|\cdot\|_\Delta)$ by $\mathcal{P}(^2\Delta)$. The following result provides a Bernstein function $\mathcal{M}_K(x,y)$ in $\mathcal{P}(^2\Delta)$ extended to the whole plane $\mathbb{R}^2$ and not just Δ. We call this function $\Psi_\Delta(x,y)$ and it has been obtained by applying (3.89) to every $(x,y) \in \mathbb{R}^2$ having in mind that if B_Δ is the closed unit ball of $\mathcal{P}(^2\Delta)$, then from Theorem 1.2.3

$$\operatorname{ext}(\mathsf{B}_\Delta) = \{\pm(1, -2-2\sqrt{2(1+t)}, t), \pm(t, -2-2\sqrt{2(1+t)}, 1) : t \in [-1,1]\}.$$

The details are shown below.

Theorem 3.2.16 (Muñoz, Révész, and Seoane [82]**)** *For each $(x,y) \in \mathbb{R}^2$, let $\Psi_\Delta(x,y)$ denote the best constant in*

$$\|DP(x,y)\|_\Delta \leq \Psi_\Delta(x,y)\|P\|_\Delta, \tag{3.91}$$

for every $P \in \mathcal{P}(^2\Delta)$, then

$$\Psi_\Delta(x,y) = \begin{cases} |2x - 6y| & \text{if } x = 0 \text{ or } x \neq 0 \text{ and } (\frac{y}{x} \leq -1 \text{ or } \frac{y}{x} \geq 2), \\ |2x + 2y + y^2/x| & \text{if } x \neq 0 \text{ and } 1 \leq \frac{y}{x} \leq 2, \\ |2x + 2y + x^2/y| & \text{if } y \neq 0 \text{ and } 1 \leq \frac{x}{y} \leq 2, \\ |6x - 2y| & \text{if } y = 0 \text{ or } y \neq 0 \text{ and } (\frac{x}{y} \leq -1 \text{ or } \frac{x}{y} \geq 2). \end{cases}$$

Notice that $\mathcal{M}_\Delta(x,y) = \Psi_\Delta(x,y)$ for all $(x,y) \in \Delta$. A graph of Ψ_Δ is shown in Figure 3.15. Moreover, $M_\Delta = 6$, meaning that if $P \in \mathcal{P}(^2\Delta)$ then

$$\max_{(x,y)\in\Delta} \|DP(x,y)\|_\Delta \leq 6\|P\|_\Delta \tag{3.92}$$

and 6 is optimal in (3.92) since equality holds for the polynomial $P(x,y) = x^2 - 6xy + y^2$.

PROOF. *For a fixed $(x,y) \in \mathbb{R}^2$ we have, by virtue of the Krein-Milman approach that*

$$\Psi_\Delta(x,y) = \sup\{\|\nabla P(x,y)\|_\Delta : P \in \operatorname{ext}(\mathsf{B}_\Delta)\}.$$

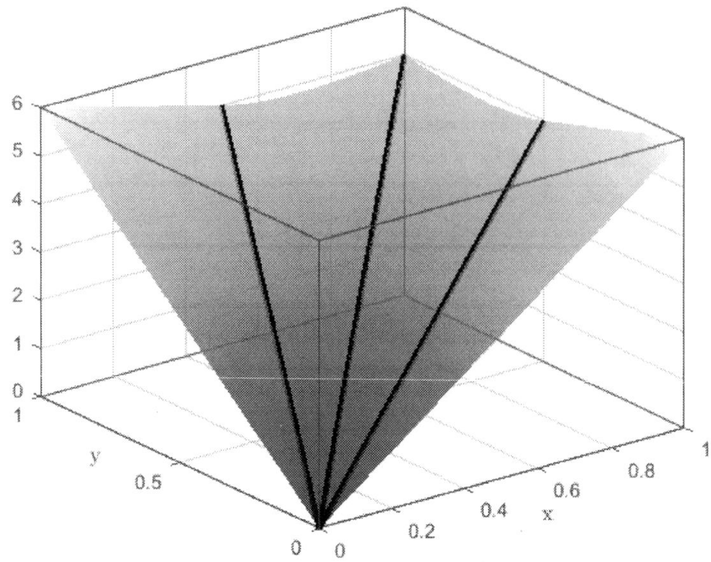

FIGURE 3.15: Representation of $\Psi_\Delta(x,y)$ in the square $[0,1]^2$.

Consider a quadratic polynomial of the form $P(\xi,\zeta) = a\xi^2 + b\xi\zeta + c\zeta^2$ *for* $(\xi,\zeta) \in \mathbb{R}^2$, *where* $a, b, c \in \mathbb{R}$. *For a fixed point* $(x,y) \in \mathbb{R}^2$, *we analyze the mapping*

$$\mathbb{R}^2 \ni (h,k) \mapsto |\nabla P(x,y) \cdot (h,k)| = |(2ax+by)h + (bx+2cy)k|.$$

Since this function is convex and Δ *is a convex body, the supremum* $\|\nabla P(x,y)\|_\Delta$ *is attained at one of the points* $(0,0)$, $(1,0)$, *or* $(0,1)$. *This leads to*

$$\begin{aligned}\|\nabla P(x,y)\|_\Delta &= \sup\{|(2ax+by)h + (bx+2cy)k| : (h,k) \in \Delta\}\\ &= \max\{|2ax+by|, |bx+2cy|\}.\end{aligned}$$

This formula further reveals that $\Psi_\Delta(x,y)$ *is a convex function in* $(x,y) \in \mathbb{R}^2$, *as it is expressed as the maximum of a convex function over* (a,b,c) *belonging to* B_Δ.

Since

$$\text{ext}(\mathsf{B}_\Delta) = \{\pm(1,-f(t),t), \pm(t,-f(t),1) : t \in [-1,1]\},$$

where $f(t) = 2 + 2\sqrt{2(1+t)}$ *for* $t \in [-1,1]$, *we obtain*

$$\begin{aligned}\Psi_\Delta(x,y) &= \sup_{s,t\in[-1,1]} \{|2x - f(t)y|, |-f(t)x + 2ty|, |2tx - f(t)y|, |-f(t)x + 2y|\}\\ &= \max\{\|F_k^{xy}\| : k = 1,2,3,4\},\end{aligned}$$

where $F_1^{xy}(t) := 2x - f(t)y$, $F_2^{xy}(t) := -f(t)x + 2ty$, $F_3^{xy}(t) := 2tx - f(t)y$, *and* $F_4^{xy}(t) := -f(t)x + 2y$ *for* $t \in [-1,1]$, *and* $\|F_k^{xy}\| := \sup\{|F_k^{xy}(t)| : -1 \le t \le 1\}$ *for* $k = 1,2,3,4$.

Now, considering the special case $x = 0$,

$$\Psi_\Delta(0,y) = \max_{t\in[-1,1]} \{|f(t)y|, |2ty|, |2y|\} = \max_{t\in[-1,1]} \{|f(t)|, 2|t|, 2\}|y| = 6|y|.$$

Similarly, for $y = 0$, *we obtain* $\Psi_\Delta(x,0) = 6|x|$.

Due to the symmetry $(x, y) \leftrightarrow (y, x)$ *and the reflection symmetry* $\Psi_\Delta(-x, -y) = \Psi_\Delta(x, y)$, *it suffices to consider the cases where* $xy \neq 0$ *and* $y > 0$. *This allows us to restrict the proof to two main cases: Case I* $(0 < x \leq y)$ *and Case II* $(0 < -x \leq y)$.

The function F_1^{xy} *is monotonic, implying*

$$\|F_1^{xy}\| = \max\{|F_1^{xy}(-1)|, |F_1^{xy}(1)|\}.$$

On the other hand, F_2^{xy} *has a unique critical point at*

$$t_{xy} = \frac{x^2 - 2y^2}{2y^2} \in [-1, 1] \quad \textit{if and only if} \quad 0 \leq \frac{x}{y} \leq 2.$$

Thus,

$$\|F_2^{xy}\| = \begin{cases} \max\{|F_2^{xy}(-1)|, |F_2^{xy}(1)|, |F_2^{xy}(t_{xy})|\} & \textit{if } 0 < x \leq y, \\ \max\{|F_2^{xy}(-1)|, |F_2^{xy}(1)|\} & \textit{if } 0 < -x \leq y. \end{cases}$$

Similarly, using symmetry properties, we obtain

$$\Psi_\Delta(x, y) = \begin{cases} \max\{6y - 2x, 2x + 2y + y^2/x\} & \textit{if } 0 < x \leq y \leq 2x, \\ 6y - 2x & \textit{if } 0 < x < y/2 \textit{ or } 0 < -x \leq y. \end{cases}$$

Since in the positive quadrant we have $x \geq (2\sqrt{2} - 2)y$, *this confirms the result across all quadrants.*

Thus, we conclude that the stated formula for $\Psi_\Delta(x, y)$ *holds in all cases.*

The proof of $M_\Delta = 6$ *follows from*

$$M_\Delta = \sup\{\Psi_\Delta(x, y) : (x, y) \in \Delta\} = 6.$$

3.2.3.2 The case where K is the unit square

In this section, we will consider the case where K is the square $[0, 1]^2$ in $\mathbb{R}^2$. This square will be represented by $\square$ and the space $(\mathcal{Q}(\mathbb{R}^2), \|\cdot\|_\square)$ by $\mathcal{P}(^2\square)$. The following result provides the Bernstein function $\mathcal{M}_\square(x, y)$ obtained by applying (3.89) to every $(x, y) \in \square$ having in mind that if $\mathsf{B}_\square$ is the closed unit ball of $\mathcal{P}(^2\square)$, then the extreme points of $\mathsf{B}_\square$ have the form

$$\pm(t, 2\sqrt{1-t}, -1) \quad \text{and} \quad \pm(-1, 2\sqrt{1-t}, t) \quad \text{with } t \in [0, 1]$$

or

$$\pm(1, -1, 1),\ \pm(1 - 3, 1),\ \pm(1, 0, 0),\ \pm(0, 0, 1).$$

Theorem 3.2.17 (Gámez, Muñoz, Sánchez, and Seoane [45]**)** *For each* $(x, y) \in \square$ *we have*

$$\mathcal{M}_\square(x, y) = \begin{cases} 3x - 2y & \textit{if } y \leq (\sqrt{2} - 1)x, \\ \frac{5}{2}x - y + \frac{y^2}{2x} & \textit{if } x \neq 0 \textit{ and } (\sqrt{2} - 1)x \leq y \leq \frac{1}{2}x, \\ 2x + \frac{y^2}{2x} & \textit{if } x \neq 0 \textit{ and } \frac{1}{2}x \leq y \leq x, \\ 2y + \frac{x^2}{2y} & \textit{if } y \neq 0 \textit{ and } x \leq y \leq 2x, \\ \frac{5}{2}y - x + \frac{x^2}{2y} & \textit{if } y \neq 0 \textit{ and } 2x \leq y \leq (\sqrt{2} + 1)x, \\ 3y - 2x & \textit{if } (\sqrt{2} + 1)x \leq y. \end{cases}$$

Moreover if $P \in \mathcal{P}(^2\square)$, *then*

$$\max_{(x,y)\in\square} \|DP(x, y)\|_\square \leq 3\|P\|_\square \tag{3.93}$$

and 3 *is optimal in* (3.93) *since equality holds for the polynomials* $P(x, y) = \pm(x^2 - 3xy + y^2)$. *In other words* $M_\square = 3$.

PROOF. *If $(x,y)\in\square$, combining (3.89) with the above description of* $\text{ext}(\mathsf{B}_\square)$ *we have that*

$$\mathcal{M}_\square(x,y)=\max\{\|DP_t(x,y)\|_\square,\|DQ_t(x,y)\|_\square,\|DR_k(x,y)\|_\square:\ t\in[0,1],\ k=1,2,3,4\},$$

where

$$\begin{aligned}P_t&=(t,2\sqrt{1-t},-1)\quad \textit{with } t\in[0,1],\\ Q_t&=(-1,2\sqrt{1-t},t)\quad \textit{with } t\in[0,1],\\ R_1&=(1,-1,1),\ R_2=(1-3,1),\ R_3=(1,0,0),\ \textit{and } R_4=(0,0,1).\end{aligned}$$

The details of the case where $x=0$ or $y=0$ are left to the reader. In the rest of the proof it will be assumed that $xy\neq 0$. We deal first with the polynomials P_t with $t\in[0,1]$. Since

$$\nabla P_t(x,y)=(2tx+2\sqrt{1-t}y,2\sqrt{1-t}x-2y),$$

and the maximum of $|DP_t(x,y)h|$ for each $h=(h_1,h_2)\in\square$ is attained at the extreme points of $\square$, we have, for $t\in[0,1]$,

$$\|DP_t(x,y)\|_\square=\max\{2tx+2\sqrt{1-t}y,|2\sqrt{1-t}x-2y|,|2tx+2\sqrt{1-t}(x+y)-2y|\}.$$

Consider

$$\begin{aligned}g_1(t)&=2tx+2\sqrt{1-t}y,\\ g_2(t)&=2\sqrt{1-t}x-2y,\\ g_3(t)&=2tx+2\sqrt{1-t}(x+y)-2y,\end{aligned}$$

with $t\in[0,1]$. The function g_1 has at most one critical point in $[0,1]$, which is $t_0=1-\left(\frac{y}{2x}\right)^2$ if and only if $y\le 2x$. Furthermore, t_0 is a local maximum. Therefore,

$$\max\{g_1(t):t\in[0,1]\}=\begin{cases}g_1(t_0) & \text{if } y\le 2x,\\ g_1(1) & \text{if } y\ge 2x,\end{cases}=\begin{cases}2x+\frac{y^2}{2x} & \text{if } y\le 2x,\\ 2y & \text{if } y\ge 2x.\end{cases}$$

The function g_2 is monotone. Consequently

$$\max\{g_2(t):t\in[0,1]\}=\max\{|g_2(0)|,|g_2(1)|\}=\max\{2y,2|x-y|\}=\begin{cases}2(x-y) & \text{if } y\le \frac{x}{2},\\ 2y & \text{if } y\ge\frac{x}{2}.\end{cases}$$

It can be easily checked that g_3 has at most one critical point in $[0,1]$, which occurs at $t_1=1-\frac{1}{4}\left(1+\frac{y}{x}\right)^2$ if and only if $y\le x$. As a matter of fact t_1 is a local maximum. Since g_3 is non-negative on $[0,1]$ when $y\le x$ we arrive at

$$\begin{aligned}\max\{|g_3(t)|:t\in[0,1]\}&=\begin{cases}|g_3(t_1)| & \text{if } y\le x,\\ \max\{|g_3(0)|,|g_3(1)|\} & \text{if } y\ge x,\end{cases}\\ &=\begin{cases}\frac{5}{2}x-y+\frac{y^2}{2x} & \text{if } y\le x,\\ \max\{2x,2(y-x)|\} & \text{if } y\ge x,\end{cases}\\ &=\begin{cases}\frac{5}{2}x-y+\frac{y^2}{2x} & \text{if } y\le x,\\ 2x & \text{if } x\le y\le 2x,\\ 2(y-x) & \text{if } y\ge 2x.\end{cases}\end{aligned}$$

Hence

$$\max\{\|DP_t(x,y)\|_{\square} : t\in[0,1]\} = \begin{cases} \max\left\{2x+\frac{y^2}{2x}, 2(x-y), \frac{5}{2}x-y+\frac{y^2}{2x}\right\} & \text{if } y\le \frac{x}{2}, \\ \max\left\{2x+\frac{y^2}{2x}, 2y, \frac{5}{2}x-y+\frac{y^2}{2x}\right\} & \text{if } \frac{x}{2}\le y\le x, \\ \max\left\{2x+\frac{y^2}{2x}, 2y, 2x\right\} & \text{if } x\le y\le 2x, \\ \max\{2y, 2(y-x)\} & \text{if } y\ge 2x, \end{cases}$$

$$= x\begin{cases} \max\left\{2+\frac{\alpha^2}{2}, 2(1-\alpha), \frac{5}{2}-\alpha+\frac{\alpha^2}{2}\right\} & \text{if } 0\le\alpha\le\frac{1}{2}, \\ \max\left\{2+\frac{\alpha^2}{2}, 2\alpha, \frac{5}{2}-\alpha+\frac{\alpha^2}{2}\right\} & \text{if } \frac{1}{2}\le\alpha\le 1, \\ \max\left\{2+\frac{\alpha^2}{2}, 2\alpha, 2\right\} & \text{if } 1\le\alpha\le 2, \\ \max\{2\alpha, 2(\alpha-1)\} & \text{if } \alpha\ge 2, \end{cases}$$

where $\alpha = y/x$. *Now it is simple to show that*

$$\begin{aligned} \frac{5}{2}-\alpha+\frac{\alpha^2}{2} &\ge \max\left\{2+\frac{\alpha^2}{2}, 2(1-\alpha)\right\} && \text{if } 0\le\alpha\le\frac{1}{2}, \\ 2+\frac{\alpha^2}{2} &\ge \max\left\{2\alpha, \frac{5}{2}-\alpha+\frac{\alpha^2}{2}\right\} && \text{if } \frac{1}{2}\le\alpha\le 1, \\ 2\alpha &\ge \max\left\{2+\frac{\alpha^2}{2}, 2\right\} && \text{if } 1\le\alpha\le 2, \\ 2\alpha &\ge 2(\alpha-1) && \text{if } \alpha\ge 2. \end{aligned}$$

Therefore,

$$\max\{\|DP_t(x,y)\|_{\square} : t\in[0,1]\} = x\begin{cases} \frac{5}{2}-\alpha+\frac{\alpha^2}{2} & \text{if } 0\le\alpha\le\frac{1}{2}, \\ 2+\frac{\alpha^2}{2} & \text{if } \frac{1}{2}\le\alpha\le 2, \\ 2\alpha & \text{if } \alpha\ge 2, \end{cases}$$

$$= \begin{cases} \frac{5}{2}x-y+\frac{y^2}{2x} & \text{if } y\le\frac{x}{2}, \\ 2x+\frac{y^2}{2x} & \text{if } \frac{x}{2}\le y\le 2x, \\ 2y & \text{if } y\ge 2x. \end{cases}$$

By symmetry

$$\max\{\|DQ_t(x,y)\|_{\square} : t\in[0,1]\} = \begin{cases} 2x & \text{if } y\le\frac{x}{2}, \\ 2y+\frac{x^2}{2y} & \text{if } \frac{x}{2}\le y\le 2x, \\ \frac{5}{2}y-x+\frac{x^2}{2y} & \text{if } y\ge 2x, \end{cases}$$

from which we obtain

$$\max\{\|DP_t(x,y)\|_{\square}, \|DQ_t(x,y)\|_{\square} : t\in[0,1]\} = \begin{cases} \frac{5}{2}x-y+\frac{y^2}{2x} & \text{if } y\le\frac{x}{2}, \\ 2x+\frac{y^2}{2x} & \text{if } \frac{x}{2}\le y\le x, \\ 2y+\frac{x^2}{2y} & \text{if } x\le y\le 2x, \\ \frac{5}{2}y-x+\frac{x^2}{2y} & \text{if } y\ge 2x. \end{cases} \tag{3.94}$$

On the other hand

$$\|DR_1(x,y)\|_{\square} = \max\{|2x-3y|, |x+y|, |-3x+2y|\} = \begin{cases} 3x-2y & \text{if } 0 \le y \le \frac{2x}{3}, \\ x+y & \text{if } \frac{2x}{3} \le y \le \frac{3x}{2}, \\ 3y-2x & \text{if } \frac{3x}{2} \le y \le 1. \end{cases}$$

$$\|DR_2(x,y)\|_{\square} = \max\{|2x-y|, |x+y|, |-x+2y|\} = \begin{cases} 2x-y & \text{if } 0 \le y \le \frac{x}{2}, \\ x+y & \text{if } \frac{x}{2} \le y \le 2x, \\ 2y-x & \text{if } 2x \le y \le 1. \end{cases}$$

$$\|DR_3(x,y)\|_{\square} = 2x,$$
$$\|DR_4(x,y)\|_{\square} = 2y.$$

it is left to the reader that

$$\max\{\|DR_k(x,y)\|_{\square} : k = 1,2,3,4\} = \begin{cases} 3x-2y & \text{if } 0 \le y \le \frac{x}{2}, \\ 2x & \text{if } \frac{x}{2} \le y \le x, \\ 2y & \text{if } x \le y \le 2x, \\ 3y-2x & \text{if } y \ge 2x. \end{cases} \tag{3.95}$$

When comparing (3.94) and (3.95) we may restrict our attention to the case $y \le x$. Observe that the surfaces $z = 3x-2y$ and $z = \frac{5}{2}x - y + \frac{y^2}{2x}$ meet along the line $y = (\sqrt{2}-1)x$. In fact we have

$$3x - 2y \ge \frac{5}{2}x - y + \frac{y^2}{2x} \qquad \text{if } 0 \le y \le (\sqrt{2}-1)x,$$
$$\frac{5}{2}x - y + \frac{y^2}{2x} \ge 3x - 2y \qquad \text{if } (\sqrt{2}-1)x \le y \le \frac{x}{2},$$

whereas

$$2x + \frac{y^2}{2x} \ge 2x \qquad \text{if } \frac{x}{2} \le y \le x.$$

The desired formula for $\mathcal{M}_{\square}(x,y)$ follows from the latter by symmetry with respect to the line $y = x$.

To obtain the Markov-type estimate, we use the convexity of $\mathcal{M}_{\square}$ on the square $\square$:

$$\begin{aligned} M_{\square} &= \max\{\mathcal{M}_{\square}(x,y) : (x,y) \in \square\} \\ &= \max\{\mathcal{M}_{\square}(0,0), \mathcal{M}_{\square}(1,0), \mathcal{M}_{\square}(0,1), \mathcal{M}_{\square}(1,1)\} \\ &= \max\left\{0, 3, \frac{5}{2}\right\} = 3. \end{aligned}$$

Sharp estimates on the length of $\nabla P(x,y)$ can also be derived using the Krein-Milman approach as shown below.

Theorem 3.2.18 (Gámez, Muñoz, Sánchez, and Seoane [45]) *If $(x,y) \in \square$ then*

$$\mathcal{M}_{\square,2}(x,y) = \begin{cases} \sqrt{\frac{24y^4+12x^2y^2+x^4+x(8y^2+x^2)^{\frac{3}{2}}}{8y^2}} & \text{if } 0 < \alpha_0 x \le y \le x, \\ \sqrt{\frac{24x^4+12x^2y^2+y^4+y(8x^2+y^2)^{\frac{3}{2}}}{8x^2}} & \text{if } 0 < x \le y \le \frac{x}{\alpha_0}, \\ \sqrt{13x^2-24xy+13y^2} & \text{otherwise,} \end{cases}$$

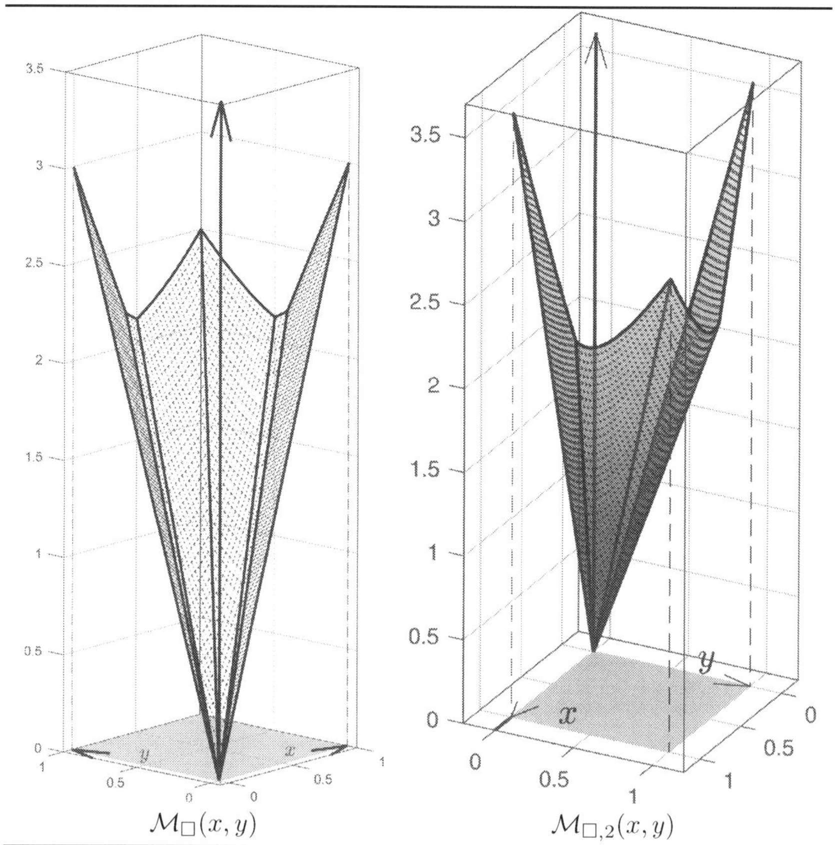

FIGURE 3.16: Representations of $\mathcal{M}_{\square}(x,y)$ and $\mathcal{M}_{\square,2}(x,y)$ on the square $[0,1]^2$.

where $\alpha_0 \approx 0.4029036618$ *is the unique root of the equation*

$$80\alpha^4 - 192\alpha^3 + 92\alpha^2 - 1 = (8\alpha^2+1)^{\frac{3}{2}}$$

in the interval $[\frac{3-\sqrt{5}}{2}, \frac{12-3\sqrt{3}}{13}]$. *Moreover,*

$$\max_{(x,y)\in\square} \|\nabla P(x,y)\|_2 \le \sqrt{13}\cdot\|P\|_{\square}$$

with equality for the polynomials $\pm(x^2-3xy+y^2)$ *at the points* $(0,1)$ *and* $(1,0)$*. In particular* $M_{\square,2} = \sqrt{13}$.

PROOF. *Applying the Krein-Milman approach, we have*

$$\mathcal{M}_{\square,2}(x,y) = \max\{\|\nabla P_t(x,y)\|_2, \|\nabla Q_t(x,y)\|_2, \|\nabla R_k(x,y)\|_2 : t\in[0,1], k=1,2,3,4\},$$

where P_t, Q_t *and the* R_k*'s are as in the proof of Theorem 3.2.17. The result is straightforward to check when either* $x=0$ *or* $y=0$*. Therefore, throughout the proof, we may assume that* $xy\neq 0$.

We begin by analyzing the behaviour of the polynomials P_t *and* Q_t *at any point* $(x,y)\in\square$*. Introducing the change of variable* $u=\sqrt{1-t}$*, which transforms* $t\in[0,1]$ *into* $u\in[0,1]$*, we can rewrite the maximum as follows:*

$$\max\{\|\nabla P_t(x,y)\|_2, \|\nabla Q_t(x,y)\|_2 : t\in[0,1]\} = \begin{cases}\max\{2\sqrt{f_1(u)} : u\in[0,1]\}, & \textit{if } y\le x,\\ \max\{2\sqrt{f_2(u)} : u\in[0,1]\}, & \textit{if } y\ge x,\end{cases}$$

where the functions f_1 *and* f_2 *are given by*

$$f_1(u) = y^2u^4 - 2xyu^3 - (y^2 - x^2)u^2 + x^2 + y^2, \quad f_2(u) = x^2u^4 - 2xyu^3 + (y^2 - x^2)u^2 + x^2 + y^2,$$

for $u \in [0, 1]$.

Due to the inherent symmetry of the problem with respect to the line $y = x$, *it suffices to analyze points* $(x, y) \in \square$ *where* $0 \leq \alpha := y/x \leq 1$. *Thus, it suffices to maximize*

$$f_1(u) = x^2[\alpha^2u^4 - 2\alpha u^3 + (1 - \alpha^2)u^2 + 1 + \alpha^2],$$

where $u \in [0, 1]$. *Since* f_1 *is non-negative on this interval, we proceed by studying its critical points. The function* $f_1(u)$ *has one stationary point at* $u = 0$ *and at most one additional critical point within* $[0, 1]$, *located at*

$$u_0 = \frac{3 - \sqrt{8\alpha^2 + 1}}{4\alpha}.$$

This critical point satisfies $u_0 \in [0, 1]$ *if and only if*

$$\frac{3 - \sqrt{5}}{2} \leq \alpha \leq 1.$$

A simple analysis reveals that f_1 *attains its maximum at* u_0 *whenever this condition holds. Otherwise,* f_1 *is increasing throughout* $[0, 1]$. *From this, we obtain*

$$\max\{2\sqrt{f_1(u)} : u \in [0, 1]\} = \begin{cases} x\sqrt{4\alpha^2 - 8\alpha + 8}, & \text{if } 0 \leq \alpha \leq \frac{3-\sqrt{5}}{2}, \\ x\sqrt{\frac{24\alpha^4+12\alpha^2+1+(8\alpha^2+1)^{\frac{3}{2}}}{8\alpha^2}}, & \text{if } \frac{3-\sqrt{5}}{2} \leq \alpha \leq 1. \end{cases}$$

We now consider the contribution from the R_k*'s. If* $(x, y) \in \square$ *with* $x \neq 0$, *we have*

$$\max\{\|\nabla R_k(x, y)\|_2 : k = 1, 2, 3, 4\} = x \max\{\sqrt{13\alpha^2 - 24\alpha + 13}, \sqrt{5\alpha^2 - 8\alpha + 5}, 2, 2\alpha\}.$$

Since $\alpha = y/x$ *as before, we analyze these terms individually. It is straightforward to check that*

- $\sqrt{13\alpha^2 - 24\alpha + 13} \geq \sqrt{5\alpha^2 - 8\alpha + 5}$ *for all* $0 \leq \alpha \leq 1$.
- *The inequality* $\sqrt{13\alpha^2 - 24\alpha + 13} \leq 2$ *holds whenever*

$$\frac{12 - 3\sqrt{3}}{13} \leq \alpha \leq 1.$$

From this, we conclude

$$\max\{\|\nabla R_k(x, y)\|_2 : k = 1, 2, 3, 4\} = \begin{cases} x\sqrt{13\alpha^2 - 24\alpha + 13}, & \text{if } 0 \leq \alpha \leq \frac{12-3\sqrt{3}}{13}, \\ 2x, & \text{if } \frac{12-3\sqrt{3}}{13} \leq \alpha \leq 1. \end{cases}$$

Noting that

$$\frac{3 - \sqrt{5}}{2} \approx 0.3820 < 0.5234 \approx \frac{12 - 3\sqrt{3}}{13},$$

we can now compare the two maximum values. For $0 \leq \alpha \leq \frac{3-\sqrt{5}}{2}$, *we establish that*

$$\sqrt{4\alpha^2 - 8\alpha + 8} \leq \sqrt{13\alpha^2 - 24\alpha + 13}.$$

Similarly, for $\frac{12-3\sqrt{3}}{13} \le \alpha \le 1$, *we confirm that*

$$2 \le \sqrt{\frac{24\alpha^4 + 12\alpha^2 + 1 + (8\alpha^2+1)^{\frac{3}{2}}}{8\alpha^2}}.$$

Since the functions $\sqrt{13\alpha^2 - 24\alpha + 13}$ *and* $\sqrt{\frac{24\alpha^4+12\alpha^2+1+(8\alpha^2+1)^{\frac{3}{2}}}{8\alpha^2}}$ *intersect at a single point* α_0 *within the interval* $\left[\frac{3-\sqrt{5}}{2}, \frac{12-3\sqrt{3}}{13}\right]$, *we conclude that for* $(x,y) \in \Box$ *with* $x \neq 0$ *and* $0 \le \alpha \le 1$

$$\mathcal{M}_{\Box,2}(x,y) = \begin{cases} x\sqrt{13\alpha^2 - 24\alpha + 13}, & \text{if } 0 \le \alpha \le \alpha_0, \\ x\sqrt{\frac{24\alpha^4+12\alpha^2+1+(8\alpha^2+1)^{\frac{3}{2}}}{8\alpha^2}}, & \text{if } \alpha_0 \le \alpha \le 1. \end{cases}$$

By symmetry, exchanging x *and* y *gives* $\mathcal{M}_{\Box,2}(x,y)$ *for* $y \ge x$.

To derive the Markov estimate, we use the convexity of $\mathcal{M}_{\Box,2}$ *in* $\Box$ *to obtain*

$$\begin{aligned}\max\{\mathcal{M}_{\Box,2}(x,y) : (x,y) \in \Box\} &= \max\{\mathcal{M}_{\Box,2}(0,0), \mathcal{M}_{\Box,2}(1,0), \mathcal{M}_{\Box,2}(0,1), \mathcal{M}_{\Box,2}(1,1)\} \\ &= \max\{0, \sqrt{13}, 2\sqrt{2}\} = \sqrt{13}.\end{aligned}$$

3.2.3.3 Bernstein and Markov inequalities for polynomials on circle sectors

It is convenient to recall that for each $\alpha, \beta \in [0, 2\pi]$ with $\alpha \le \beta$ the notation $D(\alpha, \beta)$ stands for the sector

$$D(\alpha, \beta) := \{re^{i\theta} : \ 0 \le r \le 1, \ \alpha \le \theta \le \beta\}.$$

When $\alpha = 0$, we replace $D(0,\beta)$ by $D(\beta)$ for simplicity. We represent $(\mathcal{Q}(\mathbb{R}^2), \|\cdot\|_{D(0,\beta)})$ by $\mathcal{P}(^2D(\beta))$.

The geometry of the spaces $\mathcal{P}(^2D(\beta))$ has been fully described in several works (see for instance [10, 81]) and we have presented in this book in Section 1.4, allows us to obtain sharp Bernstein and Markov inequalities in $\mathcal{P}(^2D(\beta))$ by means of (3.89) or (3.90) with $K = D(\beta)$. We will focus only on the cases $\beta = \frac{\pi}{4}$ and $\beta = \frac{\pi}{2}$.

Bernstein and Markov inequalities in $\mathcal{P}(^2D(\pi/4))$

It has been proved in [81] and we have given an explicit proof in Theorem 1.4.5 that the extreme points of the unit ball $\mathsf{B}_{D(\pi/4)}$ of $\mathcal{P}(^2D(\pi/4))$ are given by

$$\text{ext}(\mathsf{B}_{D(\pi/4)}) = \left\{\pm P_t, \pm Q_s, \pm(1,1,0) : \ -1 \le t \le 1 \text{ and } 1 \le s \le 5 + 4\sqrt{2}\right\},$$

where

$$\begin{aligned} P_t &:= (t, -2-2t-4\sqrt{1+t}, 4+t+4\sqrt{1+t}), \\ Q_s &:= (1, -2\sqrt{2(1+s)}, s). \end{aligned}$$

Using the previous characterization of $\text{ext}(\mathsf{B}_{D(\pi/4)})$ in (3.89) we obtain the following Bersntein and Markov sharp inequalities.

Theorem 3.2.19 (Araújo, Jiménez, Muñoz, and Seoane [2]) *For every* $(x,y) \in D(\frac{\pi}{4})$ *we have*

$$\mathcal{M}_{D(\frac{\pi}{4})}(x,y) = \begin{cases} \sqrt{2}\left[\left(1+2\sqrt{2}\right)x - \left(3+2\sqrt{2}\right)y\right] & \text{if } 0 \le y < \frac{2\sqrt{2}-1}{7}x, \\ \frac{\sqrt{2}(x^2+3y^2)}{2y} & \text{if } \frac{2\sqrt{2}-1}{7}x \le y < (\sqrt{2}-1)x, \\ 2\left(x + \frac{y^2}{x-y}\right) & \text{if } (\sqrt{2}-1)x \le y < \left(2-\sqrt{2}\right)x, \\ 4\left(1+\sqrt{2}\right)y - 2x & \text{if } \left(2-\sqrt{2}\right)x \le y \le x \end{cases}$$

Moreover $M_{D(\pi/4)} = 4 + \sqrt{2}$.

PROOF. *In order to calculate* $\Psi(x, y) := \sup\{\|DP(x, y)\|_{D(\frac{\pi}{4})} : \|P\|_{D(\frac{\pi}{4})}) \leq 1\}$, *by the Krein-Milman approach, it suffices to calculate*

$$\sup\{\|DP(x, y)\|_{D(\frac{\pi}{4})} : P \in ext(B_{D(\frac{\pi}{4})})\}.$$

By symmetry, we may restrict our attention to the extreme polynomials with positive sign. Let us start first with

$$P_t(x, y) = tx^2 + \left(4 + t + 4\sqrt{1+t}\right) y^2 - \left(2 + 2t + 4\sqrt{1+t}\right) xy.$$

So, we may write

$$\nabla P_t(x, y) = \left(2tx - \left(2 + 2t + 4\sqrt{1+t}\right) y,\, 2\left(4 + t + 4\sqrt{1+t}\right) y - \left(2 + 2t + 4\sqrt{1+t}\right) x\right),$$

from which

$$\begin{aligned}
\|DP_t(x, y)\|_{D(\frac{\pi}{4})} = &\sup_{0 \leq \theta \leq \frac{\pi}{4}} \Big|2\left[tx - \left(1 + t + 2\sqrt{1+t}\right) y\right] \cos\theta \\
&+ 2\left[\left(4 + t + 4\sqrt{1+t}\right) y - \left(1 + t + 2\sqrt{1+t}\right) x\right] \sin\theta\Big| \\
= &\, 2x \sup_{0 \leq \theta \leq \frac{\pi}{4}} |f_\lambda(t, \theta)|, \\
&\textit{for } f_\lambda(t, \theta) = \left[t - \left(1 + t + 2\sqrt{1+t}\right) \lambda\right] \cos\theta \\
&\qquad + \left[\left(4 + t + 4\sqrt{1+t}\right) \lambda - \left(1 + t + 2\sqrt{1+t}\right)\right] \sin\theta,
\end{aligned}$$

where $\lambda = \frac{y}{x}$, $x \neq 0$ *(the case* $x = 0$ *is trivial, since the only point in* $D(\frac{\pi}{4})$ *where* $x = 0$ *is* $(0, 0)$, *in which case* $P_t(0, 0) = \|DP_t(0, 0)\|_{D(\frac{\pi}{4})} = 0$*).*

We aim to calculate

$$\sup_{-1 \leq t \leq 1} \|DP_t(x, y)\|_{D(\frac{\pi}{4})} = 2x \sup_{\substack{0 \leq \theta \leq \frac{\pi}{4} \\ -1 \leq t \leq 1}} |f_\lambda(t, \theta)|.$$

Define $C_1 = [-1, 1] \times [0, \frac{\pi}{4}]$. *We will analyze* 5 *cases.*

(1) $(t, \theta) \in (-1, 1) \times (0, \frac{\pi}{4})$.

We are interested just in critical points. Hence,

$$\begin{aligned}
\frac{\partial f_\lambda}{\partial t}(t, \theta) = &\left[\left(1 + \frac{2}{\sqrt{1+t}}\right) \lambda - \left(1 + \frac{1}{\sqrt{1+t}}\right)\right] \sin\theta \\
&+ \left[1 - \left(1 + \frac{1}{\sqrt{1+t}}\right) \lambda\right] \cos\theta = 0,
\end{aligned} \tag{3.96}$$

$$\begin{aligned}
\frac{\partial f_\lambda}{\partial \theta}(t, \theta) = &\left[\left(1 + t + 2\sqrt{1+t}\right) \lambda - t\right] \sin\theta \\
&+ \left[\left(4 + t + 4\sqrt{1+t}\right) \lambda - \left(1 + t + 2\sqrt{1+t}\right)\right] \cos\theta = 0
\end{aligned} \tag{3.97}$$

From equation (3.97), *we have*

$$\sin\theta = \frac{\left(4 + t + 4\sqrt{1+t}\right) \lambda - \left(1 + t + 2\sqrt{1+t}\right)}{t - \left(1 + t + 2\sqrt{1+t}\right) \lambda} \cos\theta. \tag{3.98}$$

Substituting (3.98) *into* (3.96), *we obtain*

$$0 = \left\{ \left[1 - \left(1 + \frac{1}{\sqrt{1+t}}\right)\lambda\right] + \left[\left(1 + \frac{2}{\sqrt{1+t}}\right)\lambda - \left(1 + \frac{1}{\sqrt{1+t}}\right)\right] \right.$$
$$\left. \times \frac{\left(4 + t + 4\sqrt{1+t}\right)\lambda - \left(1 + t + 2\sqrt{1+t}\right)}{t - \left(1 + t + 2\sqrt{1+t}\right)\lambda} \right\} \cos\theta.$$

Since $0 < \theta < \frac{\pi}{4}$, *we can conclude*

$$0 = \left[1 - \left(1 + \frac{1}{\sqrt{1+t}}\right)\lambda\right] + \left[\left(1 + \frac{2}{\sqrt{1+t}}\right)\lambda - \left(1 + \frac{1}{\sqrt{1+t}}\right)\right]$$
$$\times \frac{\left(4 + t + 4\sqrt{1+t}\right)\lambda - \left(1 + t + 2\sqrt{1+t}\right)}{t - \left(1 + t + 2\sqrt{1+t}\right)\lambda}$$

and thus

$$\begin{aligned}
0 &= \left[1 - \left(1 + \frac{1}{\sqrt{1+t}}\right)\lambda\right] \cdot \left[t - \left(1 + t + 2\sqrt{1+t}\right)\lambda\right] \\
&\quad + \left[\left(1 + \frac{2}{\sqrt{1+t}}\right)\lambda - \left(1 + \frac{1}{\sqrt{1+t}}\right)\right] \cdot \left[\left(4 + t + 4\sqrt{1+t}\right)\lambda - \left(1 + t + 2\sqrt{1+t}\right)\right] \\
&= t - \left(1 + t + 2\sqrt{1+t}\right)\lambda - t\lambda + \left(1 + t + 2\sqrt{1+t}\right)\lambda^2 - \frac{\lambda t}{\sqrt{1+t}} \\
&\quad + \frac{\lambda^2}{\sqrt{1+t}}\left(1 + t + 2\sqrt{1+t}\right) + \left(1 + \frac{2}{\sqrt{1+t}}\right)\left(4 + t + 4\sqrt{1+t}\right)\lambda^2 \\
&\quad - \left(1 + \frac{2}{\sqrt{1+t}}\right)\left(1 + t + 2\sqrt{1+t}\right)\lambda - \left(1 + \frac{1}{\sqrt{1+t}}\right)\left(4 + t + 4\sqrt{1+t}\right)\lambda \\
&\quad + \left(1 + \frac{1}{\sqrt{1+t}}\right)\left(1 + t + 2\sqrt{1+t}\right) \\
&= t\left(1 - 2\lambda + 2\lambda^2 - 2\lambda + 1\right) + \left(-2\lambda + 2\lambda^2 + 4\lambda^2 - 2\lambda - 4\lambda + 2\right)\sqrt{1+t} \\
&\quad + \frac{t}{\sqrt{1+t}}\left(-\lambda + \lambda^2 + 2\lambda^2 - 2\lambda - \lambda + 1\right) + \frac{1}{\sqrt{1+t}}\left(\lambda^2 + 8\lambda^2 - 2\lambda - 4\lambda + 1\right) \\
&\quad + \left(-\lambda + \lambda^2 + 2\lambda^2 + 4\lambda^2 - \lambda - 4\lambda + 1 + 2 + 8\lambda^2 - 8\lambda\right) \\
&= 2t(\lambda - 1)^2 + 6\sqrt{1+t}(\lambda - 1)\left(\lambda - \frac{1}{3}\right) + 3\frac{t}{\sqrt{1+t}}(\lambda - 1)\left(\lambda - \frac{1}{3}\right) \\
&\quad + \frac{1}{\sqrt{1+t}}(3\lambda - 1)^2 + 15\left(\lambda - \frac{1}{3}\right)\left(\lambda - \frac{3}{5}\right).
\end{aligned}$$

Working with this last expression, we have

$$\begin{aligned}
0 &= 2t\sqrt{1+t}(\lambda - 1)^2 + 6(1+t)(\lambda - 1)\left(\lambda - \frac{1}{3}\right) \\
&\quad + 3t(\lambda - 1)\left(\lambda - \frac{1}{3}\right) + (3\lambda - 1)^2 \\
&\quad + 15\sqrt{1+t}\left(\lambda - \frac{1}{3}\right)\left(\lambda - \frac{3}{5}\right),
\end{aligned}$$

from which

$$\begin{aligned}
&\sqrt{1+t}\left[15\left(\lambda-\frac{1}{3}\right)\left(\lambda-\frac{3}{5}\right)+2t(\lambda-1)^2\right] \\
&\quad=-9t(\lambda-1)\left(\lambda-\frac{1}{3}\right)-15\left(\lambda-\frac{1}{3}\right)\left(\lambda-\frac{3}{5}\right). \qquad (3.99)
\end{aligned}$$

If $\lambda = 1$*, we obtain*

$$\sqrt{1+t}+1=0$$

so $\lambda \neq 1$*. The solutions of equation* (3.99) *are*

$$t_1(\lambda)=\frac{-1+2\lambda+3\lambda^2}{(\lambda-1)^2} \quad \text{and} \quad t_2(\lambda)=\frac{5\lambda^2+2\lambda-3}{4(\lambda-1)^2}.$$

From (3.96),

$$\tan\theta=\frac{\left(1+\frac{1}{\sqrt{1+t}}\right)\lambda-1}{\left(1+\frac{2}{\sqrt{1+t}}\right)\lambda-\left(1+\frac{1}{\sqrt{1+t}}\right)}.$$

In particular, evaluation at $t_1(\lambda)$ *yields*

$$\tan\theta_1=\frac{\left(1+\frac{1-\lambda}{2\lambda}\right)\lambda-1}{\left(1+\frac{1-\lambda}{\lambda}\right)\lambda-\left(1+\frac{1-\lambda}{2\lambda}\right)}=\lambda,$$

in which case

$$D_{1,1}(\lambda):=|f_\lambda(t_1,\theta_1)|=\left|-\sqrt{1+\lambda^2}\right|=\sqrt{1+\lambda^2}.$$

Regarding $t_2(\lambda)$*, we obtain*

$$\tan\theta_2=\frac{\left(1+\sqrt{\frac{4(\lambda-1)^2}{(3\lambda-1)^2}}\right)\lambda-1}{\left(1+2\sqrt{\frac{4(\lambda-1)^2}{(3\lambda-1)^2}}\right)\lambda-\left(1+\sqrt{\frac{4(\lambda-1)^2}{(3\lambda-1)^2}}\right)}.$$

Since $\theta_2 \in \left(0,\frac{\pi}{4}\right)$*, we need to guarantee that* $0<\tan\theta_2<1$*. Therefore, we must have* $0<\lambda<\frac{1}{5}$*. Hence*

$$\tan\theta_2=\frac{5\lambda-1}{7\lambda-3}$$

and in this case,

$$D_{1,2}(\lambda):=|f_\lambda(t_2,\theta_2)|=\frac{39\lambda^2-26\lambda+7}{2\sqrt{74\lambda^2-52\lambda+10}}.$$

(2) $\theta=0, -1\le t\le 1$.

We have

$$f_\lambda(t,0)=t-\left(1+t+2\sqrt{1+t}\right)\lambda.$$

Then,

$$\begin{aligned}
f_\lambda(-1,0)&=-1,\\
f_\lambda(1,0)&=1-2\left(1+\sqrt{2}\right)\lambda,
\end{aligned}$$

and hence

$$|f_\lambda(1,0)|=\begin{cases}1-2(1+\sqrt{2})\lambda & \text{if } 0\le\lambda<\frac{\sqrt{2}-1}{2},\\ 2\left(1+\sqrt{2}\right)\lambda-1 & \text{if } \frac{\sqrt{2}-1}{2}\le\lambda\le 1.\end{cases}$$

Now considering $t \in (-1,1)$, *since*

$$f'_\lambda(t,0) = 1 - \left(1 + \frac{1}{\sqrt{1+t}}\right)\lambda,$$

the critical point of $f_\lambda(t,0)$ *is*

$$t = \frac{\lambda^2}{(1-\lambda)^2} - 1.$$

Keep in mind that $-1 < t < 1$. *Therefore,*

$$\lambda < \frac{\sqrt{2}}{1+\sqrt{2}} = 2 - \sqrt{2}.$$

Substituting the critical point of $f_\lambda(t,0)$ *into* $f_\lambda(t,0)$, *we obtain*

$$f_\lambda\left(\frac{\lambda^2}{(\lambda-1)^2} - 1, 0\right) = \frac{\lambda^2}{(\lambda-1)^2} - 1 - \left[\frac{\lambda^2}{(\lambda-1)^2} + \frac{2\lambda}{1-\lambda}\right]\lambda = \frac{\lambda^2}{\lambda-1} - 1,$$

which implies

$$\left|f_\lambda\left(\frac{\lambda^2}{(\lambda-1)^2} - 1, 0\right)\right| = 1 + \frac{\lambda^2}{1-\lambda}.$$

- *Assume first* $0 \le \lambda < \frac{\sqrt{2}-1}{2}$. *Then,*

$$\sup_{-1\le t\le 1} |f_\lambda(t,0)| = \max\left\{1,\, 1 - 2\left(1+\sqrt{2}\right)\lambda,\, 1 + \frac{\lambda^2}{1-\lambda}\right\} = 1 + \frac{\lambda^2}{1-\lambda}.$$

- *Assume now* $\frac{\sqrt{2}-1}{2} \le \lambda < 2 - \sqrt{2}$. *Then,*

$$\sup_{-1\le t\le 1} |f_\lambda(t,0)| = \max\left\{1,\, 2\left(1+\sqrt{2}\right)\lambda - 1,\, 1 + \frac{\lambda^2}{1-\lambda}\right\} = 1 + \frac{\lambda^2}{1-\lambda}.$$

- *Assume finally* $2 - \sqrt{2} \le \lambda \le 1$. *Then,*

$$\sup_{-1\le t\le 1} |f_\lambda(t,0)| = \max\left\{1,\, 2\left(1+\sqrt{2}\right)\lambda - 1\right\} = 2\left(1+\sqrt{2}\right)\lambda - 1.$$

Consequently,

$$\begin{aligned}\sup_{-1\le t\le 1} |f_\lambda(t,0)| &= \begin{cases} 1 + \frac{\lambda^2}{1-\lambda} & \text{if } 0 \le \lambda < 2-\sqrt{2}, \\ (2+2\sqrt{2})\lambda - 1 & \text{if } 2-\sqrt{2} \le \lambda \le 1, \end{cases} \\ &=: \begin{cases} D_{2,1}(\lambda) & \text{if } 0 \le \lambda < 2-\sqrt{2}, \\ D_{2,2}(\lambda) & \text{if } 2-\sqrt{2} \le \lambda \le 1. \end{cases}\end{aligned}$$

(3) $\theta = \frac{\pi}{4}$ *and* $-1 \le t \le 1$.

Since

$$\begin{aligned} f_\lambda\left(t, \frac{\pi}{4}\right) &= \frac{\sqrt{2}}{2}\left[t - \left(1+t+2\sqrt{1+t}\right)\lambda + \left(4+t+4\sqrt{1+t}\right)\lambda - \left(1+t+2\sqrt{1+t}\right)\right] \\ &= \frac{\sqrt{2}}{2}\left[\left(3+2\sqrt{1+t}\right)\lambda - \left(1+2\sqrt{1+t}\right)\right], \end{aligned}$$

it follows,

$$\begin{aligned} f_\lambda\left(-1, \frac{\pi}{4}\right) &= \frac{\sqrt{2}}{2}(3\lambda - 1), \\ f_\lambda\left(1, \frac{\pi}{4}\right) &= \frac{\sqrt{2}}{2}\left[\left(3+2\sqrt{2}\right)\lambda - \left(1+2\sqrt{2}\right)\right], \\ f'_\lambda\left(t, \frac{\pi}{4}\right) &= \frac{\sqrt{2}}{2}\left[\frac{\lambda}{\sqrt{1+t}} - \frac{1}{\sqrt{1+t}}\right]. \end{aligned}$$

and $f'_\lambda(t, \frac{\pi}{4}) = 0$ *implies* $\lambda = 1$ *(in which case* $f_\lambda(t, \frac{\pi}{4}) = \sqrt{2}$ *for every* t*).*

- *Assume first* $0 \le \lambda < \frac{1}{3}$. *Then,*

$$\begin{aligned} \sup_{-1\le t\le 1} \left|f_\lambda\left(t, \frac{\pi}{4}\right)\right| &= \frac{\sqrt{2}}{2}\max\left\{\left(1+2\sqrt{2}\right) - \left(3+2\sqrt{2}\right)\lambda, 1-3\lambda\right\} \\ &= \frac{\sqrt{2}}{2}\left[\left(1+2\sqrt{2}\right) - \left(3+2\sqrt{2}\right)\lambda\right]. \end{aligned}$$

- *Assume now* $\frac{1}{3} \le \lambda < 4\sqrt{2} - 5$. *Then,*

$$\begin{aligned} \sup_{-1\le t\le 1} \left|f_\lambda\left(t, \frac{\pi}{4}\right)\right| &= \frac{\sqrt{2}}{2}\max\left\{\left(1+2\sqrt{2}\right) - \left(3+2\sqrt{2}\right)\lambda, 3\lambda - 1\right\} \\ &= \begin{cases} \frac{\sqrt{2}}{2}\left[(1+2\sqrt{2}) - (3+2\sqrt{2})\lambda\right] & \text{if } \frac{1}{3} \le \lambda < \frac{2\sqrt{2}+1}{7}, \\ \frac{\sqrt{2}}{2}(3\lambda - 1) & \text{if } \frac{2\sqrt{2}+1}{7} \le \lambda < 4\sqrt{2} - 5. \end{cases} \end{aligned}$$

- *Assume finally* $4\sqrt{2} - 5 \le \lambda \le 1$. *Then,*

$$\sup_{-1\le t\le 1} \left|f_\lambda\left(t, \frac{\pi}{4}\right)\right| = \frac{\sqrt{2}}{2}\max\left\{3\lambda - 1, \left(3+2\sqrt{2}\right)\lambda - \left(1+2\sqrt{2}\right)\right\} = \frac{\sqrt{2}}{2}(3\lambda - 1).$$

Hence,

$$\begin{aligned} \sup\nolimits_{-1\le t\le 1} \left|f_\lambda\left(t, \tfrac{\pi}{4}\right)\right| &= \begin{cases} \frac{\sqrt{2}}{2}\left[1+2\sqrt{2} - (3+2\sqrt{2})\lambda\right] & \text{if } 0 \le \lambda < \frac{2\sqrt{2}+1}{7} \\ \frac{\sqrt{2}}{2}(3\lambda - 1) & \text{if } \frac{2\sqrt{2}+1}{7} \le \lambda \le 1. \end{cases} \\ &=: \begin{cases} D_{3,1}(\lambda) & \text{if } 0 \le \lambda < \frac{2\sqrt{2}+1}{7} \\ D_{3,2}(\lambda) & \text{if } \frac{2\sqrt{2}+1}{7} \le \lambda \le 1. \end{cases} \end{aligned}$$

(4) $t = -1$, $0 \le \theta \le \frac{\pi}{4}$.

Therefore,

$$\begin{aligned} \sup_{0\le\theta\le\frac{\pi}{4}} f_\lambda(-1, \theta) &= \begin{cases} 1 & \text{if } 0 \le \lambda < \frac{1+\sqrt{2}}{3}, \\ \frac{\sqrt{2}}{2}(3\lambda - 1) & \text{if } \frac{1+\sqrt{2}}{3} \le \lambda \le 1. \end{cases} \\ &=: \begin{cases} D_{4,1}(\lambda) & \text{if } 0 \le \lambda < \frac{1+\sqrt{2}}{3}, \\ D_{4,2}(\lambda) & \text{if } \frac{1+\sqrt{2}}{3} \le \lambda \le 1. \end{cases} \end{aligned}$$

(5) $t = 1$, $0 \le \theta \le \frac{\pi}{4}$.

Now observe that that $\frac{b}{a} < 0$ *if and only if* $\lambda \in \left[0, \frac{\sqrt{2}-1}{2}\right) \cup \left(\frac{6-2\sqrt{2}}{7}, 1\right]$ *and* $\frac{b}{a} > 1$ *if and only if* $\frac{\sqrt{2}-1}{2} < \lambda < \frac{3+4\sqrt{2}}{23}$. *Therefore, putting* $a = 1 - \left(2+2\sqrt{2}\right)\lambda$ *and* $b = \left(5+4\sqrt{2}\right)\lambda - \left(2+2\sqrt{2}\right)$ *in the easily checked formula*

$$\sup_{\theta\in\left[0,\frac{\pi}{4}\right]} |a\cos\theta + b\sin\theta| = \begin{cases} \sqrt{a^2+b^2} & \text{if } 0 < \frac{b}{a} < 1, \\ \frac{\sqrt{2}}{2}|a+b| & \text{if } \left(1-\sqrt{2}\right)b < a < b \text{ or } b < a < \left(1-\sqrt{2}\right)b, \\ |a| & \text{if } -\left(1+\sqrt{2}\right)a < b < 0 \text{ or } 0 < b < -\left(1+\sqrt{2}\right)a, \end{cases} \tag{3.100}$$

we arrive at

$$\sup_{0\le\theta\le\frac{\pi}{4}} |f_\lambda(1,\theta)|$$
$$= \begin{cases} \max\left\{\left|1-\left(2+2\sqrt{2}\right)\lambda\right|, \frac{\sqrt{2}}{2}\left|\left(3+2\sqrt{2}\right)\lambda-\left(1+2\sqrt{2}\right)\right|\right\} & \text{if } 0 \le \lambda < \frac{3+4\sqrt{2}}{23}, \\ \sqrt{\left(1-\left(2+2\sqrt{2}\right)\lambda\right)^2 + \left(\left(5+4\sqrt{2}\right)\lambda - \left(2+2\sqrt{2}\right)\right)^2} & \text{if } \frac{3+4\sqrt{2}}{23} \le \lambda < \frac{6-2\sqrt{2}}{7}, \\ \max\left\{\left|1-\left(2+2\sqrt{2}\right)\lambda\right|, \frac{\sqrt{2}}{2}\left|\left(3+2\sqrt{2}\right)\lambda-\left(1+2\sqrt{2}\right)\right|\right\} & \text{if } \frac{6-2\sqrt{2}}{7} \le \lambda \le 1. \end{cases}$$

Since $0 \le \lambda < \sqrt{2}-1$ *implies* $\left|1-\left(2+2\sqrt{2}\right)\lambda\right| < \frac{\sqrt{2}}{2}\left|\left(3+2\sqrt{2}\right)\lambda - \left(1+2\sqrt{2}\right)\right|$, *it follows that*

$$\sup_{0\le\theta\le\frac{\pi}{4}} |f_\lambda(1,\theta)|$$
$$= \begin{cases} \frac{\sqrt{2}}{2}\left|\left(3+2\sqrt{2}\right)\lambda-\left(1+2\sqrt{2}\right)\right| & \text{if } 0 \le \lambda < \frac{3+4\sqrt{2}}{23} \\ \sqrt{48\sqrt{2}\lambda^2 - 56\lambda + 69\lambda^2 - 40\sqrt{2}\lambda + 8\sqrt{2} + 13} & \text{if } \frac{3+4\sqrt{2}}{23} \le \lambda < \frac{6-2\sqrt{2}}{7} \\ \left|1-\left(2+2\sqrt{2}\right)\lambda\right| & \text{if } \frac{6-2\sqrt{2}}{7} \le \lambda \le 1 \end{cases}$$
$$= \begin{cases} \frac{\sqrt{2}}{2}\left[1+2\sqrt{2}-\left(3+2\sqrt{2}\right)\lambda\right] & \text{if } 0 \le \lambda < \frac{3+4\sqrt{2}}{23} \\ \sqrt{48\sqrt{2}\lambda^2 - 56\lambda + 69\lambda^2 - 40\sqrt{2}\lambda + 8\sqrt{2} + 13} & \text{if } \frac{3+4\sqrt{2}}{23} \le \lambda < \frac{6-2\sqrt{2}}{7} \\ \left(2+2\sqrt{2}\right)\lambda - 1 & \text{if } \frac{6-2\sqrt{2}}{7} \le \lambda \le 1. \end{cases}$$
$$=: \begin{cases} D_{5,1}(\lambda) & \text{if } 0 \le \lambda < \frac{3+4\sqrt{2}}{23} \\ D_{5,2}(\lambda) & \text{if } \frac{3+4\sqrt{2}}{23} \le \lambda < \frac{6-2\sqrt{2}}{7} \\ D_{5,3}(\lambda) & \text{if } \frac{6-2\sqrt{2}}{7} \le \lambda \le 1. \end{cases}$$

According to Figures 3.17 and 3.18,

$$D_{1,1}(\lambda) \le \begin{cases} D_{2,1}(\lambda) & \text{if } 0 \le \lambda < 2-\sqrt{2}, \\ D_{2,2}(\lambda) & \text{if } 2-\sqrt{2} \le \lambda \le 1, \end{cases}$$
$$D_{1,2}(\lambda) \le D_{3,1}(\lambda) \text{ for } 0 < \lambda < \tfrac{1}{5},$$

we can rule out case (1). Since

$$D_{3,1}(\lambda) = D_{5,1}(\lambda) \quad \text{for } 0 \le \lambda \le \tfrac{3+4\sqrt{2}}{23},$$
$$D_{3,2}(\lambda) = D_{4,2}(\lambda) \quad \text{for } \tfrac{1+\sqrt{2}}{3} \le \lambda \le 1,$$

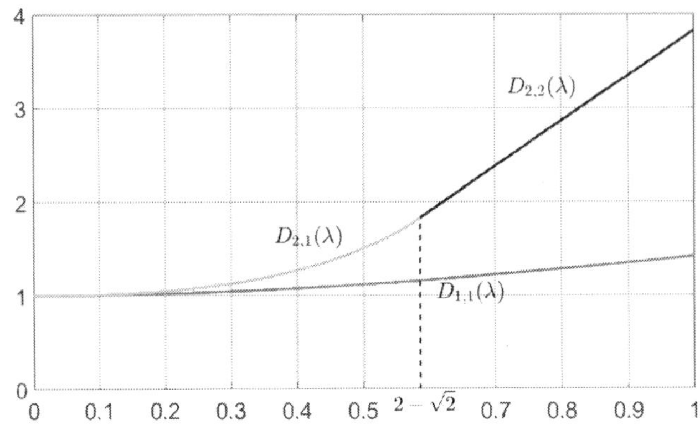

FIGURE 3.17: Graphs of the mappings $D_{1,1}(\lambda)$, $D_{2,1}(\lambda)$, and $D_{2,2}(\lambda)$.

we can directly rule out case (3). Since (see Figures 3.17 and 3.19)

$$D_{4,1}(\lambda) = 1 \leq \begin{cases} D_{2,1}(\lambda) & \textit{if } 0 \leq \lambda < 2-\sqrt{2}, \\ D_{2,2}(\lambda) & \textit{if } 2-\sqrt{2} \leq \lambda < \frac{1+\sqrt{2}}{3}, \end{cases}$$

$$D_{4,2}(\lambda) \leq D_{2,2} \ \textit{for} \ \tfrac{1+\sqrt{2}}{3} \leq \lambda \leq 1,$$

we can rule out case (4). Finally, since (see Figure 3.20)

$$D_{5,2}(\lambda) \leq D_{2,1}(\lambda) \quad \textit{for} \ \tfrac{3+4\sqrt{2}}{23} \leq \lambda < \tfrac{6-2\sqrt{2}}{7},$$

$$D_{5,3}(\lambda) = D_{2,2}(\lambda) \quad \textit{for} \ 2-\sqrt{2} \leq \lambda \leq 1,$$

we can rule out the expressions $D_{5,2}(\lambda)$ and $D_{5,3}(\lambda)$ of case (5).

Therefore, combining all the cases together it is possible to show that

$$\sup_{(t,\theta)\in C_1} |f_\lambda(t,\theta)| = \begin{cases} D_{5,1}(\lambda) & \textit{if } 0 \leq \lambda < \frac{(2-3\sqrt{2})\sqrt{4\sqrt{2}+7}+5\sqrt{2}+6}{14}, \\ D_{2,1}(\lambda) & \textit{if } \frac{(2-3\sqrt{2})\sqrt{4\sqrt{2}+7}+5\sqrt{2}+6}{14} \leq \lambda < 2-\sqrt{2}, \\ D_{2,2}(\lambda) & \textit{if } 2-\sqrt{2} \leq \lambda \leq 1, \end{cases}$$

$$= \begin{cases} \frac{\sqrt{2}}{2}\left[(1+2\sqrt{2}) - (3+2\sqrt{2})\,\lambda\right] & \textit{if } 0 \leq \lambda < \frac{(2-3\sqrt{2})\sqrt{4\sqrt{2}+7}+5\sqrt{2}+6}{14}, \\ 1 + \frac{\lambda^2}{1-\lambda} & \textit{if } \frac{(2-3\sqrt{2})\sqrt{4\sqrt{2}+7}+5\sqrt{2}+6}{14} \leq \lambda < 2-\sqrt{2}, \\ (2+2\sqrt{2})\,\lambda - 1 & \textit{if } 2-\sqrt{2} \leq \lambda \leq 1, \end{cases}$$

from which

$$\sup_{-1\leq t\leq 1} \|DP_t(x,y)\|_{D(\frac{\pi}{4})} = 2x \sup_{(t,\theta)\in C_1} |f_\lambda(t,\theta)|$$

$$= \begin{cases} \sqrt{2}\left[(1+2\sqrt{2})\,x - (3+2\sqrt{2})\,y\right] & \textit{if } 0 \leq y < \frac{(2-3\sqrt{2})\sqrt{4\sqrt{2}+7}+5\sqrt{2}+6}{14}x, \\ 2\left(x + \frac{y^2}{x-y}\right) & \textit{if } \frac{(2-3\sqrt{2})\sqrt{4\sqrt{2}+7}+5\sqrt{2}+6}{14}x \leq y < (2-\sqrt{2})\,x, \\ 4\left(1+\sqrt{2}\right)y - 2x & \textit{if } (2-\sqrt{2})\,x \leq y \leq x. \end{cases}$$

We have been considering $x \neq 0$ all the time (see Figure 3.21).

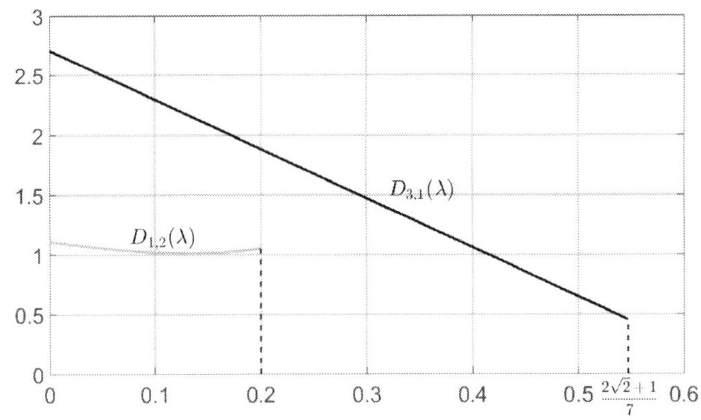

FIGURE 3.18: Graphs of the mappings $D_{1,2}(\lambda)$ and $D_{3,1}(\lambda)$.

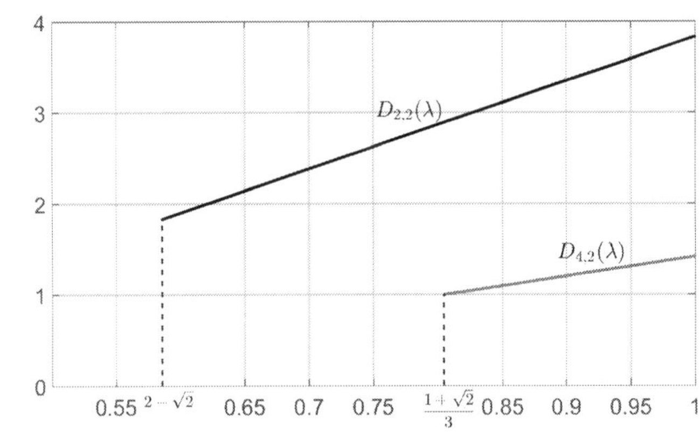

FIGURE 3.19: Graphs of the mappings $D_{2,2}(\lambda)$ and $D_{4,2}(\lambda)$.

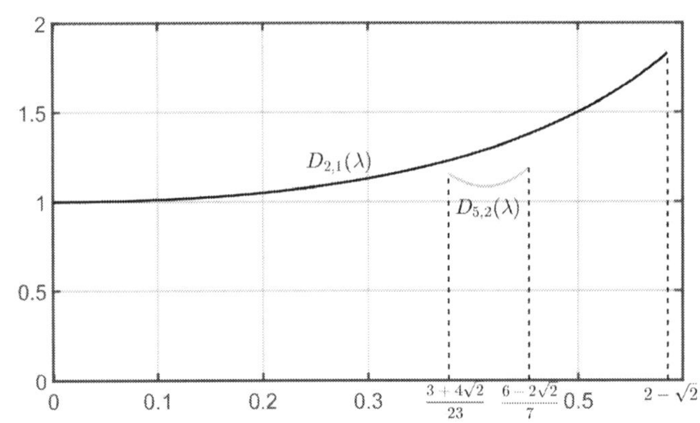

FIGURE 3.20: Graphs of the mappings $D_{2,1}(\lambda)$ and $D_{5,2}(\lambda)$.

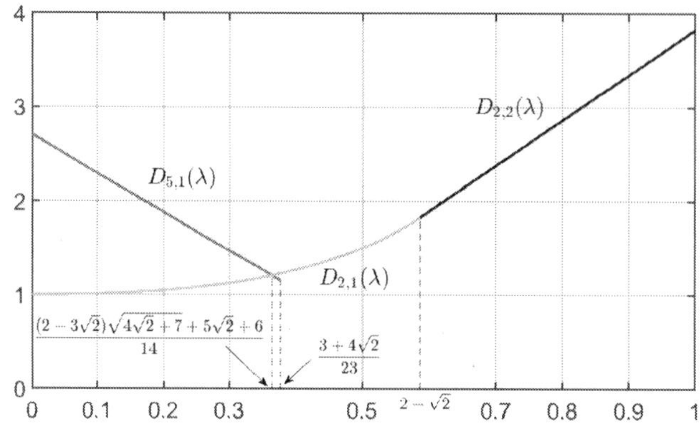

FIGURE 3.21: Graphs of the mappings $D_{2,1}(\lambda)$, $D_{2,2}(\lambda)$, and $D_{5,1}(\lambda)$.

If

$$Q_s(x,y) = x^2 + sy^2 - 2\sqrt{2(1+s)}xy, \quad 1 \le s \le 5+4\sqrt{2},$$

then

$$\nabla Q_s(x,y) = \left(2x - 2\sqrt{2(1+s)}y,\, 2sy - 2\sqrt{2(1+s)}x\right),$$

$$\|DQ_s(x,y)\|_{D(\frac{\pi}{4})} = \sup_{0\le\theta\le\frac{\pi}{4}} \left|2x\left[\left(1-\sqrt{2(1+s)}\lambda\right)\cos\theta + \left(s\lambda - \sqrt{2(1+s)}\right)\sin\theta\right]\right|,$$

and therefore

$$\sup_{1\le s\le 5+4\sqrt{2}} \|DQ_s(x,y)\|_{D(\frac{\pi}{4})} = 2x \sup_{(s,\theta)\in C_2} |g_\lambda(s,\theta)|,$$

where

$$g_\lambda(s,\theta) = \left(1-\sqrt{2(1+s)}\lambda\right)\cos\theta + \left(s\lambda - \sqrt{2(1+s)}\right)\sin\theta$$

and $C_2 = [1, 5+4\sqrt{2}] \times [0, \frac{\pi}{4}]$. *Let us decompose the problem into several cases:*

(6) $(s,\theta) \in (1, 5+4\sqrt{2}) \times (0, \frac{\pi}{4})$.

First calculate the critical points of g_λ *over* C_2.

$$\frac{\partial g_\lambda}{\partial s}(s_0,\theta_0) = \frac{-\lambda}{\sqrt{2(1+s_0)}}\cos\theta_0 + \left(\lambda - \frac{1}{\sqrt{2(1+s_0)}}\right)\sin\theta_0,$$

$$\frac{\partial g_\lambda}{\partial \theta}(s_0,\theta_0) = \left(s_0\lambda - \sqrt{2(1+s_0)}\right)\cos\theta_0 - \left(1-\sqrt{2(1+s_0)}\lambda\right)\sin\theta_0,$$

so, if $Dg_\lambda(s_0,\theta_0) = 0$, *it follows that* $\tan\theta_0 = \frac{\lambda}{\sqrt{2(1+s_0)}\lambda - 1}$ *from the first expression and* $\tan\theta_0 = \frac{s_0\lambda - \sqrt{2(1+s_0)}}{1-\sqrt{2(1+s_0)}\lambda}$ *using the second one.*

Hence,

$$\frac{s_0\lambda - \sqrt{2(1+s_0)}}{1-\sqrt{2(1+s_0)}\lambda} = \frac{\lambda}{\sqrt{2(1+s_0)}\lambda - 1}$$

from which

$$s_0 = \frac{2-\lambda^2}{\lambda^2}.$$

Then, $\tan\theta_0 = \lambda$ *and also* $\sqrt{2}-1 < \lambda < 1$ *if we want to guarantee that* $1 < s_0 < 5+4\sqrt{2}$. *In that case,* $\sin\theta_0 = \frac{\lambda}{\sqrt{1+\lambda^2}}$ *and* $\cos\theta_0 = \frac{1}{\sqrt{1+\lambda^2}}$, *and then*

$$g_\lambda(s_0,\theta_0) = \frac{-1}{\sqrt{1+\lambda^2}} + \frac{-\lambda^2}{\sqrt{1+\lambda^2}} = -\sqrt{1+\lambda^2},$$

so

$$D_6(\lambda) := |g_\lambda(s_0,\theta_0)| = \sqrt{1+\lambda^2}.$$

(7) $s = 1$, $0 \le \theta \le \frac{\pi}{4}$.

Apply now (3.100) *with* $a = 1-2\lambda$ *and* $b = \lambda - 2$. *Using* $0 \le \lambda \le 1$, *observe that we always have* $b < 0$ *and* $b \le a$. *Additionally,* $a < (1-\sqrt{2})\,b$ *if and only if* $\lambda > \frac{5-3\sqrt{2}}{7}$. *Putting it all together, we obtain*

$$\begin{aligned}\sup_{0\le\theta\le\frac{\pi}{4}} |g_\lambda(1,\theta)| &= \begin{cases} 1-2\lambda & \textit{if } 0 \le \lambda < \frac{5-3\sqrt{2}}{7}, \\ \frac{\sqrt{2}}{2}(1+\lambda) & \textit{if } \frac{5-3\sqrt{2}}{7} \le \lambda \le 1, \end{cases} \\ &= \begin{cases} D_{7,1}(\lambda) & \textit{if } 0 \le \lambda < \frac{5-3\sqrt{2}}{7}, \\ D_{7,2}(\lambda) & \textit{if } \frac{5-3\sqrt{2}}{7} \le \lambda \le 1. \end{cases}\end{aligned}$$

(8) $s = 5+4\sqrt{2}$, $0 \le \theta \le \frac{\pi}{4}$.

In this case we have to apply (3.100) *with* $a = 1 - 2\left(1+\sqrt{2}\right)\lambda$ *and* $b = \left(5+4\sqrt{2}\right)\lambda - 2\left(1+\sqrt{2}\right)$. *It can be seen easily that* $a < 0$ *if and only if* $\lambda > \frac{\sqrt{2}-1}{2}$, $b < 0$ *if and only if* $\lambda < \frac{6-2\sqrt{2}}{7}$ *and* $a < b$ *if and only if* $\lambda > \frac{3+4\sqrt{2}}{23}$. *Combining all the above, for* $\frac{3+4\sqrt{2}}{23} < \lambda < \frac{6-2\sqrt{2}}{7}$, *we have*

$$\sup_{0\le\theta\le\frac{\pi}{4}} |g_\lambda(5+4\sqrt{2},\theta)| = \sqrt{a^2+b^2} = \sqrt{13+8\sqrt{2} - \left(56+40\sqrt{2}\right)\lambda + \left(69+48\sqrt{2}\right)\lambda^2}.$$

Also, notice that, for any $\lambda \in [0,1]$, *we are going to have* $b < -\left(1+\sqrt{2}\right)a$ *and* $a < \left(1-\sqrt{2}\right)b$. *Hence,*

$$\begin{aligned}&\sup_{0\le\theta\le\frac{\pi}{4}} |g_\lambda(5+4\sqrt{2},\theta)| \\ &= \begin{cases} \frac{\sqrt{2}}{2}\left[\left(1+2\sqrt{2}\right) - \left(3+2\sqrt{2}\right)\lambda\right] & \textit{if } 0 \le \lambda < \frac{3+4\sqrt{2}}{23}, \\ \sqrt{13+8\sqrt{2} - \left(56+40\sqrt{2}\right)\lambda + \left(69+48\sqrt{2}\right)\lambda^2} & \textit{if } \frac{3+4\sqrt{2}}{23} \le \lambda < \frac{6-2\sqrt{2}}{7}, \\ 2\left(1+\sqrt{2}\right)\lambda - 1 & \textit{if } \frac{6-2\sqrt{2}}{7} \le \lambda \le 1, \end{cases} \\ &=: \begin{cases} D_{8,1}(\lambda) & \textit{if } 0 \le \lambda < \frac{3+4\sqrt{2}}{23}, \\ D_{8,2}(\lambda) & \textit{if } \frac{3+4\sqrt{2}}{23} \le \lambda < \frac{6-2\sqrt{2}}{7}, \\ D_{8,3}(\lambda) & \textit{if } \frac{6-2\sqrt{2}}{7} \le \lambda \le 1. \end{cases}\end{aligned}$$

(9) $\theta = 0$, $1 \le s \le 5+4\sqrt{2}$.

We have

$$\begin{aligned} g_\lambda(s,0) &= 1 - \sqrt{2(1+s)}\lambda, \\ g_\lambda(1,0) &= 1-2\lambda, \\ g_\lambda(5+4\sqrt{2},0) &= 1 - 2\left(1+\sqrt{2}\right)\lambda, \\ g'_\lambda(s,0) &= -\frac{\lambda}{\sqrt{2(1+s)}} \neq 0 \textit{ for } \lambda \neq 0. \end{aligned}$$

Therefore,

$$\begin{aligned}\sup_{1\le s\le 5+4\sqrt{2}} |g_\lambda(s,0)| &= \max\left\{|1-2\lambda|,\ |1-2(1+\sqrt{2})\lambda|\right\}\\ &= \begin{cases} 1-2\lambda & \text{if } 0\le\lambda<\frac{2-\sqrt{2}}{2},\\ 2\left(1+\sqrt{2}\right)\lambda-1 & \text{if } \frac{2-\sqrt{2}}{2}\le\lambda\le 1,\end{cases}\\ &=: \begin{cases} D_{9,1}(\lambda) & \text{if } 0\le\lambda<\frac{2-\sqrt{2}}{2},\\ D_{9,2}(\lambda) & \text{if } \frac{2-\sqrt{2}}{2}\le\lambda\le 1.\end{cases}\end{aligned}$$

(10) $\theta=\frac{\pi}{4}$, $1\le s\le 5+4\sqrt{2}$.

We have

$$g_\lambda\left(s,\frac{\pi}{4}\right)=\frac{\sqrt{2}}{2}\left[1+s\lambda-\sqrt{2(1+s)}(1+\lambda)\right].$$

Then

$$\begin{aligned} g_\lambda\left(1,\frac{\pi}{4}\right) &= -\frac{\sqrt{2}}{2}(1+\lambda),\\ g_\lambda\left(5+4\sqrt{2},\frac{\pi}{4}\right) &= \frac{\sqrt{2}}{2}\left[\left(3+2\sqrt{2}\right)\lambda-\left(1+2\sqrt{2}\right)\right],\\ g'_\lambda\left(s_0,\frac{\pi}{4}\right) &= 0 \text{ if and only if } s_0=\frac{(1+\lambda)^2}{2\lambda^2}-1\end{aligned}$$

and since we need to ensure that $1<s_0<5+4\sqrt{2}$, *we need* $\frac{2\sqrt{2}-1}{7}<\lambda<1$. *In that case,*

$$g_\lambda\left(s_0,\frac{\pi}{4}\right)=-\frac{\sqrt{2}(1+3\lambda^2)}{4\lambda}.$$

Hence,

$$\begin{aligned}\sup_{1\le s\le 5+4\sqrt{2}}\left|g_\lambda\left(\frac{\pi}{4}s\right)\right| &= \begin{cases}\frac{\sqrt{2}}{2}\left[\left(1+2\sqrt{2}\right)-\left(3+2\sqrt{2}\right)\lambda\right] & \text{if } 0\le\lambda<\frac{2\sqrt{2}-1}{7},\\ \frac{\sqrt{2}(1+3\lambda^2)}{4\lambda} & \text{if } \frac{2\sqrt{2}-1}{7}\le\lambda\le 1,\end{cases}\\ &=: \begin{cases} D_{10,1}(\lambda) & \text{if } 0\le\lambda<\frac{2\sqrt{2}-1}{7},\\ D_{10,2}(\lambda) & \text{if } \frac{2\sqrt{2}-1}{7}\le\lambda\le 1.\end{cases}\end{aligned}$$

Figure 3.22 let us clarify that

$$D_6(\lambda)\le\begin{cases} D_{8,2}(\lambda) & \text{if } \sqrt{2}-1<\lambda<\frac{6-2\sqrt{2}}{7},\\ D_{8,3}(\lambda) & \text{if } \frac{6-2\sqrt{2}}{7}\le\lambda<1,\end{cases}$$

so we can rule out case (6). Also Figures 3.23 and 3.24 allow us to see that

$$\begin{aligned} D_{7,1}(\lambda) &\le D_{10,1}(\lambda) \text{ for } 0\le\lambda<\frac{5-3\sqrt{2}}{7}\\ D_{7,2}(\lambda) &\le \begin{cases} D_{10,1}(\lambda) & \text{if } \frac{5-3\sqrt{2}}{7}\le\lambda<\frac{2\sqrt{2}-1}{7},\\ D_{10,2}(\lambda) & \text{if } \frac{2\sqrt{2}-1}{7}\le\lambda\le 1,\end{cases}\end{aligned}$$

so we can rule out case (7). Since

$$D_{8,1}(\lambda)=D_{10,1}(\lambda) \text{ for } 0\le\lambda<\frac{2\sqrt{2}-1}{7}$$

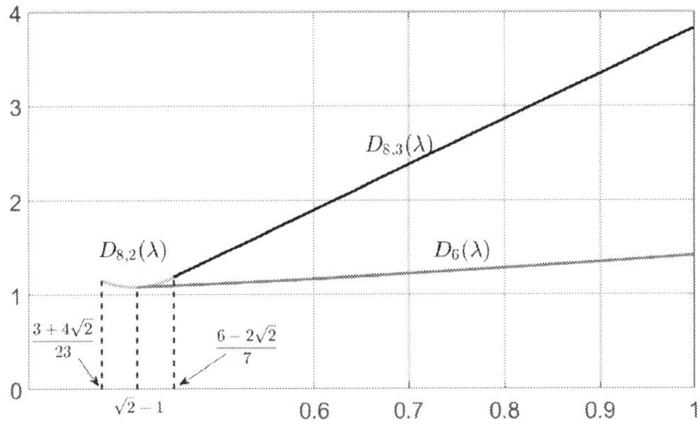

FIGURE 3.22: Graphs of the mappings $D_6(\lambda)$, $D_{8,2}(\lambda)$, and $D_{8,3}(\lambda)$.

the expression $D_{8,1}(\lambda)$ of case (8) can be ruled out as well. Since

$$D_{9,1}(\lambda) = D_{7,1}(\lambda) \quad \textit{for } 0 \le \lambda < \tfrac{5-3\sqrt{2}}{7},$$
$$D_{9,2}(\lambda) = D_{8,3}(\lambda) \quad \textit{for } \tfrac{6-2\sqrt{2}}{7} \le \lambda \le 1,$$

we can directly rule out case (9). Morerover, from Figure 3.25

$$D_{8,2}(\lambda) \le D_{10,2}(\lambda) \ \textit{for } \tfrac{3+4\sqrt{2}}{23} \le \lambda < \tfrac{6-2\sqrt{2}}{7},$$
$$D_{8,3}(\lambda) \le D_{10,2}(\lambda) \ \textit{for } \tfrac{6-2\sqrt{2}}{7} \le \lambda \le \tfrac{(4\sqrt{2}-5)\sqrt{4\sqrt{2}+7}+8-5\sqrt{2}}{7},$$

so we conclude that

$$\sup_{(s,\theta)\in C_2} |g_\lambda(s,\theta)| = \begin{cases} D_{10,1}(\lambda) & \textit{if } 0 \le \lambda < \frac{2\sqrt{2}-1}{7}, \\ D_{10,2}(\lambda) & \textit{if } \frac{2\sqrt{2}-1}{7} \le \lambda < \frac{(4\sqrt{2}-5)\sqrt{4\sqrt{2}+7}+8-5\sqrt{2}}{7}, \\ D_{8,3}(\lambda) & \textit{if } \frac{(4\sqrt{2}-5)\sqrt{4\sqrt{2}+7}+8-5\sqrt{2}}{7} \le \lambda \le 1. \end{cases}$$
$$= \begin{cases} \frac{\sqrt{2}}{2}\left[1+2\sqrt{2}-\left(3+2\sqrt{2}\right)\lambda\right] & \textit{if } 0 \le \lambda < \frac{2\sqrt{2}-1}{7}, \\ \frac{\sqrt{2}(1+3\lambda^2)}{4\lambda} & \textit{if } \frac{2\sqrt{2}-1}{7} \le \lambda < \frac{(4\sqrt{2}-5)\sqrt{4\sqrt{2}+7}+8-5\sqrt{2}}{7}, \\ 2\left(1+\sqrt{2}\right)\lambda - 1 & \textit{if } \frac{(4\sqrt{2}-5)\sqrt{4\sqrt{2}+7}+8-5\sqrt{2}}{7} \le \lambda \le 1, \end{cases}$$

and hence

$$\sup_{1\le s\le 5+4\sqrt{2}} \|DQ_s(x,y)\|_{D(\frac{\pi}{4})}$$
$$= \begin{cases} \sqrt{2}\left[\left(1+2\sqrt{2}\right)x - \left(3+2\sqrt{2}\right)y\right] & \textit{if } 0 \le y < \frac{2\sqrt{2}-1}{7}x, \\ \frac{\sqrt{2}(x^2+3y^2)}{2y} & \textit{if } \frac{2\sqrt{2}-1}{7}x \le y < \frac{(4\sqrt{2}-5)\sqrt{4\sqrt{2}+7}+8-5\sqrt{2}}{7}x, \\ 4\left(1+\sqrt{2}\right)y - 2x & \textit{if } \frac{(4\sqrt{2}-5)\sqrt{4\sqrt{2}+7}+8-5\sqrt{2}}{7}x \le y \le x. \end{cases}$$

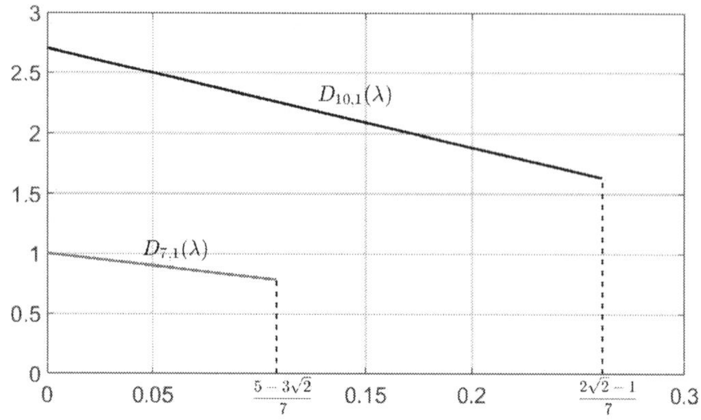

FIGURE 3.23: Graphs of the mappings $D_{7,1}(\lambda)$ and $D_{10,1}(\lambda)$.

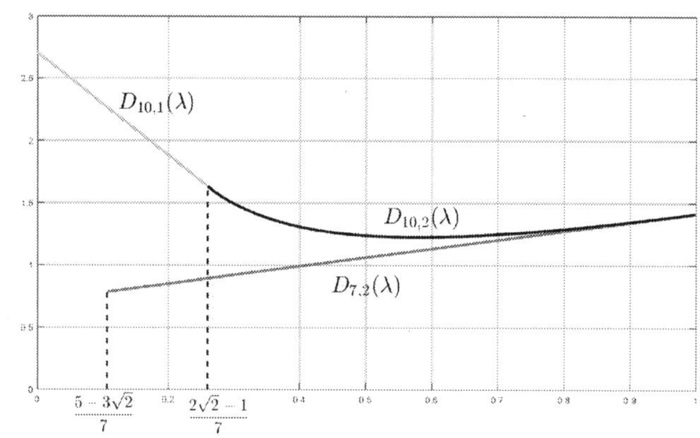

FIGURE 3.24: Graphs of the mappings $D_{7,2}(\lambda)$, $D_{10,1}(\lambda)$, and $D_{10,2}(\lambda)$.

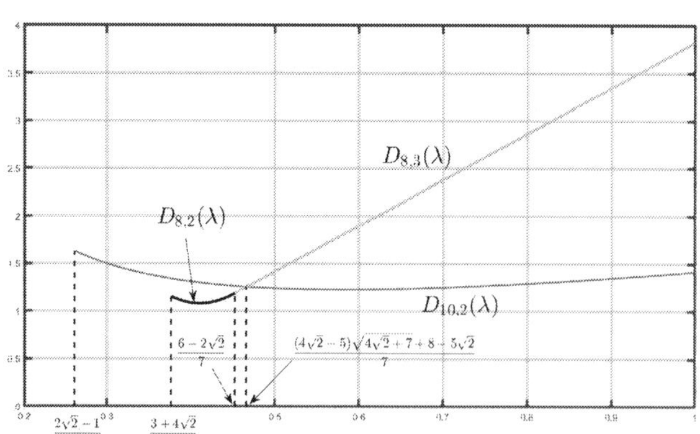

FIGURE 3.25: Graphs of the mappings $D_{8,2}(\lambda)$, $D_{8,3}(\lambda)$, and $D_{10,2}(\lambda)$.

Finally, if we compare the results obtained with P_t and Q_s, since $\frac{\sqrt{2}(1+3\lambda^2)}{4\lambda} \geq 1 + \frac{\lambda^2}{1-\lambda}$ whenever $\lambda \leq \sqrt{2}-1$, and we arrive at

$$\Phi(x,y) = \begin{cases} \sqrt{2}\left[\left(1+2\sqrt{2}\right)x - \left(3+2\sqrt{2}\right)y\right] & \text{if } 0 \leq y < \frac{2\sqrt{2}-1}{7}x, \\ \frac{\sqrt{2}(x^2+3y^2)}{2y} & \text{if } \frac{2\sqrt{2}-1}{7}x \leq y < \left(\sqrt{2}-1\right)x, \\ 2\left(x + \frac{y^2}{x-y}\right) & \text{if } \left(\sqrt{2}-1\right)x \leq y < \left(2-\sqrt{2}\right)x, \\ 4\left(1+\sqrt{2}\right)y - 2x & \text{if } \left(2-\sqrt{2}\right)x \leq y \leq x. \end{cases}$$

Plugging the polynomials from $\text{ext}(\mathsf{B}_{D(\pi/4)})$ into (3.90), the following estimates on the length of $\nabla P(x,y)$ can be obtained.

Theorem 3.2.20 (Araújo, Jiménez, Muñoz, and Seoane [2]) *For every $(x,y) \in D\left(\frac{\pi}{4}\right)$ the mapping $\mathcal{M}_{D(\pi/4),2}(x,y)$ is given by*

$$\mathcal{M}_{D(\pi/4),2}(x,y) = \begin{cases} 4\left[\left(13+8\sqrt{2}\right)x^2 + \left(69+48\sqrt{2}\right)y^2 - 2\left(28+20\sqrt{2}\right)xy\right] & \text{if } (x,y) \in A, \\ \frac{x^4}{y^2} + 4(x^2+y^2) & \text{if } (x,y) \in B, \\ \frac{\left(3x^2-2xy+3y^2\right)^2}{2(x-y)^2} & \text{if } (x,y) \in C, \end{cases}$$

where

$$A = \left\{(x,y) \in D\left(\frac{\pi}{4}\right) : 0 \leq y \leq \frac{\sqrt{2}-1}{2}x \text{ or } \left(4\sqrt{2}-5\right)x \leq y \leq x\right\},$$

$$B = \left\{(x,y) \in D\left(\frac{\pi}{4}\right) : \frac{\sqrt{2}-1}{2}x \leq y \leq \left(\sqrt{2}-1\right)x\right\},$$

$$C = \left\{(x,y) \in D\left(\frac{\pi}{4}\right) : \left(\sqrt{2}-1\right)x \leq y \leq \left(4\sqrt{2}-5\right)x\right\}.$$

Furthermore if $P \in \mathcal{P}\left(D\left(\frac{\pi}{4}\right)\right)$, then

$$\sup\left\{\|\nabla P(x,y)\|_2 : (x,y) \in D\left(\frac{\pi}{4}\right)\right\} \leq 4(13+8\sqrt{2})\|P\|_{D\left(\frac{\pi}{4}\right)},$$

with equality for the polynomials $P_1(x,y) = \pm\left(x^2 + (5+4\sqrt{2})y^2 - 2(2+2\sqrt{2})xy\right)$. In particular $M_{D(\pi/4),2} = 4(13+8\sqrt{2})$.

PROOF. *Fixing $(x,y) \in D(\pi/4)$, the Krein-Milman approach yields*

$$\mathcal{M}_{D(\pi/4),2} = \sup\left\{\|\nabla P(x,y)\|_2 : P \in \text{ext}(B_{D\left(\frac{\pi}{4}\right)})\right\}.$$

It suffices to consider the polynomials P_t and Q_s. We begin with the polynomial $P_t(x,y) = tx^2 + \left(4+t+4\sqrt{1+t}\right)y^2 - 2\left(1+t+2\sqrt{1+t}\right)xy$, $t \in [-1,1]$. Since

$$\nabla P_t(x,y) = \left(2tx - 2\left(1+t+2\sqrt{1+t}\right)y, 2\left(4+t+4\sqrt{1+t}\right)y - 2\left(1+t+2\sqrt{1+t}\right)x\right),$$

then

$$\begin{aligned} \|\nabla P_t(x,y)\|_2^2 =& 4t^2x^2 + 4\left(1+t+2\sqrt{1+t}\right)^2 y^2 - 8t\left(1+t+2\sqrt{1+t}\right)xy \\ &+ 4\left(4+t+4\sqrt{1+t}\right)^2 y^2 + 4\left(1+t+2\sqrt{1+t}\right)^2 x^2 \\ &- 8\left(4+t+4\sqrt{1+t}\right)\left(1+t+2\sqrt{1+t}\right)xy \end{aligned}$$

Introducing the change of variable $u = \sqrt{1+t} \in \left[0, \sqrt{2}\right]$, *we have*

$$\begin{aligned}\|\nabla P_u(x,y)\|_2^2 =& 8(x-y)^2u^4 + 16\left(x^2-4xy+3y^2\right)u^3 \\ &+ 8\left(x^2-10xy+13y^2\right)u^2 + 32\left(3y^2-xy\right)u + 4\left(x^2+9y^2\right),\end{aligned}$$

from which

$$\frac{\partial}{\partial u}\|\nabla P_u(x,y)\|_2^2 = 16\left(2\left(x-y\right)^2u^2 + \left(x^2-8xy+7y^2\right)u + 2y\left(3y-x\right)\right)(u+1).$$

Now the critical points of $\|\nabla P_u(x,y)\|_2^2$ *are* $u = \frac{2y}{x-y}$, $u = \frac{3y-x}{2(x-y)}$, *and* $u=-1$ *if* $x \neq y$, *and* $u = 4$ *and* $u=-1$ *if* $x=y$. *Since* $0 \leq u \leq \sqrt{2}$, *we can omit the case* $x=y$. *Thus we can write*

$$\frac{\partial}{\partial u}\|\nabla P_u(x,y)\|_2^2 = 32(x-y)^2\left(u - \frac{2y}{x-y}\right)\left(u - \frac{3y-x}{2(x-y)}\right)(u+1).$$

Let $u_1 = \frac{2y}{x-y}$ *and* $u_2 = \frac{3y-x}{2(x-y)}$. *Additionally, we have the extra conditions* $u_1 \in [0,\sqrt{2}]$ *whenever* $0 \leq y \leq \left(\sqrt{2}-1\right)x$ *and* $u_2 \in [0,\sqrt{2}]$ *whenever* $\frac{1}{3}x \leq y \leq \left(4\sqrt{2}-5\right)x$. *Taking all this into consideration, we need to compare the quantities*

$$\begin{aligned}C_1(x,y) := \|\nabla P_{u_1}(x,y)\|_2^2 = \|\nabla P_{t_1}\|_2^2 &= 4\frac{x^6-4x^5y+7x^4y^2-8x^3y^3+7x^2y^4-4xy^5+y^6}{(x-y)^4} \\ &= 4\left(x^2+y^2\right),\end{aligned}$$

for $0 \leq y \leq \left(\sqrt{2}-1\right)x$ *and* $t_1 = \frac{3y^2+2xy-x^2}{(x-y)^2}$,

$$\begin{aligned}C_2(x,y) := \|\nabla P_{u_2}(x,y)\|_2^2 &= \|\nabla P_{t_2}\|_2^2 \\ &= \frac{9x^6-30x^5y+55x^4y^2-68x^3y^3+55x^2y^4-30xy^5+9y^6}{2(x-y)^4} \\ &= \frac{\left(3x^2-2xy+3y^2\right)^2}{2(x-y)^2},\end{aligned}$$

for $\frac{1}{3}x \leq y \leq \left(4\sqrt{2}-5\right)x$ *and* $t_2 = \frac{5y^2+2xy-3x^2}{4(x-y)^2}$,

$$C_3(x,y) := \|\nabla P_{t_3=-1}\|_2^2 = 4\left(x^2+9y^2\right),$$

and

$$C_4(x,y) := \|\nabla P_{t_4=1}\|_2^2 = 4\left[\left(13+8\sqrt{2}\right)x^2 + \left(69+48\sqrt{2}\right)y^2 - 2\left(28+20\sqrt{2}\right)xy\right].$$

We focus on $Q_s = \left(1, s, -2\sqrt{2(1+s)}\right)$, $1 \leq s \leq 5+4\sqrt{2}$. *Therefore,*

$$\|\nabla Q_s(x,y)\|_2^2 = 4x^2 + 4s^2y^2 + 8(1+s)(x^2+y^2) - 8(1+s)\sqrt{2(1+s)}xy.$$

Considering the change of variable $v = \sqrt{2(1+s)} \in \left[2, 2+2\sqrt{2}\right]$, *we need to study the mapping*

$$\|\nabla Q_v(x,y)\|_2^2 = v^2\left(y^2v^2 - 4xyv + 4x^2\right) + 4\left(x^2+y^2\right).$$

Since $\|\nabla Q_v(x,y)\|_2^2 = 0$ *when* $x=y=0$, *so we will assume both* $x \neq 0$ *and* $y \neq 0$. *The critical points of* $\|\nabla Q_v(x,y)\|_2^2$ *are* $v = \frac{x}{y}$, $v = \frac{2x}{y}$ *and* $v=0$ *(but* $0 \notin [2, 2+2\sqrt{2}]$*).*

Notice that $v_1 = \frac{x}{y} \in \left[2, 2+2\sqrt{2}\right]$ *if* $\frac{\sqrt{2}-1}{2}x \le y \le \frac{1}{2}x$ *and* $v_2 = \frac{2x}{y} \in \left[2, 2+2\sqrt{2}\right]$ *if* $y \ge \left(\sqrt{2}-1\right)x$. *Consequently, we must compare the following functions as well*

$$C_5(x,y) := \|\nabla Q_{v_1}(x,y)\|_2^2 = \|\nabla Q_{s_1}(x,y)\|_2^2 = \frac{x^4}{y^2} + 4\left(x^2+y^2\right),$$

for $\frac{\sqrt{2}-1}{2}x \le y \le \frac{1}{2}x$ *and* $s_1 = \frac{x^2-2y^2}{2y^2}$,

$$C_6(x,y) := \|\nabla Q_{v_2}(x,y)\|_2^2 = \|\nabla Q_{s_2}(x,y)\|_2^2 = 4\left(x^2+y^2\right),$$

for $\left(\sqrt{2}-1\right)x \le y \le x$ *and* $s_2 = \frac{2x^2-y^2}{y^2}$, *and also*

$$C_7(x,y) := \|\nabla Q_{s_3=1}\|_2^2 = 4\left(x^2+y^2\right) + 16(x-y)^2,$$

and

$$\begin{aligned} C_8(x,y) &:= \|\nabla Q_{s_4=5+4\sqrt{2}}\|_2^2 \\ &= \left(12+8\sqrt{2}\right)\left[4x^2 + \left(12+8\sqrt{2}\right)y^2 - \left(8+8\sqrt{2}\right)xy\right] + 4\left(x^2+y^2\right) \\ &= 4\left[\left(13+8\sqrt{2}\right)x^2 + \left(69+48\sqrt{2}\right)y^2 - 2\left(28+20\sqrt{2}\right)xy\right]. \end{aligned}$$

At this point, it is helpful for the reader to consult Figures 3.26-3.28 to realize that

$$C_1(x,y), C_6(x,y) \le C_7(x,y) \le \begin{cases} C_4(x,y) & \textit{if } 0 \le y \le \frac{2-\sqrt{2}}{2}x \textit{ or } \frac{1}{2}x \le y \le x, \\ C_5(x,y) & \textit{if } \frac{\sqrt{2}-1}{2}x \le y \le \frac{1}{2}x, \end{cases}$$

$$C_3(x,y) \le \begin{cases} C_2(x,y) & \textit{if } \frac{1}{3}x \le y \le \left(4\sqrt{2}-5\right)x, \\ C_4(x,y) & \textit{if } 0 \le y \le \frac{1}{3}x \textit{ or } (4\sqrt{2}-5)x \le y \le x, \end{cases}$$

$$C_8(x,y) = C_4(x,y).$$

Hence, for $(x,y) \in D\left(\frac{\pi}{4}\right)$,

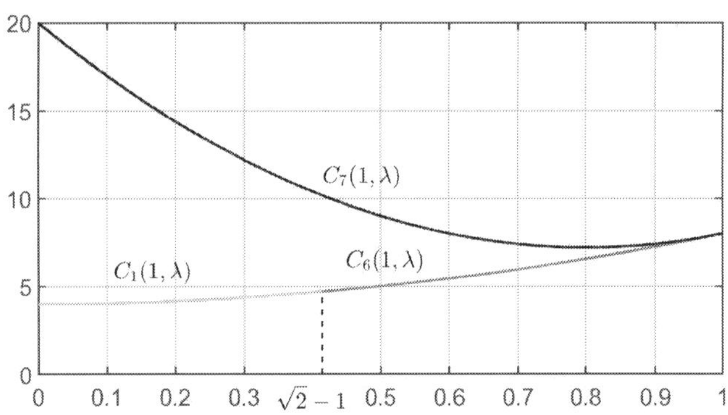

FIGURE 3.26: Graphs of the mappings $C_1(1,\lambda)$, $C_6(1,\lambda)$, and $C_7(1,\lambda)$.

$$\begin{aligned} \mathcal{M}_{D(\pi/4),2}(x,y) &= \sup\left\{\|\nabla P(x,y)\|_2 : P \in \text{ext}\left(B_{D\left(\frac{\pi}{4}\right)}\right)\right\} \\ &= \begin{cases} C_4(x,y) & \textit{if } 0 \le y \le \frac{\sqrt{2}-1}{2}x \textit{ or } (4\sqrt{2}-5)x \le y \le x, \\ C_5(x,y) & \textit{if } \frac{\sqrt{2}-1}{2}x \le y \le (\sqrt{2}-1)x, \\ C_2(x,y) & \textit{if } (\sqrt{2}-1)x \le y \le (4\sqrt{2}-5)x. \end{cases} \end{aligned}$$

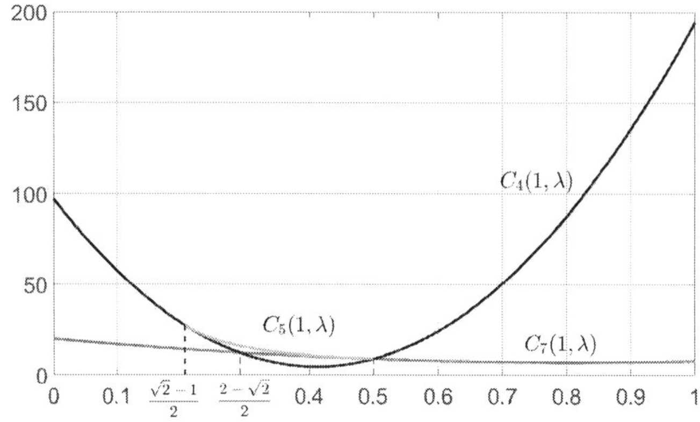

FIGURE 3.27: Graphs of the mappings $C_4(1,\lambda)$, $C_5(1,\lambda)$, and $C_7(1,\lambda)$.

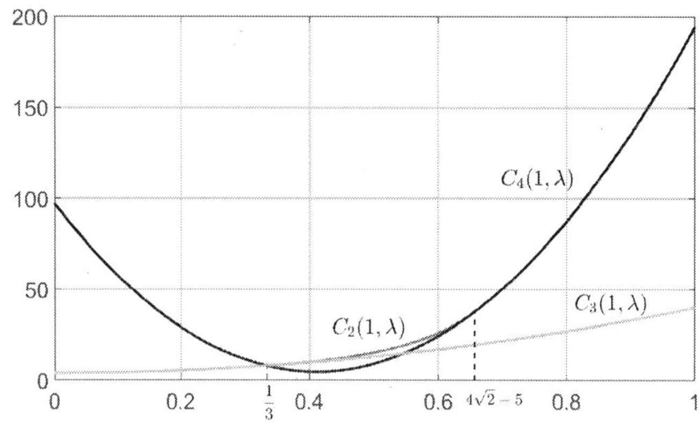

FIGURE 3.28: Graphs of the mappings $C_2(1,\lambda)$, $C_3(1,\lambda)$, and $C_4(1,\lambda)$.

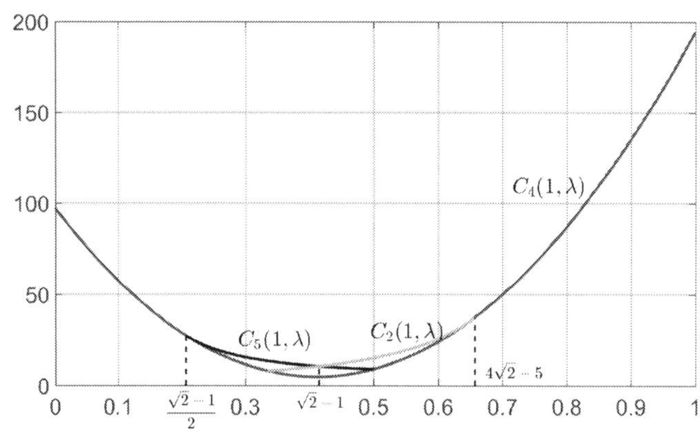

FIGURE 3.29: Graphs of the mappings $C_2(1,\lambda)$, $C_4(1,\lambda)$, and $C_5(1,\lambda)$.

The previous step is better visualized in Figure 3.29.

The Markov constant $M_{D(\pi/4),2}$ *can be obtained by maximizing* $\mathcal{M}_{D(\pi/4),2}(x,y)$ *over* $D(\pi/4)$. *We obtain* $M_{D(\pi/4),2} = 4(13+8\sqrt{2})$. *We leave the details to the reader.*

Bernstein and Markov inequalities in $\mathcal{P}(^2D(\pi/2))$

To apply (3.89) and (3.90), it is necessary to characterize $\text{ext}(\mathsf{B}(\pi/2))$. This characterization was established in [81] (see also Theorem 1.4.7). Specifically, we have that the set of extreme points of $\mathsf{B}_{D(\frac{\pi}{2})}$ is given by

$$\text{ext}\left(\mathsf{B}_{D(\frac{\pi}{2})}\right) = \{\pm P_t,\ \pm Q_t\ :\ -1 \leq t \leq 1\} \cup \{\pm(1,1,0)\},$$

where

$$P_t = (t, 1, -2\sqrt{2(1+t)}),$$
$$Q_t = (1, t, -2\sqrt{2(1+t)}).$$

Theorem 3.2.21 *For every $(x,y) \in D(\frac{\pi}{2})$ we have that*

$$\mathcal{M}_{D(\pi/2)}(x,y) = \begin{cases} 2(2x-y) & \text{if } 0 \leq y < \frac{x}{2}, \\ 2\left(y + \frac{x^2}{2y}\right) & \text{if } \frac{x}{2} \leq y < x, \\ 2\left(x + \frac{y^2}{2x}\right) & \text{if } x \leq y < 2x, \\ 2(2y-x) & \text{if } y \geq 2x. \end{cases}$$

Moreover, if $P \in \mathcal{P}\left(^2D\left(\frac{\pi}{2}\right)\right)$ then

$$\sup\{\|DP(x,y)\|_{D(\pi/2)} : (x,y) \in D(\pi/2)\} \leq 4\|P\|_{D(\frac{\pi}{2})} \tag{3.101}$$

and 4 is the best estimate in (3.101) *since equality is achieved for*

$$\pm P_1(x,y) = \pm Q_1(x,y) = \pm(x^2 + y^2 - 4xy).$$

In particular $M_{D(\pi/2)} = 4$.

PROOF. *According to* (3.89) *we have*

$$\mathcal{M}_{D(\pi/2)}(x,y) = \sup\{\|DP(x,y)\|_{D(\frac{\pi}{2})} : P \in \text{ext}(B_{D(\frac{\pi}{2})})\}$$

for each $(x,y) \in D(\pi/2)$. The case $xy = 0$ is trivial, so we may assume $x \neq 0$ and $y \neq 0$.
Since $Q_t(x,y) = x^2 + ty^2 - 2\sqrt{2(1+t)}xy$, its gradient is given by

$$\nabla Q_t(x,y) = \left(2x - 2\sqrt{2(1+t)}y, 2ty - 2\sqrt{2(1+t)}x\right).$$

Thus,

$$\|DQ_t(x,y)\|_{D(\frac{\pi}{2})} = \sup_{(h,k)\in D(\frac{\pi}{2})} \left|(2x - 2\sqrt{2(1+t)}y)h + (2ty - 2\sqrt{2(1+t)}x)k\right|.$$

By restricting our attention to extreme points of $D(\frac{\pi}{2})$ (excluding $(0,0)$), we set $(h,k) = (\cos\theta, \sin\theta)$ with $0 \leq \theta \leq \frac{\pi}{2}$ and define $\lambda = \frac{y}{x}$. We now compute

$$\sup_{-1\leq t\leq 1} \|\nabla Q_t(x,y)\|_{D(\frac{\pi}{2})} = 2x \sup_{(t,\theta)\in C} \left|\left(1 - \sqrt{2(1+t)}\lambda\right)\cos\theta + \left(t\lambda - \sqrt{2(1+t)}\right)\sin\theta\right|,$$

where $C = [-1,1] \times [0, \frac{\pi}{2}]$.

Defining

$$f_\lambda(t,\theta) = \left(1 - \sqrt{2(1+t)}\lambda\right)\cos\theta + \left(t\lambda - \sqrt{2(1+t)}\right)\sin\theta,$$

we analyze different cases based on values of λ*, solving critical point equations and comparing supremum values across boundary conditions.*

After performing a detailed case analysis and simplifications, we obtain

$$\sup_{-1\le t\le 1} \|\nabla Q_t(x,y)\|_{D(\frac{\pi}{2})} = \Psi_1(x,y),$$

where

$$\Psi_1(x,y) = \begin{cases} 2(2x-y), & 0 \le y < \frac{x}{2}, \\ 2\left(y + \frac{x^2}{2y}\right), & \frac{x}{2} \le y < \frac{1+\sqrt{2}}{2}x, \\ 2\sqrt{(2x-y)^2 + (2y-x)^2}, & \frac{1+\sqrt{2}}{2}x \le y < 2x, \\ 2(2y-x), & y \ge 2x. \end{cases}$$

Notice that $P_t(x,y) = Q_t(y,x)$ *and therefore*

$$\sup_{-1\le t\le 1} \|\nabla P_t(x,y)\|_{D(\frac{\pi}{2})} = \Psi_2(x,y) = \Psi_1(y,x).$$

Thus, we conclude that

$$\mathcal{M}_{D(\pi/2)}(x,y) = \max\left\{\Psi_1(x,y), \Psi_2(x,y)\right\}.$$

This yields

$$\mathcal{M}_{D(\pi/2)}(x,y) = \begin{cases} 2(2x-y), & 0 \le y < \frac{x}{2}, \\ 2\left(y + \frac{x^2}{2y}\right), & \frac{x}{2} \le y < x, \\ 2\left(x + \frac{y^2}{2x}\right), & x \le y < 2x, \\ 2(2y-x), & y \ge 2x. \end{cases}$$

To prove the Markov-type estimate (3.101) *we just need to take into consideration that*

$$M_{D(\pi/2)} = \sup\{\mathcal{M}_{D(\pi/2)}(x,y) : (x,y) \in D(\pi/2)\} = 4.$$

Furthermore, the previous supremum is attained at $(1,0)$ *and* $(0,1)$*. From the above proof it can be deduced easily that the extreme polynomials corresponding to* $(1,0)$ *and* $(0,1)$ *are*

$$\pm P_1(x,y) = \pm Q_1(x,y) = \pm(= x^2 + y^2 - 4xy).$$

We conclude this section by presenting sharp Bernstein and Markov estimates on the length of the gradient of polynomials in $\mathcal{P}(^2D(\pi/2))$.

Theorem 3.2.22 *For every* $(x,y) \in D\left(\frac{\pi}{2}\right)$ *we have*

$$\mathcal{M}_{D(\pi/2),2}(x,y) = \begin{cases} \sqrt{16(x-y)^2 + 4(x^2+y^2)} & \text{if } 0 \le y \le \frac{x}{2}, \\ \sqrt{\frac{x^4}{y^2} + 4(x^2+y^2)} & \text{if } 0 < \frac{x}{2} < y \le x, \\ \sqrt{\frac{y^4}{x^2} + 4(x^2+y^2)} & \text{if } 0 < x < y \le 2x, \\ \sqrt{16(y-x)^2 + 4(x^2+y^2)} & \text{if } 2x < y \le 1. \end{cases}$$

Furthermore, for every $P \in \mathcal{P}(^2D(\pi/2))$ *we have*

$$\sup\{\|\nabla P(x,y)\|_2 : (x,y) \in D(\pi/2)\} \le 2\sqrt{5}\|P\|_{D(\frac{\pi}{2})}, \tag{3.102}$$

with equality attained for $\pm P_1(x,y) = \pm Q_1(x,y) = \pm(x^2+y^2-4xy)$*. In particular* $M_{D(\pi/2)} = 2\sqrt{5}$.

PROOF. *By virtue of* (3.90), *for every* $(x, y) \in D(\pi/2)$, *we have*

$$\mathcal{M}_{D(\pi/2),2}(x,y) = \sup\left\{\|\nabla P(x,y)\|_2 \,:\, P \in \text{ext}\left(B_{D(\frac{\pi}{2})}\right)\right\}.$$

For $P = (1,1,0)$, *we compute* $\|\nabla P(x,y)\|_2^2 = 4(x^2+y^2)$.

For other polynomials, the case $xy = 0$ *is trivial and left to the reader. Assume* $x \neq 0$ *and* $y \neq 0$. *Consider first* $P_t = (t, 1, -2\sqrt{2(1+t)})$. *Then,*

$$\|\nabla P_t(x,y)\|_2^2 = 4t^2x^2 + 8(1+t)y^2 - 8t\sqrt{2(1+t)}xy + 4y^2 + 8(1+t)x^2 - 8\sqrt{2(1+t)}xy.$$

Changing variables with $u = \sqrt{2(1+t)}$ *(so* $u \in [0,2]$*), we define*

$$g_{x,y}(u) := \|\nabla P_u(x,y)\|_2^2 = x^2u^4 - 4xyu^3 + 4y^2u^2 + 4(x^2+y^2).$$

The critical points of $g_{x,y}$ *occur at* $u = 0$, $u = \frac{2y}{x}$, *and* $u = \frac{y}{x}$. *Since* $g''_{x,y}(\frac{2y}{x}) > 0$, *this point is a local minimum and is therefore excluded. Evaluating at these points, we have*

$$\begin{aligned} g_{x,y}(0) &= 4(x^2+y^2), \\ g_{x,y}\left(\frac{y}{x}\right) &= \frac{y^4}{x^2} + 4(x^2+y^2), \\ g_{x,y}(2) &= 16(x-y)^2 + 4(x^2+y^2). \end{aligned}$$

Thus,

$$\sup\left\{\|\nabla P_t(x,y)\|_2^2 \,:\, -1 \le t \le 1\right\} = \begin{cases} \max\left\{g_{x,y}\left(\frac{y}{x}\right), g_{x,y}(2)\right\} & \text{if } 0 \le \frac{y}{x} \le 2, \\ g_{x,y}(2) & \text{otherwise.} \end{cases}$$

Since $g_{x,y}(2) \le g_{x,y}\left(\frac{y}{x}\right)$ *for* $0 \le \frac{y}{x} \le 2$, *we conclude that*

$$\sup\left\{\|\nabla P_t(x,y)\|_2^2 \,:\, -1 \le t \le 1\right\} = \begin{cases} \frac{y^4}{x^2} + 4(x^2+y^2) & \text{if } 0 \le \frac{y}{x} \le 2, \\ 16(x-y)^2 + 4(x^2+y^2) & \text{otherwise.} \end{cases}$$

By symmetry, since $Q_t(x,y) = P_t(y,x)$, *we obtain*

$$\sup\left\{\|\nabla Q_t(x,y)\|_2^2 \,:\, -1 \le t \le 1\right\} = \begin{cases} \frac{x^4}{y^2} + 4(x^2+y^2) & \text{if } 0 \le \frac{x}{y} \le 2, \\ 16(x-y)^2 + 4(x^2+y^2) & \text{otherwise.} \end{cases}$$

Combining results, we conclude

$$\sup_{-1 \le t \le 1}\left\{\|\nabla P_t(x,y)\|_2^2, \|\nabla Q_t(x,y)\|_2^2\right\}$$

as

$$\begin{cases} \max\left\{\frac{y^4}{x^2} + 4(x^2+y^2), 16(x-y)^2 + 4(x^2+y^2)\right\} & \text{if } 0 \le \frac{y}{x} \le \frac{1}{2}, \\ \max\left\{\frac{y^4}{x^2} + 4(x^2+y^2), \frac{x^4}{y^2} + 4(x^2+y^2)\right\} & \text{if } \frac{1}{2} \le \frac{y}{x} \le 2, \\ \max\left\{\frac{x^4}{y^2} + 4(x^2+y^2), 16(x-y)^2 + 4(x^2+y^2)\right\} & \text{if } 2 \le \frac{y}{x}. \end{cases}$$

Consequently,

$$\mathcal{M}_{D(\pi/2),2}(x,y) = \begin{cases} \sqrt{16(x-y)^2 + 4(x^2+y^2)} & \text{if } 0 \le y \le \frac{x}{2}, \\ \sqrt{\frac{x^4}{y^2} + 4(x^2+y^2)} & \text{if } \frac{x}{2} < y \le x, \\ \sqrt{\frac{y^4}{x^2} + 4(x^2+y^2)} & \text{if } x < y \le 2x, \\ \sqrt{16(y-x)^2 + 4(x^2+y^2)} & \text{if } 2x < y \le 1. \end{cases}$$

To prove the Markov-type inequality (3.102) *it suffices to check that*

$$\sup_{(x,y)\in D\left(\frac{\pi}{2}\right)} \mathcal{M}_{D(\pi/2)}(x,y) = 2\sqrt{5},$$

being the maximum attained at the points $(1,0)$ *and* $(0,1)$. *An inspection of the above argument reveals that the extreme polynomials corresponding to* $(1,0)$ *and* $(0,1)$ *in the Bernstein function* $\mathcal{M}_{D(\pi/2),2}$ *are* $\pm P_1 = \pm Q_1$, *or in other words* $\pm(x^2+y^2-4xy)$.

3.3 Polarization constants

The concept of the polarization constant arises as a consequence of a well-known result by Martin from the 1930s, which provides an estimate (generally optimal) for the norm of a polynomial in comparison with the norm of its polar form. As is customary, if L is a multilinear mapping, $\widehat{L}$ denotes its associated homogeneous polynomial. Let us recall the relationship between the norms of L and $\widehat{L}$.

Theorem 3.3.1 (Martin [73]) *If* X *is a real or complex Banach space,* $P\in\mathcal{P}(^nX)$, *and* $L\in\mathcal{L}^s(^nX)$ *satisfies* $\widehat{L}=P$ *(L is called the polar of P), then*

$$\|L\| \le \frac{n^n}{n!}\|P\|.$$

Equality is attained for $X=\ell_1^n$ *and the polynomial* $\Phi_n(x_1,\ldots,x_n):=x_1\cdots x_n$. *This polynomial is sometimes referred to as the Nachbin Polynomial.*

The previous results give a positive answer to problem 73 by Mazur and Orlicz of the Scottish Book (see [74]). Although the constant $\frac{n^n}{n!}$ in Martin's theorem is generally optimal, it can be improved for certain spaces, leading to the following definition.

Definition 3.3.1 (Polarization Constants) *Given a Banach space* X *over* $\mathbb{K}$, *we define the* n*-th polarization constant of* X *as*

$$\mathbb{K}(n;X) := \inf\{K>0 : \|L\|\le K\|P\| : P\in\mathcal{P}(^nX) \text{ and } \widehat{L}=P\}.$$

Furthermore, we say that a polynomial $P\in\mathcal{P}(^nX)$ *is extremal if, together with its polar* $L\in\mathcal{L}^s(^nX)$, *it satisfies*

$$\|L\| = \mathbb{K}(n;X)\|P\|.$$

Evidently, from Theorem 3.3.1, we deduce that $\mathbb{K}(n,\ell_1^n)=\frac{n^n}{n!}$ and that Φ_n is an extremal polynomial in ℓ_1^n.

A related concept concerning polarization constant arises from the following result:

Theorem 3.3.2 (Harris [60]) *Let* E *be a complex Banach space, and let* $n_1,\ldots,n_k$ *be natural numbers. If* $n=n_1+\ldots+n_k$ *and* $L\in\mathcal{L}^s(^nE)$, *then*

$$\sup\{|L(x_1^{n_1},\ldots,x_k^{n_k})| : \|x_i\|=1,\ 1\le i\le k\} \le \frac{n_1!\cdots n_k!\ n^n}{n_1^{n_1}\cdots n_k^{n_k} n!}\|\widehat{L}\|,$$

where $L(x_1^{n_1},\ldots,x_k^{n_k})$ *denotes* $L(\underbrace{x_1,\ldots,x_1}_{n_1 \text{ times}},\ldots,\underbrace{x_k,\ldots,x_k}_{n_k \text{ times}})$.

The smallest constant in the above inequality is referred to as the generalized polarization constant.

Definition 3.3.2 (Generalized Polarization Constants) *Consider a Banach space E and let $n, n_1, \ldots, n_k \in \mathbb{N}$ with $n = n_1 + \ldots + n_k$. The generalized polarization constant $\mathbb{K}(n_1, \ldots, n_k; E)$ is defined as*

$$\mathbb{K}(n_1, \ldots, n_k; E) := \inf\{c > 0 : |L(x_1^{n_1}, \ldots, x_k^{n_k})| \le c\|\widehat{L}\|,\ L \in \mathcal{L}^s(^nE),\ \|x_i\| = 1\}.$$

The value of the polarization constants in any space ranges between 1 and $\frac{n^n}{n!}$, that is,

$$1 \le \mathbb{K}(n, X) \le \frac{n^n}{n!}$$

for any real or complex Banach space X. The question of which spaces attain these extreme values has generated significant interest, leading to numerous results that provide definitive answers in specific cases. The computation of $\mathbb{K}(n, \ell_p)$ and $\mathbb{K}(n, L_p)$ has been the subject of extensive research. Naturally, the study of general properties of polarization constants is also an area of frequent inquiry that has attracted considerable effort and interest. In the following sections, we present a selection of key results from each of these topics, offering the reader a clear perspective on the significance and relevance of these issues. We begin by examining the case where $\mathbb{K}(n; X) = \frac{n^n}{n!}$.

The Case Where $\mathbb{K}(n, X) = \frac{n^n}{n!}$

We anticipate that this case is closely related to ℓ_1 and the extremal polynomial Φ_n from Theorem 3.3.1. The following series of results substantiates this claim while also presenting general findings on extremal polynomials.

Theorem 3.3.3 (Sarantopoulos [98]) *If X is a complex Banach space of dimension n, then $\mathbb{C}(n, X) = \dfrac{n^n}{n!}$ if and only if X is isometrically isomorphic to ℓ_1^n. Moreover, if X is a complex Banach space and $\mathbb{C}(n, X) = \dfrac{n^n}{n!}$, then $\ell_1^n \ll X$.*

It is worth clarifying that in Theorem 3.3.3, the notation $\ell_1^n \ll X$ means that ℓ_1^n is finitely represented in X. In general, we say that a space Y is finitely represented in X if for every finite-dimensional subspace E of Y of dimension n and for every $\varepsilon > 0$, there exists a subspace F of X of dimension n such that $d(E, F) < 1 + \varepsilon$, where $d(E, F)$ denotes the Banach-Mazur distance between E and F. Recall that

$$d(E, F) = \inf\{\|T\|\|T^{-1}\| : T : E \to F \text{ is an isomorphism}\}.$$

For extremal polynomials in general, we have:

Theorem 3.3.4 (Sarantopoulos [98]) *Let X be a complex Banach space, and let $P \in \mathcal{P}(^nX)$ and $L \in \mathcal{L}^s(^nX)$ be the polar of P. Suppose that $\|L\| = \mathbb{C}(n, X)\|P\|$. Then the following conditions are equivalent*

(a) L attains its norm at $(x_1, \ldots, x_n) \in (\mathsf{B}_X)^n$, where B_X denotes the closed unit ball of X.

(b) The following two conditions hold simultaneously:

1. *P attains its norm at the points* $\dfrac{\left(e^{i\theta_1}x_1 + \cdots + e^{i\theta_n}x_n\right)}{n}$ *for any* $\theta_1, \ldots, \theta_n \in \mathbb{R}$.
2. $L\left(x_1^{m_1}, \ldots, x_n^{m_n}\right) = 0$ *for every* n*-tuple* $m = (m_1, \ldots, m_n)$ *of non-negative integers, where at least one of them is greater than 1 and* $|m| = n$.

(c) The following two conditions hold simultaneously:

1. *P attains its norm at the point* $\dfrac{\left(e^{i\theta_1}x_1 + \cdots + e^{i\theta_n}x_n\right)}{n}$ *for some choice of* $\theta_1, \ldots, \theta_n \in \mathbb{R}$.
2. $L\left(x_1^{m_1}, \ldots, x_n^{m_n}\right) = 0$ *for every* n*-tuple* $m = (m_1, \ldots, m_n)$ *of non-negative integers, where at least one of them is greater than 1 and* $|m| = n$.

Finally, by combining Theorems 3.3.3 and 3.3.4, we obtain the following result.

Theorem 3.3.5 (Sarantopoulos [98]) *Let* X *be a complex Banach space such that* $\mathbb{C}(n, X) = \dfrac{n^n}{n!}$, *and let* $P \in \mathcal{P}(^nX)$ *and* $L \in \mathcal{L}^s(^nX)$ *be the polar of* P. *Suppose that* P *is an extremal polynomial that attains its norm at the point* $(x_1, \ldots, x_n) \in (\mathsf{S}_X)^n$, *where* S_X *denotes the unit sphere of* X. *Then, there exists* $c \in \mathbb{C}$ *such that*

$$P\left(\sum_{j=1}^{n} \alpha_i x_i\right) = c\alpha_1 \cdots \alpha_n$$

for all $\alpha_1, \ldots, \alpha_n \in \mathbb{C}$.

It should be noted that the real Banach space versions of Theorems 3.3.3 and 3.3.4 have been studied and hold with some adaptations. The real case was addressed by Kirwan, Sarantopoulos, and Tonge in [65], and is based on the following result.

Theorem 3.3.6 (Kirwan, Sarantopoulos, and Tonge [65]) *Let* P *be an extremal* n*-homogeneous polynomial on a real Banach space* X *with* $\mathbb{R}(n; X) = \frac{n^n}{n!}$, *whose polar* L *satisfies*

$$\|L\| = L\left(x_1, \ldots, x_n\right)$$

for $x_1, \ldots, x_n \in X$ *with* $\|x_i\| = 1$ *for all* $i = 1, \ldots, n$. *Then, for all* $a_1, \ldots, a_n \in \mathbb{R}$, *we have*

$$\left\|\sum_{i=1}^{n} a_i x_i\right\| = \sum_{i=1}^{n} |a_i| .$$

Therefore, span $\{x_1, \ldots, x_n\} \subseteq E$ *is isometrically isomorphic to* ℓ_1^n, *and the isomorphism maps* $\{x_1, \ldots, x_n\}$ *to the canonical basis of* ℓ_1^n.

The Case Where $\mathbb{K}(n; X) = 1$

If H is a Hilbert space over $\mathbb{K}$, a classical result dating back to Banach [7] establishes that $\mathbb{K}(n; H) = 1$ for a positive integer n. The proofs for the real and complex cases are quite different. For the complex case, we recommend the proof given in [38]. For the proof of the real case, one may refer to [17].

In the real case, it is also known that if X is a real Banach space such that $\mathbb{R}(n; X) = 1$ for every $n \in \mathbb{N}$, then X must be a Hilbert space. A further refinement leads to the following result.

Theorem 3.3.7 (Benítez and Sarantopoulos [9]) *A real Banach space X is a Hilbert space if and only if $\mathbb{R}(2;X)=1$.*

However, the above characterization does not hold when $\mathbb{C}(n;E)=1$ for a complex Banach space X. In fact, if $(H,\|\cdot\|)$ is a complex Hilbert space, then the space $H\oplus_\infty\mathbb{C}$ is a non-Euclidean space for which it can be shown (see [96]) that $\mathbb{C}(n;H\oplus_\infty\mathbb{C})=1$ for all $n\in\mathbb{N}$. Another example is the complex ℓ_∞^2. Observe, however, that $\mathbb{C}(2;\ell_\infty^3)>1$ (see [104]).

Note that $H\oplus_\infty\mathbb{C}$ represents the space $H\times\mathbb{C}$ equipped with the norm

$$\|(x,\lambda)\|_\infty=\max\{\|x\|,|\lambda|\},$$

for all $x\in H$ and $\lambda\in\mathbb{C}$.

3.3.0.1 Polarization Constants of ℓ_p and L_p Spaces

The calculation of the polarization constants of specific spaces, such as the real and complex versions of ℓ_p and L^p $(1\le p\le\infty)$, has been the subject of numerous publications. In [95] and [98], the values of $\mathbb{K}(n;\ell_p)$ are provided in some cases. As an illustration, we present the following values and estimates.

Theorem 3.3.8 (Sarantopoulos [95]) *If X is an infinite-dimensional complex space $L_p(\mu)$ or $X=\ell_p(\mathbb{C})$, then*

1. *If $n=2^m$ and $1\le p\le\infty$, then $\mathbb{C}(n;X)\le\left(\frac{n^n}{n!}\right)^{\frac{|p-2|}{p}}$.*
2. *If $1\le p\le\frac{n}{n-1}$, then $\mathbb{C}(n;X)=\frac{n^{\frac{n}{p}}}{n!}$.*
3. *If $n\le p\le\infty$ and q is the conjugate exponent of p, i.e., $\frac{1}{p}+\frac{1}{q}=1$ with $q=1$ if $p=\infty$, then $\mathbb{C}(n;X)\le\frac{n^{\frac{n}{q}}}{n!}$.*

Special attention should be given to the studies on the polarization constants of spaces of type ℓ_∞. Both the real and complex cases present highly interesting open problems. In this regard, the following result provides the best-known estimate for the polarization constants of the complex version of ℓ_∞. A modern and accessible proof can be found in [38].

Theorem 3.3.9 (Harris [60])

$$\mathbb{C}(n;\ell_\infty)\le\frac{n^{\frac{n}{2}}(n+1)^{\frac{n+1}{2}}}{2^n n!}. \tag{3.103}$$

3.3.0.2 General Results on Polarization Constants

Significant effort has been devoted to establishing general properties of polarization constants. As an example of the properties we refer to, we mention the relationship between the polarization constants of a space and those of its subspaces, or the connection existing between subspaces and quotients. These and other aspects related to polarization constants are addressed with considerable generality in [38], where proofs (mostly highly non-trivial) of the following specific properties can be found.

Proposition 3.3.10 (Invariance under Isometries) *If X and Y are isometric Banach spaces, then*

$$\mathbb{K}(n,X)=\mathbb{K}(n,Y).$$

Proposition 3.3.11 *Let X be a Banach space and let Y be a closed subspace of X such that there exists a continuous projection $\pi : X \to Y$, then*

$$\mathbb{K}(n, Y) \leq \|\pi\|^n \mathbb{K}(n, X).$$

Proposition 3.3.12 *If Y is a closed subspace of a Banach space X, then*

$$\mathbb{K}(n, X/Y) \leq \mathbb{K}(n, X).$$

Proposition 3.3.13 *If X and Y are isomorphic Banach spaces, then*

$$\mathbb{K}(n, Y) \leq d(X, Y)^n \mathbb{K}(n, X).$$

We recall the following definition.

Definition 3.3.3 *Let Y be a subspace of a Banach space X. The pair $\{Y, X\}$ is said to have the Polynomial Extension Property if for each $n \in \mathbb{N}$ and for each $P \in \mathcal{P}(^nY)$, there exists $Q \in \mathcal{P}(^nX)$ such that*

$$Q|_Y = P \text{ and } \|Q\|_X = \|P\|_Y.$$

For pairs with the Polynomial Extension Property, we have:

Proposition 3.3.14 *If $\{Y, X\}$ has the Polynomial Extension Property, then*

$$\mathbb{K}(n, Y) \leq \mathbb{K}(n, X).$$

It can be proven, although not easily (see [38]), that for every Banach space X, the pair $\{X, X^{**}\}$ has the Polynomial Extension Property, leading to:

Proposition 3.3.15 *For every Banach space X, we have that $\mathbb{K}(n, X) \leq \mathbb{K}(n, X^{**})$.*

The vast amount of literature on this topic and the many interesting open questions regarding polarization constants make this problem a fertile ground for research.

All the results mentioned so far concern a notion of polarization constant based on the norm of a polynomial or a symmetric multilinear form, defined as the supremum norm over the unit ball of a Banach space. In the next section, we introduce a notion of polarization constant based on a polynomial norm defined on a convex body that is not the unit ball of any Banach space. Specifically, we consider unbalanced convex bodies.

3.3.1 Polarization constants in spaces of polynomials on unbalanced convex bodies

In this section, we will focus on calculating polarization constants in spaces of polynomials using the Krein-Milman approach. Let $K \subset \mathbb{R}^n$ be a convex body and assume that $\mathcal{P}(^nK)$ and $\mathcal{L}^s(^nK)$ stand for the spaces of n-homogeneous polynomials and symmetric n-linear forms on $\mathbb{R}^n$ endowed with the norms

$$\|P\|_K = \sup\{|P(x)| : x \in K\},$$
$$\|L\|_K = \sup\{|L(x_1, \ldots, x_n)| : x_1, \ldots, x_n \in K\},$$

for all $P \in \mathcal{P}(^nK)$ and $L \in \mathcal{L}^s(^nK)$. If K were the unit ball of a Banach space we would have

$$\|L\|_K \leq \frac{n^n}{n!} \|\widehat{L}\|_K \tag{3.104}$$

for all $L \in \mathcal{L}^s(^nK)$. However, (3.104) might not hold if K is an unbalanced convex body, as we will see later.

Definition 3.3.4 *We define the nth polarization constant of the convex body $K \subset \mathbb{R}^2$ as the best constant $\mathbb{R}(n, K)$ in the inequality*

$$\|L\|_K \leq \mathbb{R}(n, K)\|\widehat{L}\|_K,$$

for all $L \in \mathcal{L}^s(^nK)$.

In this section, we will calculate $\mathbb{R}(2, K)$ for several choices of $K \subset \mathbb{R}^2$ using the results of Section 3.2.3. In fact, the constants $\mathbb{R}(2, K)$ are closely related to the Markov constants M_K computed in Section 3.2.3. Indeed, if P is a quadratic form on $\mathbb{R}^n$ and L is its polar, then for all $x, y \in \mathbb{R}^n$, it is easy to check that

$$DP(x)(y) = 2L(x, y),$$

where $DP : \mathbb{R}^n \to \mathbb{R}^n$ is the first differential of P. The latter shows in particular that $M_K = 2\mathbb{R}(2, K)$ from which we deduce the following results.

Theorem 3.3.16 (Muñoz, Révesz, and Seoane [82]**)** *If $P \in \mathcal{P}(^2\Delta)$ and L is the polar of P, then*

$$\|L\|_\Delta \leq 3\|P\|_\Delta.$$

Moreover, equality is attained for the polynomial $P(x, y) = x^2 - 6xy + y^2$. Thus, $\mathbb{R}(2, \Delta) = 3$.

Theorem 3.3.16 illustrates how the geometric study of polynomial spaces yields highly non-trivial examples and counterexamples.

Other noteworthy examples are shown below. These involve calculating the polarization constants of $\mathcal{P}(^2\square)$, $\mathcal{P}(^2D(\frac{\pi}{4}))$, and $\mathcal{P}(^2D(\frac{\pi}{2}))$.

Theorem 3.3.17 (Gámez, Muñoz, Sánchez, and Seoane [45]**)** *If $P \in \mathcal{P}(^2\square)$ and $L \in \mathcal{L}^s(^2\square)$ is the polar of P, then*

$$\|L\|_\square \leq \frac{3}{2}\|P\|_\square. \tag{3.105}$$

Moreover, the constant $\frac{3}{2}$ is optimal in (3.105) since equality is attained for the polynomials $P(x, y) = \pm(x^2 - 3xy + y^2)$. Thus, $\mathbb{R}(2, \square) = \frac{3}{2}$.

Theorem 3.3.18 (Araújo, Jiménez, Muñoz, and Seoane [2]**)** *Let $P \in \mathcal{P}\left(^2D\left(\frac{\pi}{4}\right)\right)$ and suppose that $L \in \mathcal{L}^s\left(^2D\left(\frac{\pi}{4}\right)\right)$ is the polar of P. Then*

$$\|L\|_{D(\frac{\pi}{4})} \leq \left(2 + \frac{\sqrt{2}}{2}\right)\|P\|_{D(\frac{\pi}{4})}.$$

Moreover, equality is attained for the polynomials

$$P(x, y) = x^2 + \left(5 + 4\sqrt{2}\right)y^2 - \left(4 + 4\sqrt{2}\right)xy.$$

Therefore, $\mathbb{R}(2, D(\pi/4)) = 2 + \frac{\sqrt{2}}{2}$.

Theorem 3.3.19 *Let $P \in \mathcal{P}\left(^2D\left(\frac{\pi}{2}\right)\right)$ and suppose that $L \in \mathcal{L}^s\left(^2D\left(\frac{\pi}{2}\right)\right)$ is the polar of P. Then*

$$\|L\|_{D(\frac{\pi}{2})} \leq 2\|P\|_{D(\frac{\pi}{2})}.$$

Moreover, equality is attained for the polynomials

$$P(x, y) = \pm(x^2 + y^2 - 4xy).$$

Therefore, $\mathbb{R}(2, D(\pi/2)) = 2$.

3.4 Unconditional constants in polynomial spaces

Let us first review some commonly used concepts in the theory of Banach spaces concerning Schauder bases and unconditional bases.

Definition 3.4.1 *Let X and Y be Banach spaces over $\mathbb{K}$.*

- *We say that a set $\{e_j : j \in \mathbb{N}\} \subset X$ is a Schauder basis for the space X if for every $x \in X$, there exists a unique sequence $(a_j)_{j=1}^{\infty} \subset \mathbb{K}$ such that*

$$x = \sum_{j=1}^{\infty} a_j e_j.$$

- *We say that a sequence $(x_j) \subset X$ is a basic sequence if (x_j) is a Schauder basis of $\overline{[x_j : j \in \mathbb{N}]}$.*
- *If $\mathcal{B}_X = \{x_j : j \in \mathbb{N}\}$ is a basic sequence of X and $\mathcal{B}_Y = \{y_j : j \in \mathbb{N}\}$ is a basic sequence of Y, we say that $\mathcal{B}_X$ and $\mathcal{B}_Y$ are equivalent if the convergence of $\sum_{j=1}^{\infty} a_j x_j$ in X is equivalent to the convergence of $\sum_{j=1}^{\infty} a_j y_j$ in Y.*

The elements of spaces with Schauder bases can be represented similarly to how elements of the classical sequence spaces ℓ_p with $1 \leq p < \infty$ and c_0 are expressed. Needless to say, these spaces have a Schauder basis. Moreover, the existence of Schauder bases allows for a deeper study of the internal structure of a Banach space. If a space X contains a basic sequence that is equivalent to a Schauder basis of, for example, c_0 or ℓ_1, we can ensure that X contains c_0 or ℓ_1 as subspaces. We do not aim here to delve into the full significance of Schauder bases, so we will limit ourselves to presenting the necessary concepts to clarify the definition of the unconditional constant.

Every Schauder basis $\mathcal{B} = \{e_j : j \in \mathbb{N}\}$ in a Banach space X defines a sequence of canonical projections defined by

$$P_n(x) = \sum_{j=1}^{n} a_j e_j,$$

for all $x \in X$ with $x = \sum_{j=1}^{\infty} a_j e_j$. A classical result dating back to Banach himself states that the projections P_n are not only continuous but also have uniformly bounded norms. Specifically, the basic constant of a basis $\mathcal{B}$ is defined as

$$\mathrm{bc}(\mathcal{B}) = \sup_{n \in \mathbb{N}} \|P_n\|.$$

We now review the concept of an unconditional basis, which is central to this chapter. It is useful to recall beforehand what it means for a series $\sum_{j=1}^{\infty} x_j$ in a Banach space X to be unconditionally convergent to $x \in X$. A simple exercise is to show that the following two conditions are actually equivalent:

1. For every $\varepsilon > 0$, there exists a finite set $F_0 \subset \mathbb{N}$ such that for any finite set $F \subset \mathbb{N}$ with $F_0 \subset F$, we have

$$\left\| x - \sum_{j \in F} x_j \right\| < \varepsilon.$$

2. If $\sigma : \mathbb{N} \to \mathbb{N}$ is a bijection, then

$$\sum_{j=1}^{\infty} x_{\sigma(j)} = x.$$

A series $\sum_{j=1}^{\infty} x_j$ in a Banach space X is unconditionally convergent if it satisfies either of the two previous conditions. There are many other ways to define unconditionally convergent series (see for example the problems in Chapter 1 of [58]).

Definition 3.4.2 *Let X be a Banach space.*

(a) A Schauder basis $\{e_j : j \in \mathbb{N}\}$ of X is said to be unconditional if for each $x \in X$, its expansion $\sum_{j=1}^{\infty} a_j e_j$ converges unconditionally to x.

(b) A sequence (x_j) in X is said to be unconditional if it is an unconditional basis of $\overline{[e_j : j \in \mathbb{N}]}$.

The definition of unconditional constant is established from the following standard result in Banach space theory.

Proposition 3.4.1 *Let $(e_j)_{j=1}^{\infty}$ be a sequence in a Banach space X. The following statements are equivalent*

1. *$\{e_j : j \in \mathbb{N}\}$ is an unconditional basic sequence.*
2. *There exists a constant $K > 0$ such that for any choice of scalars $a_1, \ldots, a_m \in \mathbb{K}$ and any subset A of $\{1, \ldots, m\}$, we have*

$$\left\| \sum_{j \in A} a_j e_j \right\| \leq K \left\| \sum_{j=1}^{m} a_j e_j \right\|.$$

3. *There exists a constant $C > 0$ such that for any choice of scalars $a_1, \ldots, a_m \in \mathbb{K}$ and signs $\varepsilon_1, \ldots, \varepsilon_m \in \{-1, 1\}$, we have*

$$\left\| \sum_{j=1}^{m} \varepsilon_j a_j e_j \right\| \leq C \left\| \sum_{j=1}^{n} a_j x_j \right\|. \tag{3.106}$$

Definition 3.4.3 *Under the conditions of Proposition 3.4.1, the best constant that makes inequality (3.106) true is called the unconditional constant of $\mathcal{B} = \{e_j : j \in \mathbb{N}\}$ and we denote it by $ubc(\mathcal{B}, X)$ or simply by $ubc(\mathcal{B})$ if there is no ambiguity about what X is.*

Similarly, we define the unconditional constant of X as

$$ubc(X) = \inf\{ubc(\mathcal{B}) : \mathcal{B} \text{ is an unconditional basis of } X\}$$

if X has an unconditional basis and

$$ubc(X) = \infty$$

if X does not have an unconditional basis.

A natural and highly significant question in Banach space theory is whether every Banach space has an unconditional basic sequence. This question, called the problem of the unconditional basic sequence, was an open problem for a long time until T. Gowers and B. Maurey published in 1993 [56] the construction of a Banach space without unconditional basic sequences.

Despite the relevance of unconditional Schauder bases in Banach space theory, we will not delve too deeply in this section. The interested reader can consult the classic references by Lindenstrauss and Tzafriri [68,69] or Diestel [36], or more modern ones such as [1,43,59].

In this section, we focus on a very specific aspect of unconditional constants, as we will study this issue only in polynomial spaces. In the next section, in addition to reviewing some relevant results on unconditional constants in polynomial spaces, we will also see what connections exist between this problem and other issues of interest. Some of these motivations are perfectly explained in [24], and will be cited in detail in the next section. In particular, we will explore how unconditional constants relate to other significant problems, such as the Bohr radius problem, Gordon-Lewis constants, and Sidon constants.

3.4.1 Unconditional constants in polynomial spaces

In addition to the connections of unconditional constants in polynomial spaces with other problems, Defant, Díaz, García, and Maestre point out in [24] a motivation that makes the study of unconditional constants particularly interesting when restricted to the spaces $\mathcal{P}(^m\ell_p^n)$. Indeed (see [24, page 135]), if X or X^* have an unconditional basis, then both X and X^* contain uniformly complemented copies of one of the three sequences $(\ell_p^n)_{n=1}^\infty$ with $p = 1, 2, \infty$ (see [105, Theorem 1] and [64, Corollary 1]).

First, we will show a simple common reasoning that can be useful when calculating the unconditional constants of bases in polynomial spaces and also helps to clarify the connection between unconditional constants and other problems. We consider the real case and examine the canonical basis of the space of polynomials of degree at most n in m variables. Let $\alpha \in \mathbb{N}^n$ be a multi-index and suppose that $\mathbf{x}^\alpha$ represents $x_1^{\alpha_1} \cdots x_m^{\alpha_m}$, where $\mathbf{x} = (x_1, \dots, x_m) \in \mathbb{R}^m$ and $\alpha = (\alpha_1, \dots, \alpha_m)$ with $\alpha_k \in \mathbb{N} \cup \{0\}$, $1 \leq k \leq m$. Let

$$\mathcal{B}_n = \{\mathbf{x}^\alpha : |\alpha| \leq n\}$$

be the canonical basis of the space $\mathcal{P}_n(\mathbb{R}^m)$ of polynomials of degree less than or equal to n, and consider any subset $\mathcal{S} = \{\mathbf{x}^{\alpha_k} : |\alpha_k| \leq n, 1 \leq k \leq r\}$ of $\mathcal{B}_n$, a choice of signs $\varepsilon = (\varepsilon_1, \dots, \varepsilon_r)$, $P(\mathbf{x}) = a_{\alpha_1}\mathbf{x}^{\alpha_1} + \ldots + a_{\alpha_r}\mathbf{x}^{\alpha_r}$ and $P_\varepsilon(\mathbf{x}) = \varepsilon_1 a_{\alpha_1}\mathbf{x}^{\alpha_1} + \ldots + \varepsilon_r a_{\alpha_r}\mathbf{x}^{\alpha_r}$. Then, if B is a convex body in $\mathbb{R}^m$ such that $(x_1, \dots, x_m) \in B$ implies $(|x_1|, \dots, |x_m|) \in B$, we have

$$\|P_\varepsilon\|_{\mathsf{B}} := \sup_{\mathbf{x}\in\mathsf{B}} |P_\varepsilon(\mathbf{x})| \leq \sup_{\mathbf{x}\in\mathsf{B}} |a_{\alpha_1}||\mathbf{x}|^{\alpha_1} + \ldots + |a_{\alpha_r}||\mathbf{x}|^{\alpha_r} = \||P|\|_{\mathsf{B}},$$

where $|P|(\mathbf{x}) = |a_{\alpha_1}|\mathbf{x}^{\alpha_1} + \ldots + |a_{\alpha_r}|\mathbf{x}^{\alpha_r}$. Moreover, if $\varepsilon_k = \operatorname{sign}(a_{\alpha_k})$, then $\|P_\varepsilon\|_{\mathsf{B}} = \||P|\|_{\mathsf{B}}$. All of this shows that the unconditional constant C associated with the canonical basis of $(\mathcal{P}_n(\mathbb{R}^m), \|\cdot\|_{\mathsf{B}})$ coincides with the smallest constant $C_{\mathsf{B},\mathcal{S}}$ in

$$\||P|\|_{\mathsf{B}} \leq C_{\mathsf{B},\mathcal{S}}\|P\|_{\mathsf{B}},$$

for each P in the space generated by $\mathcal{S}$.

3.4.2 Sidon Constants

For the complex case, the Sidon constant of a set $\Lambda \subset \mathbb{Z}$ is nothing more than the unconditional constant of $\{z^\lambda : \lambda \in \Lambda\}$ in the space $\mathcal{C}(\mathbb{T})$, where $\mathbb{T} := \{z \in \mathbb{C} : |z| = 1\}$, when it exists. Similarly, the real Sidon constant of $\Lambda \subset \mathbb{N}$ is the unconditional constant of $\{x^\lambda : \lambda \in \Lambda\}$ in $\mathcal{C}([0,1])$, when it exists.

These observations highlight the connection between Sidon constants in several variables and unconditional constants in polynomial spaces. In [27], the Sidon constants $S(m,n)$ are defined as follows:

Definition 3.4.4 *Given $m, n \in \mathbb{N}$, the Sidon constant $S(m,n)$ corresponding to the set of indices*

$$\{\alpha = (\alpha_1, \dots, \alpha_n) \in \mathbb{N}_0^n : |\alpha| = m\}$$

is defined as the best constant $C > 0$ in the inequality

$$\sum_{|\alpha|=m} |a_\alpha| = |||P|||_{\mathbb{D}^n} \leq C\|P\|_{\mathbb{D}^n}$$

for any m-homogeneous polynomial $P = \sum_{|\alpha|=m} a_\alpha \mathbf{z}^\alpha$ in $\mathbb{C}$. In the real case, it is enough to change $\mathbb{C}$ to $\mathbb{R}$ and $\mathbb{D}^n$ to $[-1,1]^n$.

Note that the Sidon constants $S(m,n)$ coincide with the unconditional constant of the monomials $\mathbf{z}^\alpha$ of degree m in $\mathcal{P}(^m\ell_\infty^n(\mathbb{K}))$. These constants have been extensively studied for decades, and there is a vast literature on them. Here, we will limit ourselves to giving some significant results. For example, in [27], the authors observe that

$$S(m,n) \leq \sqrt{\binom{n+m-1}{m}}.$$

Previously, Defant, Díaz, García, and Maestre [24] had already determined the correct asymptotic growth of the unconditional constants of any unconditional basis in $\mathcal{P}(^m\ell_p^n)$, and in particular of $\mathcal{P}(^m\ell_\infty^n)$. Therefore, they also determined the growth of $S(m,n)$, which turns out to be dominated by $n^{\frac{m-1}{2}}$. All this is summarized in the following interesting result.

Theorem 3.4.2 (Defant, Díaz, García, and Maestre, [24]**)** *If $\mathcal{B}$ is any basis of $\mathcal{P}(^m\ell_p^n)$, then its unconditional constant satisfies*

$$ubc(\mathcal{B}) \overset{m}{\asymp} \begin{cases} n^{\frac{m-1}{2}} & \text{if } 1 < p \leq \infty, \\ n^{\frac{m-1}{q}} & \text{if } 1 \leq p \leq 2, \end{cases}$$

where q is the conjugate of p, that is, $\frac{1}{q} + \frac{1}{p} = 1$.

It is worth clarifying the notation $\overset{n}{\asymp}$ in the previous theorem. If $\{a_{m,n}\}_{m,n=1}^\infty$ and $\{b_{m,n}\}_{m,n=1}^\infty$ are two sequences, then

$$a_{m,n} \overset{n}{\asymp} b_{m,n}$$

means that for each n, there exist constants A_n and B_n such that for each m, we have

$$a_{m,n} \leq A_n b_{m,n} \quad \text{and} \quad b_{m,n} \leq B_n a_{m,n}.$$

3.4.3 Gordon-Lewis Constants

Given a Banach space X, a concept close to ubc(X) is that of Gordon-Lewis constants. In fact:

> "Gordon-Lewis constants are a fundamental tool for the study of unconditionality in Banach spaces,"

as noted by Defant, Díaz, García, and Maestre in [24]. This notion first appears in the classic reference [48] and is defined as follows:

Definition 3.4.5 (Gordon and Lewis [48]**)** *Let X be a Banach space.*

(a) We say that X has the Gordon-Lewis property if every 1-summing operator $T : X \to \ell_2$ (see Definition 3.4.6) admits a factorization

$$T : X \xrightarrow{R} L_1(\mu) \xrightarrow{S} \ell_2.$$

(b) If X has the Gordon-Lewis property, then there exists a constant $C > 0$ such that for every $T : X \to \ell_2$, we have

$$\inf \|R\|\|S\| \leq C\pi_1(T). \tag{3.107}$$

We define the Gordon-Lewis constant of X, and denote it by $gl(X)$, as the best constant C in (3.107)*. For X that do not have the Gordon-Lewis property, we set $gl(X) = \infty$. See Definition 3.4.6 to clarify the notation $\pi_1(T)$.*

Since summing operators appear in the previous definition, it might be of interest to recall their definition.

Definition 3.4.6 *If $1 \leq q \leq p_1, \ldots, p_m < \infty$ and $X_1, \ldots, x_m, Y$ are Banach spaces, a bounded m-linear operator $T \in \mathcal{L}(X_1 \times \cdots \times X_m; Y)$ is $(q; p_1, \ldots, p_m)$-summing if there exists a constant $C > 0$ such that for any $(x^j_{i_j})^{n_j}_{i_j=1} \subset X_j$ with $1 \leq j \leq m$ we have*

$$\left(\sum_{i_1,\ldots,i_m=1}^{n_1,\ldots,n_m} \|T(x^1_{i_1}, \ldots, x^m_{i_m})\|^q \right)^{\frac{1}{q}} \leq C \prod_{j=1}^{m} \|(x^j_{i_j})^{n_j}_{i_j=1}\|_{w,p_j}. \tag{3.108}$$

The notation $\prod^m_{(q;p_1,\ldots,p_n)}(X_1, \ldots, X_m; Y)$ stands for the class of all $(q; p_1, \ldots, p_n)$-summing multilinear operators in $\mathcal{L}(X_1 \times \cdots \times X_m; Y)$. Also the $(q; p_1, \ldots, p_n)$-summing norm in the class $\prod^m_{(q;p_1,\ldots,p_n)}(X_1, \ldots, X_m; Y)$ is defined as

$$\pi_{(q;p_1,\ldots,p_n)}(T) = \inf\{C > 0 : (3.108) \text{ is satisfied}\}.$$

When $p_1 = \ldots = p_m = p$ and $X_1 = \ldots = X_m = X$, we simply put $\prod_{(q;p)}(^mX; Y)$ and $\pi_{(q;p)}(T)$.

Unconditional constants are closely related to Gordon-Lewis constants, as evidenced by the following result of Gordon and Lewis (see also [37, 17.7]).

Theorem 3.4.3 (Gordon and Lewis [48]**)** *For any sequence (x_j) of non-zero vectors in a Banach space X, we have*

$$gl(\overline{[x_j : j \in \mathbb{N}]}) \leq C_{\mathbb{K}} ubc(X),$$

where $C_{\mathbb{R}} = 1$ and $C_{\mathbb{C}} = 2$. In particular,

$$gl(X) \leq C_{\mathbb{K}} ubc(X).$$

In [24], Defant, Díaz, García, and Maestre show a very interesting array of results that delve into the relationship between unconditional constants and Gordon-Lewis constants in the context of polynomial theory in Banach spaces. To make sense of these results, let us define

$$\begin{aligned}
\mathcal{M}(m,n) &= \{\alpha = (\alpha_1,\ldots,\alpha_m) : \alpha_1,\ldots,\alpha_m \in \{1,\ldots,n\}\},\\
\mathcal{J}(m,n) &= \{\alpha = (\alpha_1,\ldots,\alpha_m) \in \mathcal{M}(m,n) : \alpha_1 \le \ldots \le \alpha_m\},\\
\mathcal{M}(m) &= \mathbb{N}^m,\\
\mathcal{J}(m) &= \cup_{n=1}^{\infty} \mathcal{J}(m,n).
\end{aligned}$$

Suppose that the Banach space X is such that the dual X^* has a Schauder basis $\mathcal{B} = \{x_j^* : j \in \mathbb{N}\}$. Given $\alpha \in \mathcal{J}(m)$, a monomial with respect to $\mathcal{B}$ refers to any polynomial in $\mathcal{P}(^m X)$ of the form

$$P_\alpha = \prod_{j=1}^{m} x_{\alpha_j}^*.$$

An illustrative result of the type of connections between unconditional constants and Gordon-Lewis constants is the following:

Theorem 3.4.4 (Defant, Díaz, García, and Maestre [24]**)** *Let X be a Banach space and suppose that $\{x_j^* : j \in \mathbb{N}\}$ is an unconditional basis of X^*. If $\mathcal{P}_w(^m X)$ is the space of m-homogeneous polynomials in X that are weakly continuous on B_X, the following statements are equivalent*

(a) The monomials of degree m with respect to $\{x_j^* : j \in \mathbb{N}\}$ form an unconditional basis of $\mathcal{P}_w(^m X)$.

(b) $\mathcal{P}_w(^m X)$ has an unconditional basis.

(c) $\mathcal{P}_w(^m X)$ has the Gordon-Lewis property.

Moreover, if in Theorem 3.4.4 the basis $\{x_j^* : j \in \mathbb{N}\}$ is 1-unconditional, the following estimates hold:

Theorem 3.4.5 *Let X be a Banach space, $\{x_j^* : j \in \mathbb{N}\}$ a 1-unconditional basis of X^*, and suppose that the set of all monomials with respect to $\{x_j^* : j \in \mathbb{N}\}$, represented by $\mathcal{B}$, is an unconditional basis of $\mathcal{P}_w(^m X)$. Then*

$$ubc(\mathcal{B}, \mathcal{P}_w(^m X)) \le C_m gl(\mathcal{P}_w(^m X)),$$

where C_m is a constant independent of X such that

$$C_m \le \frac{2m^{4m+1}}{(m!)^2}.$$

Finally, we turn to exploring the relationship between unconditional constants in polynomial spaces and the Bohr radius problem.

3.4.4 Unconditional Constants and Bohr Radii

Determining the constants $C_{\mathsf{B},S}$ in the complex setting is closely linked to the problem of finding the n-dimensional Bohr radius. The Bohr radius problem can be traced back to 1914, stemming from the work of several mathematicians, notably H. Bohr. They proved that if the complex series $\sum_{k=0}^{\infty} c_k z^k$ satisfies $|\sum_{k=0}^{\infty} c_k z^k| < 1$ when $|z| < 1$, then $\sum_{k=0}^{\infty} |c_k z^k| < 1$

when $|z| < \frac{1}{3}$. This value, known as the Bohr radius, is in fact optimal. There is an analogue of the previous result in $\mathbb{C}^n$ (see the article [16] by Boas and Khavinson), but the exact value of the Bohr radius in the multidimensional case is unknown. For more information on problems related to Bohr radii, it is recommended to consult, for example, [12,26,28,29,34].

To illustrate the connection between Bohr radii and unconditional constants, observe that there exists a constant $C > 0$ such that

$$|||P|||_1 \leq C\|P\|_{\mathbb{D}^n},$$

for any homogeneous polynomial P of degree m in $\mathbb{C}^n$, where $\mathbb{D}^n$ is the unit polydisk in $\mathbb{C}^n$, $\|P\|_{\mathbb{D}^n}$ is the supremum norm of P over $\mathbb{D}^n$, and $|||P|||_1$ is the sum of the moduli of the coefficients of P, as usual. This result is deduced from [24]. This estimate has been used (see e.g., [8]) to prove that the n-dimensional Bohr radius is at least a constant times $\sqrt{\frac{\log n}{n}}$. Note that $|||P|||_1 = |||P|||_{\mathbb{D}^n}$, which implies that the smallest constant C in the previous inequality is actually $C_{\mathbb{B}^n,\mathcal{S}}$, where $\mathcal{S}$ is the canonical basis of the space of homogeneous polynomials in $\mathbb{C}^n$ of degree m.

3.4.5 Calculating unconditional constants using convex techniques

In this section, we will see how to use the the Krein-Milman approach to calculate unconditional constants in polynomial spaces. Two situations will be regarded. On the one hand, we consider the space $\mathcal{P}_{m,n}(\mathbb{R})$. We also pay attention to several spaces of quadratic forms on unbalanced convex bodies.

3.4.5.1 Unconditional constants in $\mathcal{P}_{m,n}(\mathbb{R})$

Let $C(m,n)$ be the unconditional constant of $\mathcal{P}_{m,n}(\mathbb{R})$ with respect to its canonical basis. Then $C(m,n)$ is also the smallest constant in $|||P|||_{m,n} \leq C(m,n)\|P\|_{m,n}$ for each $P \in \mathcal{P}_{m,n}(\mathbb{R})$. Observe that

$$C(m,n) = \sup\{|||P|||_{m,n} : P \in \operatorname{ext}(\mathsf{B}_{m,n})\}. \tag{3.109}$$

Recall that $\mathsf{B}_{m,n}$ is the closed unit ball of $\mathcal{P}_{m,n}(\mathbb{R})$. Also, if $P(x) = ax^m + bx^n + c$ then $|P|(x) = |a|x^m + |b|x^n + |c|$. The following can be proven:

Theorem 3.4.6 (Grecu, Muñoz, and Seoane [55]**)** *If $m,n \in \mathbb{N}$ with $m > n$, then*

$$C(m,n) = \begin{cases} 3 & \text{if } m \text{ and } n \text{ have different parity,} \\ 1 + \frac{4}{m-n}\left(\frac{m^m}{n^n}\right)^{\frac{1}{m-n}} & \text{if } m,n \text{ are even,} \\ \dfrac{n-\lambda_0 m}{n+\lambda_0 m} & \text{if } m,n \text{ are odd,} \end{cases}$$

where λ_0 is given by Lemma 3.2.8 (and also earlier in this text).

PROOF. *It is straightforward that $|||\,|a|x^m + |b|x^n + |c|\,|||_{m,n} = |a| + |b| + |c|$ so (3.109) turns into*

$$C(m,n) = \sup\{|a| + |b| + |c| : ax^m + bx^n + c \in \operatorname{ext}(\mathsf{B}_{m,n})\}. \tag{3.110}$$

The proof is divided into the following three cases:

(a) $C(m,n) = 3$ for every $m,n \in \mathbb{N}$ with different parity.

(b) $C(m,n) = 1 + \frac{4}{m-n}\left(\frac{m^m}{n^n}\right)^{\frac{1}{m-n}}$ for every $m,n \in \mathbb{N}$ even.

(c) $C(m,n) = \dfrac{n-\lambda_0 m}{n+\lambda_0 m}$ *for every* $m,n \in \mathbb{N}$ *odd.*

If m is odd then [88, Theorem 3.4] *states that the extreme polynomials of* $\mathsf{B}_{m,n}$ *are*

$$\{\pm(2x^n-1), \pm(x^m+x^n-1), \pm(-x^m+x^n-1), \pm 1\}.$$

Since the sum of the absolute values of the coefficients of $\pm(2x^n-1), \pm(x^m+x^n-1)$, *and* $\pm(-x^m+x^n-1)$ *is* 3, *the result follows.*

Now, if m is even and n is odd, then ([88, Theorem 4.6]*) the extreme polynomials of* $\mathsf{B}_{m,n}$ *are*

$$\left\{\pm\left(tx^m \pm \Gamma(t)x^n + 1 - t - \Gamma(t)\right), \pm 1 : \frac{n}{m} \leq t \leq 2\right\},$$

where $\Gamma(t)$ *is the curve defined implicitly by the equation*

$$\frac{(m-n)t}{n}\left(\frac{nx}{mt}\right)^{\frac{m}{m-n}} = 2 - t - x, \tag{3.111}$$

Then,

$$C(m,n) = \max\left\{|t| + |\Gamma(t)| + |1-t-\Gamma(t)|, 1 : \frac{n}{m} \leq t \leq 2\right\}.$$

We need to find first the critical points of $\Phi(t) = t + \Gamma(t)$ *within the interval* $[\frac{n}{m}, 2]$. *Differentiating implicitly*

$$\Gamma'(t) = \frac{\Gamma(t)}{t} \cdot \frac{2n - mt - n\Gamma(t)}{2m - mt - n\Gamma(t)}. \tag{3.112}$$

Therefore, if t_0 *is a critical point,* $\Phi'(t_0) = 0$ *is equivalent to*

$$mt_0(2-t_0) + (2n - nt_0 - mt_0)\Gamma(t_0) - n\Gamma(t_0)^2 = 0,$$

from which either $\Gamma(t_0) = 2 - t_0$ *or* $\Gamma(t_0) = -\frac{mt_0}{n}$. *The latter solution is not useful since* $\Gamma(t) \geq 0$ *over the considered interval. Thus* $\Gamma(t_0) - 2 + t_0 = 0$ *and, by equation* (3.111) *we have*

$$\frac{(m-n)t_0}{n}\left(\frac{n\Gamma(t_0)}{mt_0}\right)^{\frac{m}{m-n}} = 0.$$

we arrive then at $\Gamma(t_0) = 0$, *from which* $t_0 = 2$. *Therefore, if* $t \in \left[\frac{n}{m}, 2\right]$,

$$t + \Gamma(t) = \Phi(t) \geq \min\left\{\Phi(t) : \frac{n}{m} \leq t \leq 2\right\} = \min\left\{\Phi\left(\frac{n}{m}\right), \Phi(2)\right\} = 1 + \frac{n}{m} > 1.$$

Thus, for all $t \in \left[\frac{n}{m}, 2\right]$ *we have* $1 - t - \Gamma(t) < 0$, *from which*

$$C(m,n) = \max\left\{2t + 2\Gamma(t) - 1 : \frac{n}{m} \leq t \leq 2\right\} = \max\left\{2\Phi(t) - 1 : t \in \left[\frac{n}{m}, 2\right]\right\}.$$

Since the mappings $2\Phi - 1$ *and* Φ *have the same critical points we obtain*

$$C(m,n) = max\left\{2\Phi\left(\frac{n}{m}\right) - 1, 2\Phi(2) - 1\right\} = 3.$$

Next, if m and n are both even, then [88, Theorem 5.5] *provides the extreme polynomials of* $\mathsf{B}_{m,n}$. *In particular*

$$\{\pm\left(sx^m + \Lambda(s)x^n - 1 - s - \Lambda(s)\right), \pm\left(tx^m - \Upsilon(t)x^n + 1\right), \pm 1\},$$

where $s \in [\gamma_0, -2]$, $t \in [-\gamma_1, -\gamma_0]$, $\gamma_0 = -\frac{2}{m-n} \cdot \left(\frac{m^m}{n^n}\right)^{\frac{1}{m-n}}$, $\gamma_1 = \frac{-2n}{m-n}$ *and* Λ *and* Υ *are defined by*

$$\Lambda(s) = -\Gamma(|s|) \quad and \quad \Upsilon(t) = \left(\frac{2m}{m-n}\right)^{\frac{m-n}{m}} \cdot \left(\frac{mt}{n}\right)^{\frac{n}{m}}.$$

On the one hand

$$max\{|t| + |\Upsilon(t)| + 1 : -\gamma_1 \leq t \leq -\gamma_0\} = max\{t + \Upsilon(t) + 1 : -\gamma_1 \leq t \leq -\gamma_0\} = -2\gamma_0 + 1,$$

and on the other hand

$$max\{|s| + |\Lambda(s)| + |1 + s + \Lambda(s)| : \gamma_0 \leq s \leq -2\} = \begin{cases} -1 - 2s & if\, 1 + s + \Lambda(s) \leq 0, \\ 1 + 2\Lambda(s) & otherwise. \end{cases}$$

As proved in [88], *we have*

$$\Lambda(\gamma_0) = -\gamma_0$$

and since Λ *is a decreasing function within the specified interval, it follows from the above equation that*

$$max\{|s| + |\Lambda(s)| + |1 + s + \Lambda(s)| : \gamma_0 \leq s \leq -2\} \leq -2\gamma_0 + 1.$$

Consequently

$$C(m,n) = -2\gamma_0 + 1 = 1 + \frac{4}{m-n}\left(\frac{m^m}{n^n}\right)^{\frac{1}{m-n}}.$$

To prove (c) notice that from [88, Theorem 2.6], *the extreme polynomials of* $\mathsf{B}_{m,n}$ *are the following:*

$$\left\{\pm\left(t \cdot x^m - \frac{mt^{\frac{n}{m}}}{(m-n)^{\frac{m-n}{m}} n^{\frac{n}{m}}} \cdot x^n\right), \pm 1 : \frac{n}{m-n} \leq t \leq \frac{n}{n + m\lambda_0}\right\}. \tag{3.113}$$

Hence

$$\begin{aligned} C(m,n) &= max\left\{|t| + \left|\frac{mt^{\frac{n}{m}}}{(m-n)^{\frac{m-n}{m}} n^{\frac{n}{m}}}\right|, 1 : \frac{n}{m-n} \leq t \leq \frac{n}{n + m\lambda_0}\right\} \\ &= max\left\{t + \frac{mt^{\frac{n}{m}}}{(m-n)^{\frac{m-n}{m}} n^{\frac{n}{m}}} : \frac{n}{m-n} \leq t \leq \frac{n}{n + m\lambda_0}\right\}. \end{aligned}$$

And, since $\Psi(t) = t + \dfrac{mt^{\frac{n}{m}}}{(m-n)^{\frac{m-n}{m}} n^{\frac{n}{m}}}$ *is an increasing function, we will have that*

$$C(m,n) = \Psi\left(\frac{n}{n + m\lambda_0}\right).$$

The desired result follows from $|n + m\lambda_0| = (m-n)|\lambda_0|^{\frac{m}{m-n}}$.

3.4.5.2 Unconditional constants in polynomial spaces over convex sets

We will calculate the unconditional constant $C(K)$ of the canonical basis $\{x^2, xy, y^2\}$ of the spaces $\mathcal{P}(^2K)$ for several choices of the convex body $K \subset \mathbb{R}^2$. Notice that

$$C(K) = \sup\{|||P|||_K : P \in \text{ext}(\mathsf{B}_K)\}, \tag{3.114}$$

where, as usual, B_K is the closed unit ball of $\mathcal{P}(^2K)$ and $|P|(x,y) = |a|x^2 + |b|xy + |c|y^2$ whenever $P(x,y) = ax^2 + bxy + cy^2$.

For polynomials in the simplex, that is, for $\mathcal{P}(^2\Delta)$, the unconditional constant of the canonical basis is given in the following result.

Theorem 3.4.7 (Muñoz, Révesz, and Seoane [82]) *If $P \in \mathcal{P}(^2\Delta)$ then*

$$\||P|\|_\Delta \leq 2\|P\|_\Delta,$$

with 2 being optimal since equality is attained for the polynomials $\pm(x^2 - 6xy + y^2)$. Consequently, the unconditional constant $C(\Delta)$ of the canonical basis of $\mathcal{P}(^2\Delta)$ is 2.

PROOF. *We just have to apply (3.114) taking into consideration that, by Theorem 1.2.3.(b),*

$$\operatorname{ext}(\mathsf{B}_\Delta) = \{\pm(1, -2 - 2\sqrt{2(1+t)}, t), \pm(t, -2 - 2\sqrt{2(1+t)}, 1) : t \in [-1,1]\}.$$

Thus, using the formula for $\|\cdot\|_\Delta$ in (3.114) we obtain

$$\begin{aligned} C(\Delta) &= \sup\{\|(1, 2 + 2\sqrt{2(1+t)}, |t|)\|_\Delta, \|(|t|, 2 + 2\sqrt{2(1+t)}, 1)\|_\Delta : t \in [-1,1]\} \\ &= \sup_{t\in[-1,1]} \frac{\left[1 + \sqrt{2(1+t)}\right]^2 - |t|}{1 - |t| + 2\sqrt{2(1+t)}} = 2 \end{aligned}$$

Notice that the last supremum is attained at $t = 1$.

For polynomials in the square $\square$ the following holds.

Theorem 3.4.8 (Gámez, Muñoz, Sánchez, and Seoane [45]) *If $P \in \mathcal{P}(^2\square)$ then*

$$\||P|\|_\square \leq 5\|P\|_\square.$$

Equality is achieved for the polynomials $P(x,y) = \pm(x^2 - 3xy + y^2)$. Therefore, the unconditional constant $C(\square)$ of the canonical basis of $\mathcal{P}(^2\square)$ is 5.

PROOF. *The result follows from (3.114) and the following description of* $\operatorname{ext}(\mathsf{B}_\square)$, *given in Theorem 1.3.3.(b),*

$$\begin{aligned} \operatorname{ext}(\mathsf{B}_\square) = \{ &\pm(t, 2\sqrt{1-t}, -1), \pm(-1, 2\sqrt{1-t}, t) : t \in [0,1]\} \\ &\cup \{\pm(1,-1,1), \pm(1-3,1), \pm(1,0,0), \pm(0,0,1)\}. \end{aligned}$$

Indeed

$$C(\square) = \max_{|t|\leq 1}\{1 + |t| + 2\sqrt{1-t}, 3, 5, 1, 1\} = 5.$$

and $\pm(x^2 - 3xy + y^2)$ are the extreme polynomials for which $C(\square)$ is attained.

For homogeneous polynomials over circular sectors, we highlight the following result:

Theorem 3.4.9 (Araújo, Jimémez, Muñoz, and Seoane [2]) *The inequality*

$$\||P|\|_{D(\frac{\pi}{4})} \leq (5 + 4\sqrt{2})\|P\|_{D(\frac{\pi}{4})},$$

holds for all $P \in \mathcal{P}\left(^2D\left(\frac{\pi}{4}\right)\right)$ and it is optimal since equality is achieved for the polynomials $P(x,y) = \pm\left[x^2 + (5+4\sqrt{2})y^2 - (4+4\sqrt{2})xy\right]$. In other words, The unconditional constant of the canonical basis of $\mathcal{P}\left(^2D\left(\frac{\pi}{4}\right)\right)$ is $5 + 4\sqrt{2}$.

PROOF. *By (3.114) we have*

$$C\left(D\left(\frac{\pi}{4}\right)\right) = \sup\left\{\||P\||_{D(\frac{\pi}{4})} : P \in \operatorname{ext}\left(B_{D(\frac{\pi}{4})}\right)\right\}.$$

Recall from Theorem1.4.5 that

$$\operatorname{ext}\left(B_{D(\frac{\pi}{4})}\right) = \left\{\pm P_t, \pm Q_s, \pm(1,1,0) : -1 \le t \le 1 \text{ and } 1 \le s \le 5+4\sqrt{2}\right\},$$

where

$$\begin{aligned} P_t &:= (t, 4+t+4\sqrt{1+t}, -2-2t-4\sqrt{1+t}),\\ Q_s &:= (1, s, -2\sqrt{2(1+s)}). \end{aligned}$$

To compute the supremum above, we use the previously described extreme polynomials and the formula for $\|\cdot\|_{D(\frac{\pi}{4})}$ in Theorem 1.4.1. If we consider first the polynomials P_t, then $|P_t| = \left(|t|, 4+t+4\sqrt{1+t}, 2+2t+4\sqrt{1+t}\right)$ and we have

$$\begin{aligned} \sup_{-1\le t\le 1} \||P_t\||_{D(\frac{\pi}{4})} &= \sup_{-1\le t\le 1} \max\left\{|t|, \frac{1}{2}\left(|t|+4+t+4\sqrt{1+t}+2+2t+4\sqrt{1+t}\right)\right\}\\ &= \sup_{-1\le t\le 1} \frac{1}{2}\left(|t|+6+3t+8\sqrt{1+t}\right) = 5+4\sqrt{2}. \end{aligned}$$

Notice that the above supremum is attained at $t=1$. On the other hand, if we consider the polynomials Q_s, we have $|Q_s| = \left(1, s, 2\sqrt{2(1+s)}\right)$. Using the formula for $\|\cdot\|_{D(\frac{\pi}{4})}$ once more we arrive at

$$\begin{aligned} \sup_{1\le s\le 5+4\sqrt{2}} \||Q_s\||_{D(\frac{\pi}{4})} &= \sup_{1\le s\le 5+4\sqrt{2}} \max\left\{1, \frac{1}{2}\left(1+s+2\sqrt{2(1+s)}\right)\right\}\\ &= \sup_{1\le s\le 5+4\sqrt{2}} \frac{1}{2}\left(1+s+2\sqrt{2(1+s)}\right) = 5+4\sqrt{2}. \end{aligned}$$

Observe that the last supremum is now attained at $s = 5+4\sqrt{2}$.

Theorem 3.4.10 *The unconditional constant of the canonical basis of the space $\mathcal{P}(^2D(\frac{\pi}{2}))$ is 3. In other words, the inequality*

$$\||a|x^2+|b|y^2+|c|xy\||_{D(\frac{\pi}{2})} \le 3\|ax^2+by^2+cxy\|_{D(\frac{\pi}{2})},$$

holds for all $a,b,c \in \mathbb{R}$ and 3 is optimal since equality is achieved for the polynomials $\pm(x^2+y^2-4xy)$.

PROOF. *The extreme polynomials in the unit ball of $\mathcal{P}(^2D(\frac{\pi}{2}))$ are*

$$\begin{aligned} P_t(x) &= tx^2+y^2-2\sqrt{2(1+t)}xt,\\ Q_t(x) &= x^2+ty^2-2\sqrt{2(1+t)}xt, \end{aligned}$$

with $t \in [-1,1]$ (see Theorem 1.4.7). If we plug these into (3.85), due to the symmetry of the problem we end up with the maximum of

$$\max\left\{|t|, 1, \frac{1}{2}\left||t|+1+\sqrt{(|t|-1)^2+8(1+t)}\right|\right\}.$$

The latter function attains its maximum at 1 and turns out to be 3.

3.5 The Bohnenblust-Hille inequality

3.5.1 Some background one needs to know

Recall, as we already covered at the beginning of this text, that any homogeneous polynomial in $\mathbb{K}^n$ ($\mathbb{K} = \mathbb{R}$ or $\mathbb{C}$) of degree $m \in \mathbb{N}$ can be expressed as

$$P(x) = \sum_{|\alpha|=m} a_\alpha x^\alpha, \tag{3.115}$$

where $x = (x_1, \ldots, x_n) \in \mathbb{K}^n$, $\alpha = (\alpha_1, \ldots, \alpha_n) \in (\mathbb{N} \cup \{0\})^n$, with $|\alpha| = \alpha_1 + \cdots + \alpha_n$, and $x^\alpha = x_1^{\alpha_1} \cdots x_n^{\alpha_n}$, where each coefficient a_α belongs to $\mathbb{K}$. The space $\mathcal{P}(^m\mathbb{K}^n)$ denotes the finite-dimensional linear space of all homogeneous polynomials of degree m on $\mathbb{K}^n$.

Given a norm $\|\cdot\|$ on $\mathbb{K}^n$, the function

$$\|P\| := \sup\{|P(x)| : x \in \mathsf{B}_X\},$$

for every $P \in \mathcal{P}(^m\mathbb{K}^n)$, where B_X represents the unit ball of the Banach space $X = (\mathbb{K}^n, \|\cdot\|)$, defines a norm on $\mathcal{P}(^m\mathbb{K}^n)$, commonly referred to as the polynomial norm. The space $\mathcal{P}(^m\mathbb{K}^n)$ equipped with the polynomial norm induced by X is denoted by $\mathcal{P}(^mX)$.

In addition to the polynomial norm, another commonly used norm in $\mathcal{P}(^m\mathbb{K}^n)$ is the ℓ_p norm of the coefficients. Specifically, for a polynomial P of the form (3.115) and any $p \geq 1$, the norm

$$|P|_p := \left(\sum_{|\alpha|=m} |a_\alpha|^p \right)^{\frac{1}{p}},$$

defines an alternative norm on $\mathcal{P}(^m\mathbb{K}^n)$. It is worth noting that while the polynomial norm is often challenging to compute, the ℓ_p norm of the coefficients is relatively straightforward to determine. Since $\mathcal{P}(^m\mathbb{K}^n)$ is finite-dimensional, the polynomial norm $\|\cdot\|$ and the ℓ_p norm $|\cdot|_p$ (for $p \geq 1$) are equivalent, meaning that there exist constants $k(m,n), K(m,n) > 0$ such that

$$k(m,n)|P|_p \leq \|P\| \leq K(m,n)|P|_p, \tag{3.116}$$

for all $P \in \mathcal{P}(^m\mathbb{K}^n)$. These inequalities provide useful estimates for $\|P\|$, particularly when the sharp values of the best possible constants $k(m,n)$ and $K(m,n)$ are known.

The previous problem is an extension of the well-known polynomial Bohnenblust-Hille inequality. H. F. Bohnenblust and E. C. Hille (see Figure 3.30) proved, in 1931 [15], that there exists a constant $D_m > 0$ such that for every $P \in \mathcal{P}(^m\ell_\infty^n(\mathbb{K}))$ we have

$$|P|_{\frac{2m}{m+1}} \leq D_m \|P\|, \tag{3.117}$$

where $\ell_\infty^n(\mathbb{R})$ and $\ell_\infty^n(\mathbb{C})$ are, respectively, the real and complex versions of ℓ_∞^n. Observe that (3.117) coincides with the first inequality in (3.116) for $p = \frac{2m}{m+1}$ except for the fact that D_m in (3.117) can be chosen independently of the dimension n. Actually, Bohnenblust and Hille showed that $\frac{2m}{m+1}$ is optimal in (3.117) in the sense that for $p < \frac{2m}{m+1}$, any constant D fitting in the inequality

$$|P|_p \leq D\|P\|,$$

for all $P \in \mathcal{P}(^m\ell_\infty^n(\mathbb{K}))$ must necessarily depend on n.

The best constants in (3.117) depend considerably on whether we consider the real or the complex version of ℓ_∞^n, which motivates the following definition.

FIGURE 3.30: H. F. Bohnenblust (1906–2000) and E. C. Hille (1894–1980).

Definition 3.5.1 *The polynomial Bohnenblus-Hille constant for polynomials of degree m is defined as*

$$D_{\mathbb{K},m} := \inf\left\{D > 0 : |P|_{\frac{2m}{m+1}} \leq D\|P\|,\ \textit{for all } n \in \mathbb{N} \textit{ and } P \in \mathcal{P}(^m\ell_\infty^n(\mathbb{K}))\right\}.$$

If we restrict attention to a certain subset E of $\mathcal{P}(^m\ell_\infty^n(\mathbb{K}))$ for some $n \in \mathbb{N}$, then we define

$$D_{\mathbb{K},m}(E) := \inf\left\{D > 0 : |P|_{\frac{2m}{m+1}} \leq D\|P\| \textit{ for all } P \in E\right\}.$$

For simplicity, we will often use the notation $D_{\mathbb{K},m}(n)$ instead of $D_{\mathbb{K},m}(\mathcal{P}(^m\ell_\infty^n(\mathbb{K})))$. Observe that

$$1 \leq D_{\mathbb{K},m}(n) \leq D_{\mathbb{K},m},$$

for all $m, n \in \mathbb{N}$.

To better understand the asymptotic behaviour of the constants $D_{\mathbb{K},m}$ and $D_{\mathbb{K},m}(n)$, we introduce the following key concept.

Definition 3.5.2 *The asymptotic hypercontractivity constant associated with the polynomial Bohnenblust-Hille inequality is given by*

$$H_{\mathbb{K},\infty} := \limsup_{m\to\infty} \sqrt[m]{D_{\mathbb{K},m}}.$$

Similarly, when considering polynomials with a fixed number of variables n, we define:

$$H_{\mathbb{K},\infty}(n) := \limsup_{m\to\infty} \sqrt[m]{D_{\mathbb{K},m}(n)}.$$

These constants satisfy the inequality $1 \leq H_{\mathbb{K},\infty}(n) \leq H_{\mathbb{K},\infty}$ for every natural number n.

The authors of [27] established that the complex polynomial Bohnenblust-Hille inequality is, at most, hypercontractive. Refinements of the upper bound for $D_{\mathbb{C},m}$ were later obtained in [8], demonstrating that for every $B > 1$, there exists a constant $A > 0$ such that

$$D_{\mathbb{C},m} \leq Ae^{B\sqrt{m\log m}}.$$

As a direct consequence, it follows that $H_{\mathbb{C},\infty}(n) = H_{\mathbb{C},\infty} = 1$ for all $n \in \mathbb{N}$.

Regarding the real setting, recent findings in [19] have demonstrated that $H_{\mathbb{R},\infty} = 2$. However, determining the exact values of $D_{\mathbb{K},m}(n)$ remains a challenging task. This motivates our investigation into explicit and numerical approximations of these constants, leveraging recent developments in the geometric analysis of polynomial spaces.

This section is divided into two principal parts. Initially, we explore the geometric structure of polynomial spaces to determine the exact value of $D_{\mathbb{C},2}(2)$. We then extend this approach to compute $D_{\mathbb{R},2}(2)$. Furthermore, by utilizing numerical techniques, we establish lower bounds for $D_{\mathbb{R},m}(2)$ and $H_{\mathbb{R},\infty}(2)$, demonstrating the effectiveness of computational methods for estimating these constants.

The Bohnenblust-Hille inequality plays a significant role in various areas of mathematics and physics. Since the pioneering proof of its multilinear counterpart in 1931, the inequality has been extensively studied (see [8, 11, 13, 15, 18, 25, 28, 30, 32, 33, 35, 40, 41, 78, 80, 90–93, 99, 100], among others).

3.5.2 The exact value of $D_{\mathbb{C},2}(2)$

As it has been a common fashion in this text, and throughout this section, we will often identify any two-variable polynomial $az^2 + bwz + cw^2$ or any one-variable polynomial $a\lambda^2 + b\lambda + c$, for $a, b, c \in \mathbb{K}$, with the vector $(a, b, c) \in \mathbb{K}^3$. Also, we use the standard notation $\|az^2 + bwz + cw^2\|_{\mathbb{D}}$ for the supremum of $|az^2 + bwz + cw^2|$ for z, w in the unit disk $\mathbb{D}$ of $\mathbb{C}$. Similarly, $\|a\lambda^2 + b\lambda + c\|_{\mathbb{D}}$ stands for the supremum of $|a\lambda^2 + b\lambda + c|$ for $\lambda \in \mathbb{D}$. Observe that

$$\|az^2 + bwz + cw^2\|_{\mathbb{D}} = \|a\lambda^2 + b\lambda + c\|_{\mathbb{D}} = \max_{|\lambda|=1} |a\lambda^2 + b\lambda + c|,$$

being the last of the latter identities due to the Maximum Modulus Principle.

The main result of this section depends upon the following lemma, which is of independent interest and, although "innocent" at first sight, it is of a highly complex nature (see Lemma 2.1 in [61]).

Lemma 3.5.1 *Let $a, b, c \in \mathbb{C}$. There exist $a', b', c' \in \mathbb{R}$ such that*

$$\|az^2 + bwz + cw^2\|_{\mathbb{D}} \geq \|a'z^2 + b'zw + c'w^2\|_{\mathbb{D}} \quad \text{and} \quad \|(a, b, c)\|_{\frac{4}{3}} = \|(a', b', c')\|_{\frac{4}{3}}.$$

PROOF. *If we perform the change of variables*

$$z \mapsto ze^{-\frac{i\arg(a)}{2}} \quad \text{and} \quad w \mapsto we^{-\frac{i\arg(c)}{2}},$$

in $\|az^2 + bzw + cw^2\|_{\mathbb{D}}$, we can assume (without loss of generality) that $a, c \geq 0$. We can also assume that $a \geq c$ by swapping z and w. We have

$$\begin{aligned}
|a\lambda^2 + b\lambda + c|^2 &= \left(a\lambda^2 + b\lambda + c\right)\left(a\overline{\lambda^2} + \overline{b\lambda} + c\right) \\
&= a^2 + ac\lambda^2 + a\bar{b}\lambda + ac\overline{\lambda^2} + c^2 + c\overline{b\lambda} + ab\bar{\lambda} + bc\lambda + |b|^2 \\
&= a^2 + c^2 + |b|^2 + 2\left[acRe(\lambda^2) + aRe(\bar{b}\lambda) + cRe(b\lambda)\right] \\
&= a^2 + c^2 + |b|^2 + 2\left[acRe(\lambda^2) + (a + c)Re(b)Re(\lambda) + (a - c)Im(b)Im(\lambda)\right].
\end{aligned}$$

Similarly, if a', b', c' are real numbers, then

$$|a'\lambda^2 + b'\lambda + c'|^2 = a'^2 + c'^2 + |b'|^2 + 2\left[a'c'Re(\lambda^2) + (a' + c')b'Re(\lambda)\right].$$

1. *Assume first* $a \geq c \geq |b|$. *Then, choose*

$$a' = \frac{(c^{\frac{4}{3}} + |b|^{\frac{4}{3}})^{\frac{3}{4}}}{2^{\frac{3}{4}}},\ c' = -a',\ b' = a.$$

Then, $\|(a,b,c)\|_{\frac{4}{3}} = \|(a',b',c')\|_{\frac{4}{3}}$. *On the other hand,*

$$a'^2 + c'^2 + |b'|^2 + 2\left[a'c'Re(\lambda^2) + (a'+c')b'Re(\lambda)\right] = 2a'^2 + a^2 - 2a'^2 Re(\lambda^2),$$

so that it is easy to see

$$\|(a',b',c')\|_{\mathbb{D}}^2 = 4a'^2 + a^2 = \sqrt{2}(c^{\frac{4}{3}} + |b|^{\frac{4}{3}})^{\frac{3}{2}} + a^2.$$

Also, giving the value $\lambda = 1$ *(if* $Re(b) \geq 0$*) or* $\lambda = -1$ *(if* $Re(b) \leq 0$*), we can see that*

$$\|az^2 + cw^2 + bzw\|_{\mathbb{D}}^2 \geq a^2 + c^2 + |b|^2 + 2ac.$$

Now, we want

$$\sqrt{2}(c^{\frac{4}{3}} + |b|^{\frac{4}{3}})^{\frac{3}{2}} + a^2 \leq a^2 + c^2 + |b|^2 + 2ac,$$

that is,

$$\frac{c^2 + |b|^2 + 2ac}{\sqrt{2}(c^{\frac{4}{3}} + |b|^{\frac{4}{3}})^{\frac{3}{2}}} \geq 1.$$

Divide both, numerator and denominator, by a^2 *in order to convert the problem in having to achieve*

$$\frac{x^2 + y^2 + 2x}{\sqrt{2}(x^{\frac{4}{3}} + y^{\frac{4}{3}})^{\frac{3}{2}}} \geq 1,$$

for $0 \leq y \leq x \leq 1$.

2. *Assume next* $a \geq |b| \geq c$. *In this second part of the proof, we shall need to employ a couple of real valued functions that will come in handy to achieve our purpose. Let us first focus our attention on the choice of the constants* a', b', c', *as before,*

$$a' = \frac{\left(|c|^{\frac{4}{3}} + |b|^{\frac{4}{3}}\right)^{\frac{3}{4}}}{2^{\frac{3}{4}}},\ c' = -a',\ b' = a.$$

In this case, choose

$$\lambda = sign(Re(b))\sqrt{\frac{1}{2}} + i sign(Im(b))\sqrt{\frac{1}{2}}.$$

Then,

$$\begin{aligned}\|(a,b,c)\|_{\mathbb{D}}^2 &\geq a^2 + c^2 + |b|^2 + 2\left[\sqrt{\frac{1}{2}}(a+c)|Re(b)| + \sqrt{\frac{1}{2}}(a-c)|Im(b)|\right]\\ &\geq a^2 + c^2 + |b|^2 + \sqrt{2}|b|(a-c).\end{aligned}$$

Hence, we will achieve the desired result if we can guarantee

$$\sqrt{2}\left(|c|^{\frac{4}{3}} + |b|^{\frac{4}{3}}\right)^{\frac{3}{2}} + a^2 \leq a^2 + c^2 + |b|^2 + \sqrt{2}(a-c)|b|,$$

in other words,

$$1 \leq \Phi_1(x, y) := \frac{x^2 + y^2 + \sqrt{2}(1 - x)y}{\sqrt{2}\left(x^{\frac{4}{3}} + y^{\frac{4}{3}}\right)^{\frac{3}{2}}},$$

where $0 \leq x \leq y \leq 1$.

Let us focus now in another choice of constants a', b', c':

$$a' = \frac{\left(|a|^{\frac{4}{3}} + |c|^{\frac{4}{3}} + |b|^{\frac{4}{3}}\right)^{\frac{3}{4}}}{(2 + k^{4/3})^{\frac{3}{4}}}, \; c' = -a', \; b' = ka,$$

where k *has been chosen so that* $\|(a, b, c)\|_{\frac{4}{3}} = \|(a', b', c')\|_{\frac{4}{3}}$. *It can be proved that* $k \approx 2.828$. *Still giving the value* $\lambda = sign(Re(b))$

$$\|(a, b, c)\|_{\mathbb{D}}^2 \geq a^2 + c^2 + |b|^2 + 2ac,$$

and again we guarantee that we achieve what we are searching for if we get

$$4a'^2 + b'^2 = \frac{\left(|a|^{\frac{4}{3}} + |c|^{\frac{4}{3}} + |b|^{\frac{4}{3}}\right)^{\frac{3}{2}}}{(2 + k^{4/3})^{\frac{3}{2}}}(4 + k^2) \leq a^2 + c^2 + |b|^2 + 2ac.$$

That is, equivalently,

$$1 \leq \Psi_1(x, y) := \frac{\left(1 + x^2 + y^2 + 2yx\right)(2 + k^{4/3})^{\frac{3}{2}}}{(4 + k^2)\left(y^{\frac{4}{3}} + x^{\frac{4}{3}} + 1\right)^{\frac{3}{2}}},$$

for all $0 \leq x \leq y \leq 1$.
The reader can check using elementary calculus that, if

$$H(x, y) := \max\{\Phi_1(x, y), \; \Psi_1(x, y)\},$$

then

$$1 \leq H(x, y) \text{ for every } \quad 0 \leq x \leq y \leq 1.$$

3. *Assume finally* $|b| \geq a \geq c$. *Then, we may choose*

$$a' = \frac{\left(|a|^{\frac{4}{3}} + |c|^{\frac{4}{3}} + |b|^{\frac{4}{3}}\right)^{\frac{3}{4}}}{(2 + k^{4/3})^{\frac{3}{4}}}, \; c' = -a', \; b' = ka,$$

where k *is chosen as in the previous case. For* $\lambda = sign(Re(b))$, *we still need to make sure that*

$$1 \leq \Phi_2(x, y) := \frac{(1 + x^2 + y^2 + 2xy)(2 + k^{4/3})^{\frac{3}{2}}}{(4 + k^2)(x^{4/3} + y^{4/3} + 1)^{\frac{3}{2}}}.$$

For $\lambda = sign(Re(b))\sqrt{\frac{1}{2}} + i sign(Im(b))\sqrt{\frac{1}{2}}$, *we need to make sure that*

$$1 \leq \Psi_2(x, y) := \frac{(1 + x^2 + y^2 + \sqrt{2}(y - x)(2 + k^{4/3})^{\frac{3}{2}}}{(4 + k^2)(x^{4/3} + y^{4/3} + 1)^{\frac{3}{2}}}.$$

Next, choose

$$a' = \frac{(|a|^{4/3} + |c|^{4/3})^{3/4}}{2^{3/4}}, \quad c' = -a' \quad \text{and} \quad b' = |b|,$$

such that

$$\|(a', b', c')\|_{\mathbb{D}}^2 = 4\frac{(|a|^{4/3} + |c|^{4/3})^{3/2}}{2^{3/2}} + |b|^2,$$

and

$$\lambda = sign(Re(b))\sqrt{1 - \left(\frac{a-c}{|b|}\right)^2} + i\, sign(Im(b))\frac{a-c}{|b|}.$$

In that case,

$$\begin{aligned}\|(a,b,c)\|_{\mathbb{D}}^2 &\geq a^2 + c^2 + |b|^2 + 2\left[acRe(\lambda^2) + (a+c)Re(b)Re(\lambda) + (a-c)Im(b)Im(\lambda)\right] \\ &= a^2 + c^2 + |b|^2 + 2\left[ac\left(1 - 2\left(\frac{a-c}{|b|}\right)^2\right) + (a+c)|Re(b)|\sqrt{1 - \left(\frac{a-c}{|b|}\right)^2}\right. \\ &\quad \left. + (a-c)|Im(b)|\frac{a-c}{|b|}\right].\end{aligned}$$

Assume first $|Im(b)| \geq \frac{\sqrt{2}}{2}$. *Then,*

$$\begin{aligned}\|(a,b,c)\|_{\mathbb{D}}^2 &\geq a^2 + c^2 + |b|^2 \\ &\quad + 2\left[ac\left(1 - 2\left(\frac{a-c}{|b|}\right)^2\right) + (a-c)\frac{\sqrt{2}}{2}|b|\frac{a-c}{|b|}\right].\end{aligned}$$

Hence, we will achieve what we are searching for if we can assure that

$$1 \leq \Omega_2^{(1)}(x,y) := \frac{x^2 + y^2 + 2xy(1 - 2(y-x)^2) + \sqrt{2}(y-x)^2}{\sqrt{2}(x^{4/3} + y^{4/3})^{\frac{3}{2}}}.$$

On in on, we need to prove

$$1 \leq \max\{\Phi_2(x,y),\ \Psi_2(x,y),\ \Omega_2^{(1)}(x,y)\},$$

for $0 \leq x \leq y \leq 1$. *This can be done by means of elementary calculus, and we leave it as an exercise to the reader.*

On the other hand if, instead, we have $|Re(b)| \geq \frac{\sqrt{2}}{2}$, *then*

$$\begin{aligned}\|(a,b,c)\|_{\mathbb{D}}^2 &\geq a^2 + c^2 + |b|^2 \\ &\quad + 2\left[ac\left(1 - 2\left(\frac{a-c}{|b|}\right)^2\right) + (a+c)\frac{\sqrt{2}}{2}|b|\sqrt{1 - \left(\frac{a-c}{|b|}\right)^2}\right],\end{aligned}$$

and (in this case) we will be working with the condition

$$1 \leq \Omega_2^{(2)}(x,y) := \frac{x^2 + y^2 + 2xy(1 - 2(y-x)^2) + \sqrt{2}(y+x)\sqrt{1 - (y-x)^2}}{\sqrt{2}(x^{4/3} + y^{4/3})^{\frac{3}{2}}},$$

and, in conclusion, we shall need to guarantee that

$$1 \leq \max\{\Phi_2(x,y),\ \Psi_2(x,y),\ \Omega_2^{(2)}(x,y)\},$$

for $0 \leq x \leq y \leq 1$, *which we also leave as an exercise to the reader.*

And, with this last case, the proof is complete.

In order to prove that $D_{\mathbb{C},2}(2) = \sqrt[4]{\frac{3}{2}}$ we will also need the following description of the extreme points of the unit ball of $\mathbb{R}^3$ endowed with the norm

$$\|(a,b,c)\|_{\mathbb{D}} := \sup\{|az^2 + bz + c| : |z| \leq 1\}$$

for $a, b, c \in \mathbb{R}$. This norm has been studied by Aron and Klimek in [6], where they denote it by $\|\cdot\|_{\mathbb{C}}$. Observe, once again that

$$\|(a,b,c)\|_{\mathbb{D}} = \|az^2 + bwz + cz^2\|_{\mathbb{D}}.$$

Theorem 3.5.2 (Aron and Klimek [6]**)** *Let $E_{\mathbb{R}}$ be the real subspace of $\mathcal{P}(^2\ell_\infty^2(\mathbb{C}))$ given by $\{az^2 + bwz + cw^2 : \ (a,b,c) \in \mathbb{R}^3\}$. Then*

$$\mathrm{ext}(\mathsf{B}_{E_{\mathbb{R}}}) = \left\{ \left(s, \sqrt{4|s||t|\left(\frac{1}{(|s|+|t|)^2} - 1\right)}, t \right) : (s,t) \in G \right\},$$

where $\mathrm{ext}(\mathsf{B}_{E_{\mathbb{R}}})$ is the set of extreme points of the unit ball of $E_{\mathbb{R}}$, namely $\mathsf{B}_{E_{\mathbb{R}}}$ and $G = \{(s,t) \in \mathbb{R}^2 : |s| + |t| < 1 \text{ and } |s+t| \leq (s+t)^2\} \cup \{\pm(1,0), \pm(0,1)\}$.

Theorem 3.5.3 (Jiménez, Muñoz, Murillo, and Seoane [61]**)** *The optimal complex polynomial Bohnenblust-Hille constant for polynomials in $E_{\mathbb{R}}$, which we denote by $D_{\mathbb{C},2}(E_{\mathbb{R}})$, is given by $D_{\mathbb{C},2}(E_{\mathbb{R}}) = \sqrt[4]{\frac{3}{2}}$. Moreover,*

$$D_{\mathbb{C},2}(2) = \sqrt[4]{\frac{3}{2}} \approx 1.1066.$$

PROOF. *Using convexity we have*

$$\begin{aligned} D_{\mathbb{C},2}(E_{\mathbb{R}}) &= \sup\{\|(a,b,c)\|_{\frac{4}{3}} : \|az_1^2 + bz_1z_2 + cz_2^2\| \leq 1\} \\ &= \sup\{\|(a,b,c)\|_{\frac{4}{3}} : \|(a,b,c)\|_{\mathbb{C}} \leq 1\} \\ &= \sup\{\|(a,b,c)\|_{\frac{4}{3}} : (a,b,c) \in \mathrm{ext}(\mathsf{B}_{E_{\mathbb{R}}})\}, \end{aligned}$$

Hence

$$D_{\mathbb{C},2}(E_{\mathbb{R}}) = \sup\left\{ \left(|s|^{\frac{4}{3}} + |t|^{\frac{4}{3}} + \left[4|s||t|\left(\frac{1}{(|s|+|t|)^2} - 1\right)\right]^{\frac{2}{3}} \right)^{\frac{3}{4}} : (s,t) \in G \right\}.$$

If $\Phi(s,t) = \left(|s|^{\frac{4}{3}} + |t|^{\frac{4}{3}} + \left[4|s||t|\left(\frac{1}{(|s|+|t|)^2} - 1\right)\right]^{\frac{2}{3}} \right)^{\frac{3}{4}}$ for $(s,t) \in G$, one can prove using elementary calculus that Φ attains its maximum on G at $\pm\left(\frac{\sqrt{3}}{6}, -\frac{\sqrt{3}}{6}\right)$ and $\Phi\left(\frac{\sqrt{3}}{6}, -\frac{\sqrt{3}}{6}\right) = \sqrt[4]{\frac{3}{2}}$. Finally, from Lemma 3.5.1 we also obtain that $D_{\mathbb{C},2}(2) = \sqrt[4]{\frac{3}{2}} \approx 1.1066$.

Observe that the fact that $D_{\mathbb{C},2}(2) \geq \sqrt[4]{\frac{3}{2}}$ was already proved in [90]. The reader can find a sketch of the graph of Φ on the part of G contained in the second quadrant in Figure 3.31.

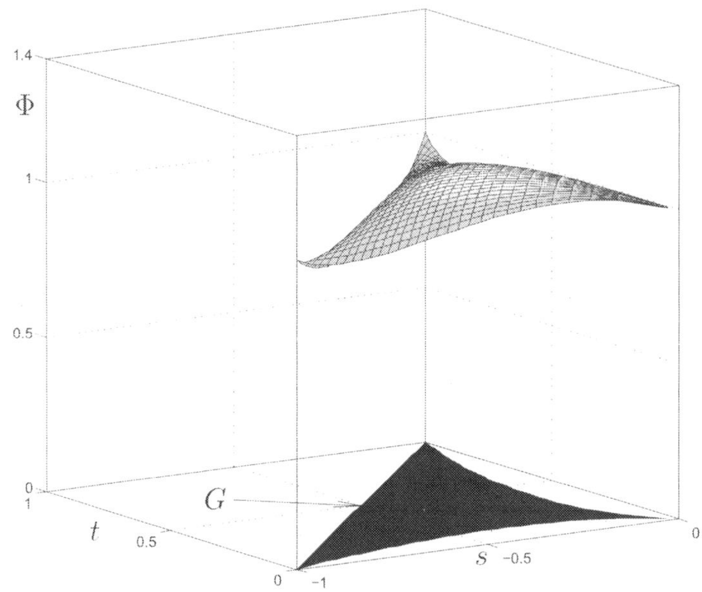

FIGURE 3.31: Graph of Φ on G intersected with the second quadrant.

3.5.3 The exact value of $D_{\mathbb{R},2}(2)$ and lower bounds for $D_{\mathbb{R},m}(2)$

In [19] it is proved that the asymptotic hypercontractivity constant of the real polynomial Bohnenblust-Hille inequality is exactly 2. Is it true that $H_{\mathbb{R},\infty}(2) = 2$? The results presented here suggest that, perhaps $H_{\mathbb{R},\infty}(2) < 2$. In this section, as in the previous section, we identify polynomials with their coefficient vectors.

Remark 3.5.4 *Throughout this section, we will compute several times norms of polynomials on the real line numerically (see [61]). This is done by using Matlab. In particular, if $P(x)$ is a real polynomial on $\mathbb{R}$, we apply the predefined Matlab function* `roots.m` *to P' in order to obtain an approximation of all the critical points of P. If $x_1, \ldots, x_k$ are all the roots of P' in $[-1, 1]$, then we approach the norm of P as*

$$\|P\| := \max\{|P(x)| : x \in [-1, 1]\} = \max\{|P(x_i)|, |P(\pm 1)| : i = 1, \ldots, k\}.$$

Another Matlab predefined function, namely `conv.m`*, is used in order to multiply polynomials. This is done to obtain Figure 3.34.*

3.5.4 The exact calculation of $D_{\mathbb{R},2}(2)$

The value of the constant $D_{\mathbb{R},2}(2)$ can be obtained using the geometry of the unit ball of $\mathcal{P}(\ell_\infty^2(\mathbb{R}))$ described in [20]. We state the result we need for completeness:

Theorem 3.5.5 (Choi and Kim [20]) *The set $\text{ext}(\mathsf{B}_{\mathcal{P}(^2\ell_\infty^2(\mathbb{R}))})$ of extreme points of the unit ball of $\mathcal{P}(^2\ell_\infty^2(\mathbb{R}))$ is given by*

$$\text{ext}(\mathsf{B}_{\mathcal{P}(^2\ell_\infty^2(\mathbb{R}))}) = \{\pm x^2,\ \pm y^2,\ \pm(tx^2 - ty^2 \pm 2\sqrt{t(1-t)}xy) :\ t \in [1/2, 1]\}.$$

As a consequence of the previous result, we obtain the following:

Theorem 3.5.6 (Jiménez, Muñoz, Murillo, and Seoane [61]) *Let f be the real valued function given by*

$$f(t) = \left[2t^{\frac{4}{3}} + (2\sqrt{t(1-t)})^{\frac{4}{3}}\right]^{\frac{3}{4}}.$$

We have that $D_{\mathbb{R},2}(2) = f(t_0) \approx 1.837373$, where

$$t_0 = \frac{1}{36}\left(2\sqrt[3]{107 + 9\sqrt{129}} + \sqrt[3]{856 - 72\sqrt{129}} + 16\right) \approx 0.867835.$$

The exact value of $f(t_0)$ is given by

$$f(t_0) = \sqrt[4]{\frac{1}{9}\left(\sqrt[3]{40465 + 387\sqrt{129}} + \sqrt[3]{40465 - 387\sqrt{129}} + 34\right)}.$$

Moreover, the following normalized polynomials are extreme for this problem

$$P_2(x, y) = \pm(t_0 x^2 - t_0 y^2 \pm 2\sqrt{t_0(1-t_0)}xy).$$

PROOF. *Let*

$$f(t) = \left[2t^{\frac{4}{3}} + (2\sqrt{t(1-t)})^{\frac{4}{3}}\right]^{\frac{3}{4}}.$$

We just have to notice that due to the convexity of the ℓ_p-norms and Theorem 3.5.5 we have

$$\begin{aligned} D_{\mathbb{R},2}(2) &= \sup\{|\mathbf{a}|_{\frac{4}{3}} : \mathbf{a} \in \mathsf{B}_{\mathcal{P}(^2\ell_\infty^2,\mathbb{R})}\} \\ &= \sup\{|\mathbf{a}|_{\frac{4}{3}} : \mathbf{a} \in \mathrm{ext}(\mathsf{B}_{\mathcal{P}(^2\ell_\infty^2,\mathbb{R})})\} = \sup_{t\in[1/2,1]} f(t). \end{aligned}$$

Some calculations will show that the last supremum is attained at $t = t_0$, which completes the proof.

Now, if $\mathbf{a}_n$ is the vector of the coefficients of P_2^n for each $n \in \mathbb{N}$, then we know that

$$D_{\mathbb{R},2n}(2) \geq \frac{|\mathbf{a}_n|_{\frac{4n}{2n+1}}}{\|P_2\|^n}. \tag{3.118}$$

Since $\|P_2\| = 1$, then (3.118) with $n = 300$ (see also Figure 3.34) proves that

$$D_{\mathbb{R},600}(2) \geq (1.36117)^{600},$$

providing numerical evidence showing that

$$H_{\mathbb{R},\infty}(2) \geq 1.36117.$$

3.5.5 Educated guess for the exact calculation of $D_{\mathbb{R},3}(2)$

To the best of our knowledge, determining the exact value of $\|P\|$ is generally a challenging task. Nevertheless, in certain specific cases, there exists a method for its computation. For example, Grecu, Muñoz, and Seoane derived the following formula in [55, Lemma 3.12]:

Lemma 3.5.7 *If for every $a, b \in \mathbb{R}$ we define $P_{a,b}(x,y) = ax^3 + bx^2y + bxy^2 + ay^3$ then*

$$\|P_{a,b}\| = \begin{cases} \left| a - \frac{b^2}{3a} + \frac{2b^3}{27a^2} + \frac{2a}{27}\left(-\frac{3b}{a} + \frac{b^2}{a^2}\right)^{\frac{3}{2}} \right| & \text{if } a \neq 0 \text{ and } b_1 < \frac{b}{a} < 3 - 2\sqrt{3}, \\ |2a + 2b| & \text{otherwise,} \end{cases}$$

where

$$b_1 = \frac{3}{7}\left(3 - \frac{2\sqrt[3]{9}}{\sqrt[3]{-12 + 7\sqrt{3}}} + 2\sqrt[3]{-36 + 21\sqrt{3}}\right) \approx -1.6692.$$

From Lemma 3.5.7 we have the following sharp polynomial Bohnenblust-Hille type constant:

Theorem 3.5.8 (Jiménez, Muñoz, Murillo, and Seoane [61]) *Let $P_{a,b}(x,y) = ax^3 + bx^2y + bxy^2 + ay^3$ for $a, b \in \mathbb{R}$ and consider the subset of $\mathcal{P}(^3\ell_\infty^2(\mathbb{R}))$ given by $E = \{P_{a,b} : a, b \in \mathbb{R}\}$. Then*

$$\frac{|(a,b,b,a)|_{\frac{3}{2}}}{\|P_{a,b}\|} = \begin{cases} \frac{27a^2\left(2|a|^{\frac{3}{2}} + 2|b|^{\frac{3}{2}}\right)^{\frac{2}{3}}}{\left|27a^3 - 9ab^2 + 2b^3 + 2\,\mathrm{sign}(a)(-3ab + b^2)^{\frac{3}{2}}\right|}, & \text{if } a \neq 0 \text{ and } b_1 < \frac{b}{a} < 3 - 2\sqrt{3}, \\ \frac{\left(2|a|^{\frac{3}{2}} + 2|b|^{\frac{3}{2}}\right)^{\frac{2}{3}}}{2|a+b|}, & \text{otherwise} \end{cases}$$

where b_1 was defined in Lemma 3.5.7. Moreover, the above function attains its maximum when $\frac{b}{a} = b_1$, which implies that

$$D_{\mathbb{R},3}(E) = \frac{\left(2 + 2|b_1|^{\frac{3}{2}}\right)^{\frac{2}{3}}}{2|1 + b_1|} \approx 2.5525$$

The authors have numerical evidence to state that

$$D_{\mathbb{R},3}(2) = D_{\mathbb{R},3}(E).$$

Moreover, one polynomial for which $D_{\mathbb{R},3}(2)$ would be attained is

$$P_3(x,y) = x^3 + b_1x^2y + b_1xy^2 + y^3,$$

where $b_1 \approx -1.6692$ is as in Lemma 3.5.7. It can be proved from Lemma 3.5.7 that

$$\|P_3\| \approx 1.33848,$$

up to 5 decimal places. If $\mathbf{a_n}$ is the vector of the coefficients of $P_3(x,y)^n$ and we use the fact that

$$D_{\mathbb{R},3n}(2) \geq \frac{|\mathbf{a}_n|_{\frac{6n}{3n+1}}}{\|P_3\|^n}, \tag{3.119}$$

then putting $n = 200$ in (3.119) we obtain, for instance,

$$D_{\mathbb{R},600}(2) \geq (1.42234)^{600},$$

which provides numerical evidence showing that

$$H_{\mathbb{R},\infty}(2) \geq 1.42234.$$

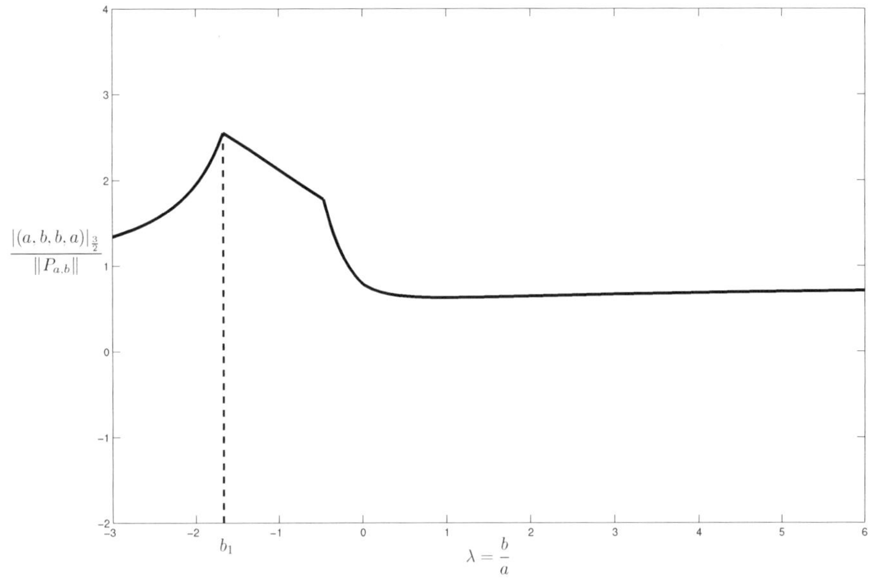

FIGURE 3.32: Graph of the quotient $\frac{|(a,b,b,a)|_{\frac{3}{2}}}{\|P_{a,b}\|}$ as a function of $\lambda = \frac{a}{b}$.

3.5.6 Numerical calculation of $D_{\mathbb{R},5}(2)$

Let us define the polynomial

$$P_5(x,y) = ax^5 - bx^4y - cx^3y^2 + cx^2y^3 + bxy^4 - ay^5,$$

with

$$\begin{aligned} a &= 0.19462, \\ b &= 0.66008, \\ c &= 0.97833. \end{aligned}$$

The norm of P_5 can be calculated numerically (using Remark 3.5.4), and it turns out to be

$$\|P_5\| = 0.28617,$$

up to 5 decimal places. The authors have numerical evidence showing that

$$D_{\mathbb{R},5}(2) \approx 6.83591.$$

In any case we have

$$D_{\mathbb{R},5}(2) \geq \frac{|(a,-b,-c,c,b,-a)|_{\frac{5}{3}}}{\|P_5\|} \approx 6.83591.$$

It is interesting to observe that we can improve numerically the estimate $H_{\infty,\mathbb{R}}(2) \geq \sqrt[8]{27} \approx 1.50980$ (see [19, Theorem 4.2]) by considering polynomials of the form P_5^n. Indeed, if $\mathbf{a}_n$ is the vector of the coefficients of P_5^n for each $n \in \mathbb{N}$, then we know that

$$D_{\mathbb{R},5n}(2) \geq \frac{|\mathbf{a}_n|_{\frac{10n}{5n+1}}}{\|P_5\|^n}, \tag{3.120}$$

Using (3.120) with $n = 120$ we obtain, in particular (see also Figure 3.34)

$$D_{\mathbb{R},600}(2) \geq (1.54987)^{600},$$

providing numerical evidence showing that

$$H_{\mathbb{R},\infty}(2) \geq 1.54987.$$

3.5.7 Educated guess for the exact calculation of $D_{\mathbb{R},6}(2)$

The authors have numerical evidence pointing to the fact that an extreme polynomial in the Bohnenblust-Hille inequality for polynomials in $\mathcal{P}(^6\ell_\infty^2(\mathbb{R}))$ may be of the form

$$Q_{a,b}(x,y) = ax^5y + bx^3y^3 + axy^5.$$

This motivates a more detailed analysis of such polynomials, as carried out in the following theorem.

Theorem 3.5.9 (Jiménez, Muñoz, Murillo, and Seoane [61]**)** *Let*

$$Q_{a,b}(x,y) = ax^5y + bx^3y^3 + axy^5$$

for $a, b \in \mathbb{R}$ and consider the subspace of $\mathcal{P}(^6\ell_\infty^2(\mathbb{R}))$ given by $F = \{Q_{a,b} : a, b \in \mathbb{R}\}$. Suppose $\lambda_0 < \lambda_1$ are the only two roots of the equation

$$\frac{|3\lambda^2 - 20 + \lambda\sqrt{9\lambda^2 - 20}|}{25}\sqrt{\frac{-3\lambda - \sqrt{9\lambda^2 - 20}}{10}} = |2 + \lambda|.$$

Then if $\lambda = \frac{b}{a}$ we have

$$\frac{|(0,a,0,b,0,a,0)|_{\frac{12}{7}}}{\|Q_{a,b}\|} = \begin{cases} \dfrac{25\sqrt{10}\left(2+|\lambda|^{\frac{12}{7}}\right)^{\frac{12}{7}}}{|3\lambda^2-20+\lambda\sqrt{9\lambda^2-20}|\sqrt{-3\lambda-\sqrt{9\lambda^2-20}}}, & \text{if } a \neq 0 \text{ and } \lambda_0 < \frac{b}{a} < \lambda_1, \\ \dfrac{\left(2+|\lambda|^{\frac{12}{7}}\right)^{\frac{7}{12}}}{|2+\lambda|}, & \text{otherwise.} \end{cases}$$

Observe that $\lambda_0 \approx -2.2654$, $\lambda_1 \approx -1.6779$ and the above function attains its maximum when $\frac{b}{a} = \lambda_0$ (see Figure 3.33), which implies that

$$D_{\mathbb{R},6}(F) = \frac{\left(2 + |\lambda_0|^{\frac{12}{7}}\right)^{\frac{12}{7}}}{|2 + \lambda_0|} \approx 10.7809.$$

PROOF. *We do not lose generality by considering only polynomials of the form $Q_{1,\lambda}$, in which case*

$$\|Q_{1,\lambda}\| = \sup\{|x^5 + \lambda x^3 + x| : x \in [0,1]\}.$$

The polynomial $q_\lambda(x) := x^5 + \lambda x^3 + x$ has no critical points if $\lambda > -\frac{2\sqrt{5}}{3}$, otherwise it has the following critical points in $[0,1]$

$$x_0 := \sqrt{\frac{-3\lambda - \sqrt{9\lambda^2 - 20}}{10}} \quad \text{and} \quad x_1 := \sqrt{\frac{-3\lambda + \sqrt{9\lambda^2 - 20}}{10}} \quad \text{if } -2 \leq \lambda \leq -\frac{2\sqrt{5}}{3},$$

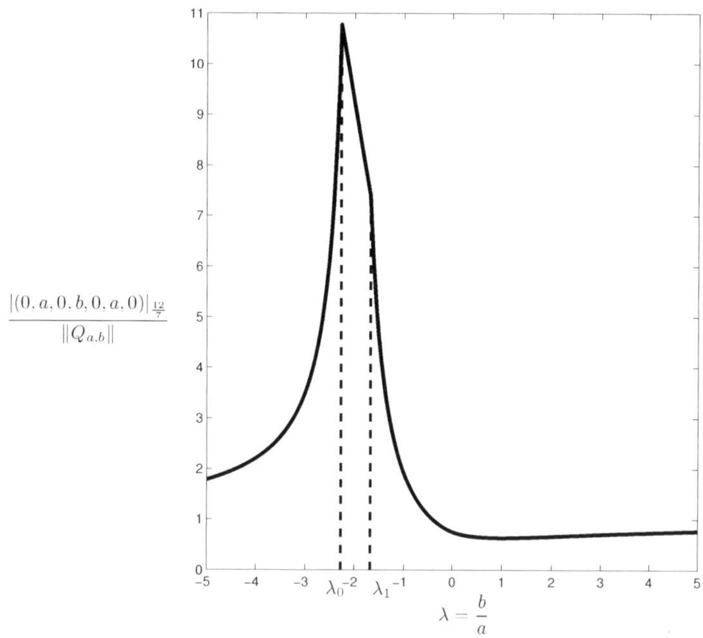

FIGURE 3.33: Graph of the quotient $\frac{|(0,a,0,b,0,a,0)|_{\frac{12}{7}}}{\|Q_{a,b}\|}$ as a function of $\lambda = \frac{b}{a}$.

and x_0 if $\lambda \leq -2$. Notice that

$$q_\lambda(x_0) = \frac{-3\lambda^2 + 20 - \lambda\sqrt{9\lambda^2 - 20}}{20} x_0$$
$$q_\lambda(x_1) = \frac{-3\lambda^2 + 20 + \lambda\sqrt{9\lambda^2 - 20}}{20} x_1.$$

It is easy to check that $|q_\lambda(x_0)| \geq |q_\lambda(x_1)|$ for $-2 \leq \lambda \leq -\frac{2\sqrt{5}}{3}$, which implies that

$$\|Q_{1,\lambda}\| = \begin{cases} \max\{|2+\lambda|, |q_\lambda(x_0)|\} & \text{if } -2 \leq \lambda \leq -\frac{2\sqrt{5}}{3}, \\ |2+\lambda| & \text{otherwise.} \end{cases}$$

The equation $|2+\lambda| = |q_\lambda(x_0)|$ turns out to have only two roots, namely $\lambda_0 \approx -2.2654$ and $\lambda_1 \approx -1.6779$. By continuity, it is easy to prove that $|2+\lambda| \leq |q_\lambda(x_0)|$ only if $-2 \leq \lambda \leq -\frac{2\sqrt{5}}{3}$, which concludes the proof.

As mentioned above, we have numerical evidence showing that

$$D_{\mathbb{R},6}(2) = D_{\mathbb{R},6}(F) = \frac{\left(2 + |\lambda_0|^{\frac{12}{7}}\right)^{\frac{12}{7}}}{|2+\lambda_0|} \approx 10.7809.$$

In any case we do have that

$$D_{\mathbb{R},6}(2) \geq 10.7809.$$

As we did in the previous cases, it would be interesting to know if we can improve numerically our best lower bound on $H_{\mathbb{R},\infty}$ by considering powers of

$$P_6(x,y) = Q_{1,\lambda_0}(x,y) = x^5y + \lambda_0 x^3y^3 + xy^5,$$

where λ_0 is the constant defined in Theorem 3.5.9 ($\lambda_0 \approx -2.2654$). If $\mathbf{a}_n$ is the vector of the coefficients of P_6^n for each $n \in \mathbb{N}$, then we know that

$$D_{\mathbb{R},6n}(2) \geq \frac{|\mathbf{a}_n|_{\frac{12n}{6n+1}}}{\|P_6\|^n}. \tag{3.121}$$

Using (3.121) with $n = 100$ and estimating $\|P_6\|$ according to Remark 3.5.4 we obtain

$$D_{\mathbb{R},600}(2) \geq (1.58432)^{600},$$

which suggests that (see Figure 3.34)

$$H_{\infty,\mathbb{R}}(2) \geq 1.58432.$$

3.5.8 Numerical calculation of $D_{\mathbb{R},7}(2)$

Let us define the polynomial

$$P_7(x,y) = -ax^7 + bx^6y + cx^5y^2 - dx^4y^3 - dx^3y^4 + cx^2y^5 + bxy^6 - ay^7,$$

with

$$\begin{aligned} a &= 0.05126, \\ b &= 0.22070, \\ c &= 0.50537, \\ d &= 0.71044. \end{aligned}$$

It can be proved numerically (using Remark 3.5.4) that

$$\|P_7\| \approx 0.07138,$$

up to 5 decimal places. The authors have numerical evidence showing that

$$D_{\mathbb{R},7}(2) \approx \frac{|(-a,b,c,-d,-d,c,b,-a)|_{\frac{7}{4}}}{\|P_7\|} \approx 19.96308.$$

If $\mathbf{a}_n$ is the vector of the coefficients of P_7^n for each $n \in \mathbb{N}$, then we know that

$$D_{\mathbb{R},7n}(2) \geq \frac{|\mathbf{a}_n|_{\frac{14n}{7n+1}}}{\|P_7\|^n}. \tag{3.122}$$

Moreover, if we put $n = 86$ in (3.122) we obtain

$$D_{\mathbb{R},602}(2) \geq (1.61725)^{602},$$

suggesting that

$$H_{\mathbb{R},\infty}(2) \geq 1.61725.$$

3.5.9 Numerical calculation of $D_{\mathbb{R},8}(2)$

Let us define the polynomial

$$P_8(x,y) = -ax^7y + bx^5y^3 - bx^3y^5 + axy^7,$$

with

$$a = 0.15258,$$
$$b = 0.64697.$$

It can be established numerically (see Remark 3.5.4) that

$$\|P_8\| \approx 0.02985,$$

up to 5 decimal places. The authors have numerical evidence showing that

$$D_{\mathbb{R},8}(2) \approx \frac{|(0, -a, 0, b, 0, -b, 0, a, 0)|_{\frac{16}{9}}}{\|P_8\|} \approx 33.36323.$$

If $\mathbf{a}_n$ is the vector of the coefficients of P_8^n for each $n \in \mathbb{N}$, then we know that

$$D_{\mathbb{R},8n}(2) \geq \frac{|\mathbf{a}_n|_{\frac{16n}{8n+1}}}{\|P_8\|^n}. \tag{3.123}$$

Moreover, using (3.123) with $n = 75$ we obtain

$$D_{\mathbb{R},600}(2) \geq (1.64042)^{600},$$

which suggests that

$$H_{\mathbb{R},\infty}(2) \geq 1.64042.$$

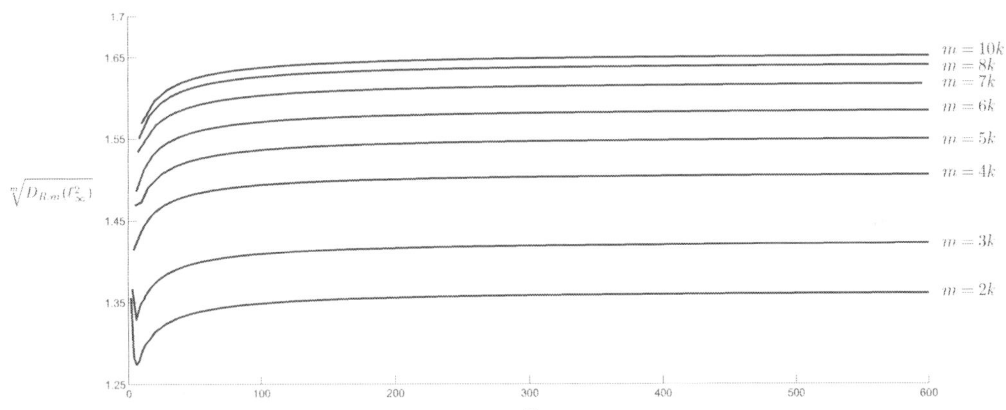

FIGURE 3.34: Graphs of the estimates on $\sqrt[m]{D_{\mathbb{R},m}(2)}$ obtained by using (3.118) through (3.124).

3.5.10 Numerical calculation of $D_{\mathbb{R},10}(2)$

In this case our numerical estimates show that there exists an extreme polynomial in the Bohnenblust-Hille polynomial inequality in $\mathcal{P}({}^{10}\ell_\infty^2(\mathbb{R}))$ of the form

$$P_{10}(x, y) = ax^9y + bx^7y^3 + x^5y^5 + bx^3y^7 + axy^9,$$

with

$$\begin{aligned} a &= 0.0938, \\ b &= -0.5938. \end{aligned}$$

It can be computed numerically (see Remark 3.5.4) that

$$\|P_{10}\| \approx 0.01530,$$

up to 5 decimal places. The authors have numerical evidence showing that

$$D_{\mathbb{R},10}(2) \approx \frac{|(0,a,0,b,0,1,0,b,0,a,0)|_{\frac{20}{11}}}{\|P_{10}\|} \approx 90.35556.$$

If $\mathbf{a}_n$ is the vector of the coefficients of P_{10}^n for each $n \in \mathbb{N}$, then we know that

$$D_{\mathbb{R},10n}(2) \geq \frac{|\mathbf{a}_n|_{\frac{20n}{10n+1}}}{\|P_{10}\|^n}. \tag{3.124}$$

If we set $n = 60$ in (3.124) then we obtain

$$D_{\mathbb{R},600}(2) \geq (1.65171)^{600},$$

which suggests that

$$H_{\mathbb{R},\infty}(2) \geq 1.65171.$$

We have sketched in Figure 3.34 a summary of the numerical results obtained in this section.

Bibliography

[1] F. Albiac and N. J. Kalton, *Topics in Banach space theory*, 2nd ed., Graduate Texts in Mathematics, vol. 233, Springer, [Cham], 2016.

[2] G. Araújo, P. Jiménez-Rodríguez, G. A. Muñoz-Fernández, and J. B. Seoane-Sepúlveda, *Polynomial inequalities on the $\pi/4$-circle sector*, J. Convex Anal. **24** (2017), no. 3, 927–953.

[3] ———, *Equivalent norms in polynomial spaces and applications*, J. Math. Anal. Appl. **445** (2017), no. 2, 1200–1220, DOI 10.1016/j.jmaa.2016.03.039.

[4] G. Araújo, G. A. Muñoz-Fernández, D. L. Rodríguez-Vidanes, and J. B. Seoane-Sepúlveda, *Sharp Bernstein inequalities using convex analysis techniques*, Math. Inequal. Appl. **23** (2020), no. 2, 725–750, DOI 10.7153/mia-2020-23-61.

[5] G. Araújo, P. H. Enflo, G. A. Muñoz-Fernández, D. L. Rodríguez-Vidanes, and J. B. Seoane-Sepúlveda, *Quantitative and qualitative estimates on the norm of products of polynomials*, Israel J. Math. **236** (2020), no. 2, 727–745.

[6] R. M. Aron and M. Klimek, *Supremum norms for quadratic polynomials*, Arch. Math. (Basel) **76** (2001), no. 1, 73–80.

[7] S. Banach, *Über homogene polynome in (l^2)*, Studia Mathematica **7** (1938), 36-44, DOI 10.4064/sm-7-1-36-44.

[8] F. Bayart, D. Pellegrino, and J. B. Seoane-Sepúlveda, *The Bohr radius of the n-dimensional polydisk is equivalent to $\sqrt{(\log n)/n}$*, Adv. Math. **264** (2014), 726–746, DOI 10.1016/j.aim.2014.07.029.

[9] C. Benítez and Y. Sarantopoulos, *Characterization of real inner product spaces by means of symmetric bilinear forms*, J. Math. Anal. Appl. **180** (1993), no. 1, 207–220, DOI 10.1006/jmaa.1993.1396.

[10] L. Bernal-González, G. A. Muñoz-Fernández, D. L. Rodríguez-Vidanes, and J. B. Seoane-Sepúlveda, *A complete study of the geometry of 2-homogeneous polynomials on circle sectors*, Rev. R. Acad. Cienc. Exactas Fís. Nat. Ser. A Mat. RACSAM **114** (2020), no. 3, Paper No. 160, 31.

[11] O. Blasco, *The Bohr radius of a Banach space*, Vector measures, integration and related topics, Oper. Theory Adv. Appl., vol. 201, Birkhäuser Verlag, Basel, 2010, pp. 59–64.

[12] H. P. Boas, *Majorant series*, J. Korean Math. Soc. **37** (2000), no. 2, 321–337. Several complex variables (Seoul, 1998).

[13] R. P. Boas, *Inequalities for the derivatives of polynomials*, Math. Mag. **42** (1969), 165–174.

[14] H. P. Boas, *The football player and the infinite series*, Notices Amer. Math. Soc. **44** (1997), no. 11, 1430–1435.

[15] H. F. Bohnenblust and E. Hille, *On the absolute convergence of Dirichlet series*, Ann. of Math. **32** (1931), no. 3, 600–622.

[16] H. P. Boas and D. Khavinson, *Bohr's power series theorem in several variables*, Proc. Amer. Math. Soc. **125** (1997), 2975–2979.

[17] J. Bochnak and J. Siciak, *Polynomials and multilinear mappings in topological vector spaces*, Studia Math. **39** (1971), 59–76, DOI 10.4064/sm-39-1-59-76.

[18] E. Bombieri and J. Bourgain, *A remark on Bohr's inequality*, Int. Math. Res. Not. **80** (2004), 4307–4330.

[19] J. R. Campos, P. Jiménez-Rodríguez, G. A. Muñoz-Fernández, D. Pellegrino, and J. B. Seoane-Sepúlveda, *On the real polynomial Bohnenblust-Hille inequality*, Linear Algebra Appl. **465** (2015), 391–400, DOI 10.1016/j.laa.2014.09.040.

[20] Y. S. Choi and S. G. Kim, *The unit ball of $\mathcal{P}(^2l_2^2)$*, Arch. Math. (Basel) **71** (1998), no. 6, 472–480.

[21] ———, *Smooth points of the unit ball of the space $\mathcal{P}(^2l_1)$*, Results Math. **36** (1999), no. 1-2, 26–33.

[22] ———, *Exposed points of the unit balls of the spaces $\mathcal{P}(^2l_p^2)$ $(p = 1, 2, \infty)$*, Indian J. Pure Appl. Math. **35** (2004), no. 1, 37–41.

[23] Y. S. Choi, S. G. Kim, and H. Ki, *Extreme polynomials and multilinear forms on l_1*, J. Math. Anal. Appl. **228** (1998), no. 2, 467–482.

[24] A. Defant, J. C. Díaz, D. García, and M. Maestre, *Unconditional basis and Gordon-Lewis constants for spaces of polynomials*, J. Funct. Anal. **181** (2001), no. 1, 119–145, DOI 10.1006/jfan.2000.3702.

[25] A. Defant and L. Frerick, *Hypercontractivity of the Bohnenblust-Hille inequality for polynomials and multidimensional Bohr radii*, arXiv:0903.3395.

[26] A. Defant and L. Frerick, *A logarithmic lower bound for multi-dimensional Bohr radii*, Israel J. Math. **152** (2006), 17–28.

[27] A. Defant, L. Frerick, J. Ortega-Cerdà, M. Ounaïes, and K. Seip, *The Bohnenblust-Hille inequality for homogeneous polynomials is hypercontractive*, Ann. of Math. **174** (2011), no. 1, 485–497.

[28] A. Defant, D. García, and M. Maestre, *Bohr's power series theorem and local Banach space theory*, J. Reine Angew. Math. **557** (2003), 173–197.

[29] A. Defant, D. García, and M. Maestre, *Estimates for the first and second Bohr radii of Reinhardt domains*, J. Approx. Theory **128** (2004), no. 1, 53–68.

[30] A. Defant, D. García, M. Maestre, and P. Sevilla-Peris, *Bohr's strips for Dirichlet series in Banach spaces*. part 2, Funct. Approx. Comment. Math. **44** (2011), no. part 2, 165–189.

[31] A. Defant, D. García, M. Maestre, and P. Sevilla-Peris, *Dirichlet series and holomorphic functions in high dimensions*, New Mathematical Monographs, vol. 37, Cambridge University Press, Cambridge, 2019.

[32] A. Defant, M. Maestre, and C. Prengel, *The arithmetic Bohr radius*, Q. J. Math. **59** (2008), no. 2, 189–205.

[33] A. Defant, M. Maestre, and U. Schwarting, *Bohr radii of vector valued holomorphic functions*, Adv. Math. **231** (2012), no. 5, 2837–2857.

[34] A. Defant and C. Prengel, *Harald Bohr meets Stefan Banach*, Methods in Banach space theory, 2006, pp. 317–339.

[35] A. Defant and P. Sevilla-Peris, *A new multilinear insight on Littlewood's 4/3-inequality*, J. Funct. Anal. **256** (2009), no. 5, 1642–1664.

[36] J. Diestel, *Sequences and series in Banach Spaces*, Graduate Texts in Mathematics, vol. 92, Springer-Verlag, Berlin, 1984.

[37] J. Diestel, H. Jarchow, and A. Tonge, *Absolutely summing operators*, Cambridge Studies in Advanced Mathematics, vol. 43, Cambridge University Press, Cambridge, 1995.

[38] S. Dineen, *Complex analysis on infinite-dimensional spaces*, Springer Monographs in Mathematics, Springer-Verlag London, Ltd., London, 1999.

[39] S. Dineen, *Extreme integral polynomials on a complex Banach space*, Math. Scand. **92** (2003), no. 1, 129–140.

[40] D. Diniz, G. A. Muñoz-Fernández, D. Pellegrino, and J. B. Seoane-Sepúlveda, *Lower bounds for the constants in the Bohnenblust-Hille inequality: the case of real scalars*, Proc. Amer. Math. Soc. **142** (2014), no. 2, 575–580.

[41] D. Diniz, G. A. Muñoz-Fernández, D. Pellegrino, and J. B. Seoane-Sepúlveda, *The asymptotic growth of the constants in the Bohnenblust-Hille inequality is optimal*, J. Funct. Anal. **263** (2012), no. 2, 415–428, DOI 10.1016/j.jfa.2012.04.014.

[42] R. Duffin and A. C. Schaeffer, *Some properties of functions of exponential type*, Bull. Amer. Math. Soc. **44** (1938), no. 4, 236–240.

[43] M. Fabian, P. Habala, P. Hájek, V. Montesinos, J. Pelant, and V. Zizler, *Functional analysis and infinite-dimensional geometry*, CMS Books in Mathematics/Ouvrages de Mathématiques de la SMC, vol. 8, Springer-Verlag, New York, 2001.

[44] J. Ferrer, D. García, M. Maestre, G. A. Muñoz-Fernández, D. L. Rodríguez-Vidanes, and J. B. Seoane-Sepúlveda, *Geometry of the Unit Sphere in Polynomial Spaces*, SpringerBriefs in Mathematics, Springer, 2023.

[45] J. L. Gámez-Merino, G. A. Muñoz-Fernández, V. M. Sánchez, and J. B. Seoane-Sepúlveda, *Inequalities for polynomials on the unit square via the Krein-Milman theorem*, J. Convex Anal. **20** (2013), no. 1, 125–142.

[46] D. García, B. C. Grecu, and M. Maestre, *Geometry in preduals of spaces of 2-homogeneous polynomials on Hilbert spaces*, Monatsh. Math. **157** (2009), no. 1, 55–67, DOI 10.1007/s00605-008-0017-7.

[47] D. García, G. A. Muñoz-Fernández, and J. B. Seoane-Sepúlveda, *Extreme homogeneous trinomials on the unit square*, J. Math. Anal. Appl. **15** (2023), no. 4, Paper No. 127134, DOI 10.1007/s43037-021-00144-8.

[48] Y. Gordon and D. R. Lewis, *Absolutely summing operators and local unconditional structures*, Acta Math. **133** (1974), 27–48, DOI 10.1007/BF02392140.

[49] B. C. Grecu, *Geometry of homogeneous polynomials on two-dimensional real Hilbert spaces*, J. Math. Anal. Appl. **293** (2004), no. 2, 578–588.

[50] ———, *Extreme 2-homogeneous polynomials on Hilbert spaces*, Quaest. Math. **25** (2002), no. 4, 421–435.

[51] ———, *Geometry of 2-homogeneous polynomials on l_p spaces*, $1 < p < \infty$, J. Math. Anal. Appl. **273** (2002), no. 2, 262–282.

[52] ———, *Smooth 2-homogeneous polynomials on Hilbert spaces*, Arch. Math. (Basel) **76** (2001), no. 6, 445–454.

[53] B. C. Grecu, *Geometry of three-homogeneous polynomials on real Hilbert spaces*, J. Math. Anal. Appl. **246** (2000), no. 1, 217–229.

[54] B. C. Grecu, G. A. Muñoz-Fernández, and J. B. Seoane-Sepúlveda, *The unit ball of the complex* $\mathcal{P}(^3H)$, Math. Z. **263** (2009), no. 4, 775–785.

[55] B. C. Grecu, G. A. Muñoz-Fernández, and J. B. Seoane-Sepúlveda, *Unconditional constants and polynomial inequalities*, J. Approx. Theory **161** (2009), no. 2, 706–722.

[56] W. T. Gowers and B. Maurey, *The unconditional basic sequence problem*, J. Amer. Math. Soc. **6** (1993), no. 4, 851–874.

[57] V. A. Gusev, *Functionals of the derivatives of an algebraic polynomial and a theorem of V. A. Markov*, Izv. Akad. Nauk SSSR Ser. Mat. **25** (1961), 367–384 (Russian).

[58] P. Habala, P. Hájek, and V. Zizler, *Introduction to Banach Spaces I and II*, Matfyzpress, Univerzity Karlovy, 1996.

[59] P. Hájek, V. Montesinos, J. Vanderwerff, and V. Zizler, *Biorthogonal Systems in Banach Spaces*, CMS Books in Mathematics, vol. 26, Springer, 2008.

[60] L. A. Harris, *A proof of Markov's theorem for polynomials on Banach spaces*, Journal of Mathematical Analysis and Applications **368** (2010), no. 1, 374–381.

[61] P. Jiménez-Rodríguez, G. A. Muñoz-Fernández, M. Murillo-Arcila, and J. B. Seoane-Sepúlveda, *Sharp values for the constants in the polynomial Bohnenblust-Hille inequality*, Linear Multilinear Algebra **64** (2016), no. 9, 1731–1749, DOI 10.1080/03081087.2015.1115810.

[62] P. Jiménez-Rodríguez, G. A. Muñoz-Fernández, D. Pellegrino, and J. B. Seoane-Sepúlveda, *Bernstein-Markov type inequalities and other interesting estimates for polynomials on circle sectors*, Math. Inequal. Appl. **20** (2017), no. 1, 285–300, DOI 10.7153/mia-20-21.

[63] P. Jiménez-Rodríguez, G. A. Muñoz-Fernández, and D. L. Rodríguez-Vidanes, *Geometry of spaces of homogeneous trinomials on* $\mathbb{R}^2$, Banach J. Math. Anal. **15** (2021), no. 4, Paper No. 61, 22, DOI 10.1007/s43037-021-00144-8.

[64] W. Johnson and L. Tzafriri, *On the local structures of subspaces of Banach lattices*, Israel J. Math. **20** (1975), 292–299.

[65] P. Kirwan, Y. Sarantopoulos, and A. M. Tonge, *Extremal homogeneous polynomials on real normed spaces*, J. Approx. Theory **97** (1999), no. 2, 201–213, DOI 10.1006/jath.1996.3273.

[66] A. G. Konheim and T. J. Rivlin, *Extreme points of the unit ball in a space of real polynomials*, Amer. Math. Monthly **73** (1966), 505–507.

[67] M. Krein and D. Milman, *On extreme points of regular convex sets*, Studia Math. **9** (1940), 133–138, DOI 10.4064/sm-9-1-133-138 (English, with Ukrainian summary).

[68] J. Lindenstrauss and L. Tzafriri, *Classical Banach spaces. I. Sequence spaces*, Ergebnisse der Mathematik und ihrer Grenzgebiete, vol. 92, Springer-Verlag, Berlin-New York, 1977.

[69] ———, *Classical Banach spaces. II. Function spaces*, Ergebnisse der Mathematik und ihrer Grenzgebiete [Results in Mathematics and Related Areas], vol. 97, Springer-Verlag, Berlin-New York, 1979.

[70] J. E. Littlewood, *On bounded bilinear forms in an infinite number of variables*, Quart. J. Math. Oxford **1** (1930), 164–174.

[71] A. Markov, *On a problem of D. I. Mendeleev*, Zapiski Imp. Akad. Nauk **62** (1889), 1–24 (Russian).

[72] ———, *On a problem of D. I. Mendeleev*, Electronic article to be downloaded from http://www.math.technion.ac.il/hat/papers.html.

[73] R. S. Martin, *Thesis*, California Inst, of Tech. (1982).

[74] R. D. Mauldin (ed.), *The Scottish Book*, Birkhäuser, Boston, MA, 1981. Mathematics from the Scottish Café; Including selected papers presented at the Scottish Book Conference held at North Texas State University, Denton, Tex., May 1979; Edited by R. Daniel Mauldin.

[75] L. Milev and N. Naidenov, *Indefinite extreme points of the unit ball in a polynomial space*, Acta Sci. Math. (Szeged) **77** (2011), no. 3-4, 409–424.

[76] ———, *Strictly definite extreme points of the unit ball in a polynomial space*, C. R. Acad. Bulgare Sci. **61** (2008), no. 11, 1393–1400.

[77] ———, *Semidefinite extreme points of the unit ball in a polynomial space*, J. Math. Anal. Appl. **405** (2013), no. 2, 631–641.

[78] A. Montanaro, *Some applications of hypercontractive inequalities in quantum information theory*, J. Math. Physics **53** (2012).

[79] V. Montesinos, P. Zizler, and V. Zizler, *An introduction to modern analysis*, Springer, Cham, 2015.

[80] G. A. Muñoz-Fernández, D. Pellegrino, and J. B. Seoane-Sepúlveda, *Estimates for the asymptotic behaviour of the constants in the Bohnenblust-Hille inequality*, Linear Multilinear Algebra **60** (2012), no. 5, 573–582, DOI 10.1080/03081087.2011.613833.

[81] G. A. Muñoz-Fernández, D. Pellegrino, J. B. Seoane-Sepúlveda, and A. Weber, *Supremum norms for 2-homogeneous polynomials on circle sectors*, J. Convex Anal. **21** (2014), no. 3, 745–764.

[82] G. A. Muñoz-Fernández, S. Gy. Révész, and J. B. Seoane-Sepúlveda, *Geometry of homogeneous polynomials on non symmetric convex bodies*, Math. Scand. **105** (2009), no. 1, 147–160.

[83] G. A. Muñoz-Fernández, S. Ruiz, and J. B. Seoane-Sepúlveda, *Geometry of trinomials revisited*, Geometry of Banach spaces and related fields, Proc. Sympos. Pure Math., vol. 106, Amer. Math. Soc., Providence, RI, [2024] ©2024, pp. 231–249.

[84] G. A. Muñoz-Fernández, V. M. Sánchez, and J. B. Seoane-Sepúlveda, *Estimates on the derivative of a polynomial with a curved majorant using convex techniques*, J. Convex Anal. **17** (2010), no. 1, 241–252.

[85] ———, *L^p-analogues of Bernstein and Markov inequalities*, Math. Inequal. Appl. **14** (2011), no. 1, 135–145, DOI 10.7153/mia-14-11.

[86] G. A. Muñoz and Y. Sarantopoulos, *Bernstein and Markov-type inequalities for polynomials on real Banach spaces*, Math. Proc. Cambridge Philos. Soc. **133** (2002), no. 3, 515–530.

[87] G. A. Muñoz-Fernández, Y. Sarantopoulos, and J. B. Seoane-Sepúlveda, *An application of the Krein-Milman theorem to Bernstein and Markov inequalities*, J. Convex Anal. **15** (2008), no. 2, 299–312.

[88] G. A. Muñoz-Fernández and J. B. Seoane-Sepúlveda, *Geometry of Banach spaces of trinomials*, J. Math. Anal. Appl. **340** (2008), no. 2, 1069–1087.

[89] S. Neuwirth, *The maximum modulus of a trigonometric trinomial*, J. Anal. Math. **104** (2008), 371–396.

[90] D. Núñez-Alarcón, *A note on the polynomial Bohnenblust-Hille inequality*, J. Math. Anal. Appl. **407** (2013), no. 1, 179–181, DOI 10.1016/j.jmaa.2013.05.008.

[91] D. Nuñez-Alarcón, D. Pellegrino, and J. B. Seoane-Sepúlveda, *On the Bohnenblust-Hille inequality and a variant of Littlewood's 4/3 inequality*, J. Funct. Anal. **264** (2013), 326–336.

[92] D. Nuñez-Alarcón, D. Pellegrino, J. B. Seoane-Sepúlveda, and D. M. Serrano-Rodríguez, *There exist multilinear Bohnenblust-Hille constants $(C_n)_{n=1}^{\infty}$ with $\lim_{n\to\infty}(C_{n+1} - C_n) = 0$*, J. Funct. Anal. **264** (2013), no. 2, 429–463, DOI 10.1016/j.jfa.2012.11.006.

[93] D. Pellegrino and J. B. Seoane-Sepúlveda, *New upper bounds for the constants in the Bohnenblust-Hille inequality*, J. Math. Anal. Appl. **386** (2012), no. 1, 300–307, DOI 10.1016/j.jmaa.2011.08.004.

[94] R. A. Ryan and B. Turett, *Geometry of spaces of polynomials*, J. Math. Anal. Appl. **221** (1998), no. 2, 698–711.

[95] I. Sarantopoulos, *Estimates for polynomial norms on $L^p(\mu)$ spaces*, Math. Proc. Cambridge Philos. Soc. **99** (1986), no. 2, 263–271, DOI 10.1017/S0305004100064185. MR0817668

[96] ———, *Polynomials on certain Banach spaces*, Bull. Soc. Math. Grèce (N.S.) **28** (1987), 89–102.

[97] Y. Sarantopoulos, *Bounds on the derivatives of polynomials on Banach spaces*, Math. Proc. Camb. Phil. Soc. **110** (1991), 307–312.

[98] I. Sarantopoulos, *Extremal multilinear forms on Banach spaces*, Proc. Amer. Math. Soc. **99** (1987), no. 2, 340–346, DOI 10.2307/2046637. MR0870797

[99] K. Seip, *Estimates for Dirichlet polynomials*, EMS Lecturer, CRM (20), http://www.euro-math-soc.eu/system/files/Seip_CRM.pdf.

[100] J. B. Seoane-Sepúlveda, *Sharp generalizations of the multilinear Bohnenblust-Hille inequality*, Oberwolfach Reports (European Mathematical Society) **11** (2014), no. 1, 378–381, DOI 10.4171/OWR/2014/06.

[101] V. I. Skalyga, *Analogues of the Markov and Bernstein inequalities for polynomials in Banach spaces*, Izv. Math. **61** (1998), 143–159.

[102] ———, *Analogues of the Markov and Bernstein inequalities on convex bodies in Banach spaces*, Izv. Math. **62** (1998), 375–397.

[103] E. Steinitz, *Bedingt konvergente Reihen und konvexe Systeme*, J. Reine Angew. Math. **143** (1913), 128–176, DOI 10.1515/crll.1913.143.128 (German).

[104] A. Tonge, *The failure of Bernstein's theorem for polynomials on $C(K)$ spaces*, J. Approx. Theory **51** (1987), no. 2, 160–162, DOI 10.1515/crll.1913.143.128.

[105] L. Tzafriri, *On Banach spaces with unconditional basis*, Israel J. Math. **17** (1974), 84–93.

[106] E. V. Voronovskaja, *The method of functionals as applied to problems of Zolotarev-Pšeborskiĭ type*, Dokl. Akad. Nauk SSSR **194** (1970), 20–23 (Russian).

[107] D. R. Wilhelmsen, *A Markov inequality in several dimensions*, J. Approx. Theory **11** (1974), 216–220.

Index

Made in the USA
Las Vegas, NV
19 September 2026

ceec2f78-a017-4518-a49a-2a02c6984349R01